AF832795

ROUTLEDGE LIBRARY EDITIONS:
ETHICS

Volume 41

PRACTICAL ETHICS

PRACTICAL ETHICS

A Sketch of the Moral Structure of Society

MARY STURT AND MARGARET HOBLING

Routledge
Taylor & Francis Group
LONDON AND NEW YORK

First published in 1949 by Routledge & Kegan Paul Ltd

This edition first published in 2021
by Routledge
2 Park Square, Milton Park, Abingdon, Oxon OX14 4RN

and by Routledge
52 Vanderbilt Avenue, New York, NY 10017

Routledge is an imprint of the Taylor & Francis Group, an informa business

© 1949 Mary Sturt and Margaret Hobling

All rights reserved. No part of this book may be reprinted or reproduced or utilised in any form or by any electronic, mechanical, or other means, now known or hereafter invented, including photocopying and recording, or in any information storage or retrieval system, without permission in writing from the publishers.

Trademark notice: Product or corporate names may be trademarks or registered trademarks, and are used only for identification and explanation without intent to infringe.

British Library Cataloguing in Publication Data
A catalogue record for this book is available from the British Library

ISBN: 978-0-367-85624-3 (Set)
ISBN: 978-1-00-305260-9 (Set) (ebk)
ISBN: 978-0-367-50291-1 (Volume 41) (hbk)
ISBN: 978-1-00-304939-5 (Volume 41) (ebk)

Publisher's Note
The publisher has gone to great lengths to ensure the quality of this reprint but points out that some imperfections in the original copies may be apparent.

Disclaimer
The publisher has made every effort to trace copyright holders and would welcome correspondence from those they have been unable to trace.

PRACTICAL ETHICS

A SKETCH OF THE MORAL STRUCTURE OF SOCIETY

by

MARY STURT, M.A.

and

MARGARET HOBLING, M.A.

LONDON

ROUTLEDGE & KEGAN PAUL LIMITED

BROADWAY HOUSE: 68–74 CARTER LANE, E.C.4

And what is freedom? Rightly understood
A universal licence to be good.

HARTLEY COLERIDGE

CONTENTS

A SKETCH OF THE MORAL STRUCTURE OF SOCIETY

PRACTICAL ETHICS

A SKETCH OF THE MORAL STRUCTURE OF SOCIETY

Introduction

Ethics is generally defined as the theoretic study of human morality; but the word is also used to cover the scheme of behaviour approved by a community or civilization and the attempt to make individual members live in accordance with it. There is thus a practical as well as a theoretic side to the study—not only an attempt to understand the best course of action but also the effort to live in the light of this understanding and to induce others to do so too. We can distinguish, in reality, three branches of the subject: the theoretic discussion of the "Good", the effort to achieve personal virtue, and the attempt to secure virtue in others. Each of these three branches can be treated separately.

The theoretic discussion of morality has not been very productive of useful results. To decide which is the highest type of life may be the first step to living it, but too often it is also the last. Moreover there has not been unanimity among the theorists. The Good which they have imagined has differed markedly on different occasions: from the Epicureans, who asserted that pleasure is the End, to the sterner Christian theologians of whom it can be said, "Following the New Testament other terms have been employed: 'self-denial', 'repentance', 'crucifixion of the flesh'. Thus awakening the feeling that abnegation of the natural life is the essence of Christian ethics".[1]

These extreme divergences in the stream of European thought serve to indicate the different sources from which that thought springs. We should become bewildered if we tried to find a common doctrine among all these varieties. If we looked farther afield we should find the divergence between the standards accepted in different cultures even greater. Anthropology has taught us that customs and beliefs differ profoundly from one culture group to another; and that, whatever may be imagined to be the ultimate good—and in most cases no ultimate good has been analysed out from the customs that prevail—it is certainly sought by methods very alien to those that we cherish. With so much uncertainty we shall not devote this book to an attempt to discuss the various types of ultimate Good. And yet it

[1] Haering, *Ethics of the Christian Life*, p. 127.

is impossible to ignore the question altogether, however practical we hope to be. We shall, therefore, a little later in this chapter, say what the ultimate good seems to us to be; and we shall leave it at that, and not try to defend our point of view against all the objections that could be raised against it.

Of the two types of practical ethics, the achievement of personal virtue is very difficult to discuss. Each man's moral life is to him a series of problems, some small, some great, that he must solve in accordance with his general standards. Those whose business it is to exhort us to virtue are apt to assume that a good will is all that is required; and that if we would surrender ourselves to virtue, there would be no further problem. In actual experience that is not so. The best will in the world will not tell us clearly how to act on this occasion or in these circumstances. If it is assumed that certain general principles always hold good and can be applied like formulae in algebra, there still remains the problem of which principle these particular circumstances call for. Do we, on this occasion, show self-sacrifice and self-control, or do we fight for what we believe to be the truth and right? To take a single example, the upbringing of a child presents us with an endless series of problems, if we take our duties seriously; and these are problems which can only be solved in relation to particular circumstances. Our dealings with our friends or our husbands may have to be adjusted even more carefully to special circumstances, and they cannot be profitably discussed except in the intimacy of individual cases.

What is left is the study of the conditions in which a general knowledge of and inclination to virtue can be imparted. Plato maintained that this was everyone's business, and, in consequence, was little regarded and less analysed. It has been considered the duty of parents, of schools, of the Church, and all have tried their best; but in the past many of the conditions of success have been lacking. There was not, in the first place, sufficient psychological knowledge, and many of the methods intended to promote virtue had an opposite effect. Secondly, virtue was looked upon as a strictly individual affair. "The individual", says Kierkegaard, "is in truth the only subject of ethics". Now this is true only if we realize that ethics concerns the behaviour of the individual in his social setting; and not individuals only, but whole groups or communities can have an ethical character. One of the chief elements in determining a man's moral standards is the group of which he forms part. Morals arise in the group, are cared for by the group, and have importance in relation to the group. The individual, apart from the group, is not an ethical being, though he may be a religious one. Therefore ethics is not only a study of the individual but of his group and his whole culture. It is this point of view which gives this book its character. So long as man's environment was ignored, and only his individual acts studied, it was hard to understand *why* he behaved as he did, or to devise means of altering his behaviour. When it is realized that man and his environment must be looked at together, then the reasons for his acts and the methods of controlling them are simple to understand.

The individualistic attitude towards ethics is largely due to Christian theology. Each soul is supremely precious. Each man is the guardian of his own soul, and each soul must be judged individually and bear the responsibility for its actions.

For the sin ye do by two and two, ye must pay for one by one.[1]

We shall return to this point several times. It is enough here to say that this particular conception has rendered much ethical teaching fruitless, mainly because it is opposed to very strong human impulses. This, then, for us, is the meaning of practical ethics, the study of man, his nature and his social environment, so that we can both understand the reason for his acts and devise means by which he can be made better and led to a more satisfactory way of life.

Although we shall not discuss them in great detail, it is necessary to say something of the origin of the particular moral ideas we hold to-day. Our ideas of what is right or wrong greatly influence our conduct, and it is important to consider how we came by the very diverse set of opinions that we hold. The fact that very different types of behaviour can all be considered right by their perpetrators might have led moralists to decide that morals had reference only to the group; and that *right* was simply a matter of convention within a society. Instead each group has generally assumed that its own standards are *absolutely* right, and has sought, in religion or metaphysics, a universal justification for them. From the strength of this position they have then been able to condemn all those who differed from them. This is particularly easy and certain when the authority claimed for the rules is divine; and the great majority of older systems did make this claim to divine authority.

Of the older systems of morality the best-known to most people is that embodied in the Old Testament. The Jews were not very different in some of their ideas from other peoples in the same stage of development, and some of the decrees of Jehovah were very like some of the details in the code of Hammurabi and not unlike some

[1] Kipling.

of the demands of Zeus; but whereas the primitive ideas of the Old Testament are almost universally known, and have had the greatest direct effect on European thought, the Greek ideas that have influenced Europe are of a later date, and it is Plato and Aristotle, not Homer and Hesiod, that for us represent Greek ethics.

Two things are clear from the Old Testament. One is that morals are a social thing within the group; the other that they rest on the decrees of God, thus being removed from question or explanation. The Ten Commandments, sent down direct from Mount Sinai, are as it were the minimum requirements for a harmonious theocratic, property-owning group. They prescribe on the one hand the positive outlines of worship, and on the other the mainly negative observations of civil life. All commands alike, religious or secular, are enjoined by divine authority, and must be obeyed under threat of divine displeasure. As the stories of the early history of the Israelites show, this displeasure was very real, and a breach of any divine ordinance would be followed by extremely serious consequences. Again and again in the Old Testament we get this made clear. The story of the crime and destruction of Achan illustrates both points. The sin against a divine command is instantly punished, and the punishment falls on the group as well as the individual.

Thus we find as one of the firmly fixed bases of morality the "Will of God". The Good Man is he who submits to this will or strives actively to fulfil it. Sin consists in opposing it, and neither reason nor expediency nor natural affection must hinder the exact performance of what is demanded.

This belief in the necessity of following the will of God exists to-day. The Swiss theologian Brunner, in his book *The Divine Imperative*, attempts to show how God's will, deducible from the Bible, should guide our actions. His discussion of the attitude of the Christian to the economic order of our time illustrates the ideas and methods. "The duty", he says, "of thinking out what obedience to God's will in the economic order really means has been laid upon us and cannot be evaded". The difficulty arises from the fact that, though economic activity is necessary for life, it is yet, according to this school of thought, intrinsically wicked:

"Since God has created man as a corporeal being, He has also created him as a being needing an economic order and capable of creating an economic order. Thus the economic order forms part of the original Divine order in creation; as such, like all Divine order, it is at the same time a Divine law. Man is commanded to create an economic order. The apostolic injunction, 'If any will not work, neither let him eat,' expresses this Divine order in creation. . . .

"But this is only one side of the question, and it is the bright side. The other side, the dark side, can never be separated from the real economic order in which we live, that is, from the curse of sin which broods over it. It is not possible for the individual to engage in economic activity without sin. Nowhere does the doctrine of Original Sin become so palpably evident as in this sphere. Each of us has an active and passive share in the universal sinfulness of the economic order."

The solution of this dilemma is not very clear. We must work to live; it is useless to think we can devise a sinless economic order; we must do our best with what we have, bearing in mind that it should never be taken for anything of supreme importance.

Such a system of ethical thought can be carried into every department of human life. This was done fairly frequently in the past, it is sometimes done to-day, though it needs great subtlety and ingenuity of mind to fit the Biblical injunctions to a state of society that has changed so greatly since they were written.

It is also clear from the Old Testament story that the commandments that were the cement of the social group did not necessarily apply outside it. The Tribe, whose members were forbidden to steal from one another, had but recently plundered the Egyptians with full divine approval; and this tribal aspect of morality has clearly descended to us to-day. It was, of course, widespread and not exclusively Jewish, and was not maintained by the Jews all through their history; yet because men were familiar with the Biblical passages which express this tribal limitation of morality, they were able to quote isolated passages in support of this human, all-too-human, tendency, ignoring the higher universalistic tendency of the prophets. Thus the Covenanters could quote verses like "and Phinehas arose and executed judgment" in support of their most bloodthirsty excesses. Nations have succeeded tribes, and "national interest" has become the excuse for all kinds of behaviour that would be rejected inside the group. A king can seize a neighbouring province, a trading nation insist on the right to introduce opium to a China that does not want it, and the enslavement of negroes can be quite legal so long as it happens outside the territorial limits of the home group.

Because an idea is felt at any time to be supported by divine authority it does not follow that that idea will remain unchanged through the ages or that it will be unaffected by social conditions. Whatever idea exists at any moment, if it is sufficiently well established, can claim to be part of the divine plan. The same Biblical precepts may be reinterpreted many times. A very obvious example is the Christian attitude to property. The poverty of the early Christian groups led to a form of religious communism, that, if not complete, was at least considerable. The well-endowed Anglican Church of the eighteenth and nineteenth centuries, functioning as part of the capitalist and aristocratic social order, saw in private property, and the inequality of rich and poor, an ordinance of God as firmly fixed as any other.

> Whene'er I take my walks abroad,
> How many poor I see;
> What have I done that God should give
> So many gifts to me?
>
> Nor more than others I deserve,
> Yet God has given me more.
> I have my food and they have none,
> And beg from door to door.[1]

It was not only economic ideas that varied from time to time, other principles were modified from one age or sect to another; but in every case the sanction was the same, and virtue consisted in obedience to the will of God. In many cases, of course, there was no direct prescription: not all cases had been foreseen. The divine will must be learnt through the authority of the priest or of the scribe, learned in the law and its interpretation, or by a consideration of one's own conscience. In some churches the priest was the accepted mouthpiece, and through his sermons or by individual advice he helped his flock to know what was right. In other cases the individual made his own appeal for guidance. With the open book and the light that God gave he found out the truth. As the proud motto of Oxford University says, *Dominus illuminatio mea.* Although the conflict between authority and individual judgment has rent Europe by wars, the essential basis of belief has always been the same. There is a right course that is the will of God, and man in some way or other can learn what this course is, and then must follow it. If he does follow it God will reward him, if he fails to do so there are the penalties of divine anger. This type of ethical theory may be called religious ethics and is based on the religious and metaphysical belief in a personal God who is directly concerned with the affairs of men.

"And are not two sparrows sold for a farthing and one of them shall not fall to the ground without your Father. Fear ye not therefore ye are of more value than many sparrows."

[1] Dr. Isaac Watts.

Of a totally different kind is the ethical theory represented for most of us by Aristotle. Of all the Greek writers, he had most effect on the thought of the Middle Ages, and so has become most embedded in European thought; though, of course, he could never compare in authority or vividness with the Bible. Aristotle started, not from a God who declared his will and was prepared to enforce it, but from a consideration of the actual facts of man's life and speech. For him ethics was a part of politics, the art by which men live together in society. Ethics is the consideration of man's ultimate good and the virtues that help him to achieve it. It is generally agreed that the thing absolutely desired for itself alone is Happiness. Therefore the supreme aim of man is the achievement of Happiness, and this is defined as "the active exercise of the soul's faculties in conformity with excellence or virtue"; and virtue, moral virtue, is later defined as a "settled disposition of the mind as regards the choice of action and feelings, consisting essentially in the observance of the mean relative to us, this being determined by principle, that is, as the prudent man would determine it". As a final example of perfect Happiness he would give the life of philosophic contemplation; though all men, naturally, would not attain to it.

Now this type of ethical thought is obviously quite different from the religious one. The Athenian gentleman, living in an intellectual society, with a religion that was mainly ceremonial and which had long ceased to interfere much in the conduct of daily life, could start an empirical inquiry unhampered by divine prescription; and could arrive at a conclusion impossible to the man under the necessity of daily toil. Also, since the social and political life of the state was so close and continuous, he would naturally see ethics as a social matter and not as the relation of the individual to God. Moreover, political change was sudden and often disastrous in Greek states, and within a very few miles of each other Athens and Sparta had organized social systems that were very diverse. It was impossible, therefore, to found an ethical system on the will of God or immemorial custom. It was all too clear that social systems were man-made and that human reason must take control. It follows from this that the only way of arriving at a concept of the Good is to consider the thought and practice of mankind. There is no supreme external authority which dictates the content of right and wrong. Thus Aristotle asks explicitly, what do men commonly think, and how do they actually conduct their lives. From these facts he constructs the basis of his theory.

Later writers have to some extent divided themselves into two camps. There are those who really belong to the religious group, even if they do not mention God. They hold that we know what is right *intuitively* by the action of *conscience* or some other God-given faculty. This power tells us definitively, without the possibility of question, that, to give Whewell's list, benevolence, justice, truth, purity and order are the qualities that one must pursue; and they must be realized in actions conformable to the social canons of the day. On the other hand, the more Aristotelian, of whom Bentham may be taken as an example, ask what does man in fact desire, and decide, as their master before them, that he does in fact desire Happiness. But Bentham was writing in an age of democratic principles, while Aristotle belonged to an aristocratic, slave-owning community, so that Bentham took as the supreme good the greatest happiness of the greatest number, each to count for one, and no one for more than one, while Aristotle is concerned with the development of the highest type of individual.

It is important to remember in studying ethical writers that they are human, and their own characters and the circumstances of their lives deeply influence their theories. Kant, writing in the infancy of the authoritarian Prussian State, saw in Duty an absolutely compelling force. It was a little hard to define Duty, and it was not very clear what it enjoined; but Duty was supreme. Acts were only good when they were done for its sake, and lost quality when done for any other end. Such a doctrine could never be the basis for a general theory of conduct. Its applications were too difficult, and too indeterminate. It can only be used, as it was perhaps intended to be, to give philosophical support to a code promulgated by authority. Hegel, who lived when Prussian absolutism had developed farther, declared that law and the customary morals of the community provided all that man needed to know of the "universal will", and that he must follow this law even against the dictates of his own conscience. In England we can trace a similar connection between ethical systems and the conditions in which their authors lived. The traditional doctrines of Anglican Christianity, as interpreted in a capitalistic society, clearly supply the content of the moral laws these writers discover. It is this dependence of theory on special circumstances that makes theoretic ethics so difficult to treat satisfactorily.

In this book we do not want to provide a treatise of this kind. Yet we cannot discuss practical ethics—the methods necessary to achieve the good life—without to some extent indicating our conception of the Good. For this we propose to go back to Aristotle and take a few examples of actions and practices commonly esteemed good or bad and then try and arrive from them at the very vaguest criterion that seems generally to be applied. To start overseas: Europeans in China have universally condemned the binding of women's feet, and one of the first efforts of missionaries and others has been directed to combating the custom. Now foot-binding is not contrary to any European code, religious or legal. It has been condemned because of the pain and misery it inflicts on the victims, because of the loss of economic power and of freedom. Round about 1870 and 1880 large areas of Birmingham and other towns were built with houses facing on to the street, and then houses behind, back to back, facing inwards on little courts. Many of these houses were built by good men and zealous citizens. They are now held in universal horror, and every effort is being made to replace them by a different type of dwelling. The reason is that life in such slums is far less pleasant and healthy than in houses with water laid on, sanitation, a garden and fresh air. In slum houses the domestic virtues flourish with difficulty, pleasant activities are impossible and physical growth is stunted. Large amounts have been spent by U.N.R.R.A. in keeping the population of several European countries from dying of starvation. We do not ask why. We *know* that it is "bad" for people to die of starvation, especially children. On the other hand we need no explanation why healthy children, comfortable houses, pretty clothes, flowering trees and such like are "good". It is quite obvious.

If we try and gather the general reason behind these judgments of "good" and "bad" we can say roughly with Aristotle that Happiness is the supreme good, and that what makes people happy is good, and what makes them unhappy is bad. We can also say with Bentham that the more people who are happy the better it is, and the fewer who are miserable the better it is also. We thus call an act good if it conduces to the general happiness, and say that it has ethical or moral value when it is done with that aim in view.

It is in our definition of Happiness that we part from our ethical authorities. Happiness is a psychological concept and should be defined in psychological terms. William McDougall has said that Happiness is a state in which the chief human impulses are receiving an harmonious satisfaction. This, like all McDougall's definitions, is concise and each word is important. Happiness is a permanent state and is connected with the deep urges of our nature. It is totally different from a temporary state of pleasure or the excitement that comes from some immediate stimulation. It arises, moreover, when there is harmony in the soul; when the different and potentially contradictory human impulses have been brought into accord with each other, and a state of tranquillity has been reached with a consequent absence of tension. In later chapters we shall consider the nature of man's impulses and show how they can be made to accord. It is enough now to say that as our criterion of the Good we take this conception of happiness for all, or as many as possible, and approve or condemn acts or social arrangements as they lead to or away from this condition.

It is possible to ask whether we know the conditions that make for happiness. There is a certain type of conservative who asserts that the poor man in his hovel is probably happier than the rich man in his luxury flat; and argues thence to the position that the poor man had best be left alone without a piped water supply lest his joy in life should vanish with a tiled bathroom. This argument, patently absurd as it is, does raise a real point. How far will good conditions automatically produce happiness or virtue? The answer is simply that happiness, as we have said, is a psychological state and material conditions can only facilitate it, never be its sole cause. Thus, whether we try to achieve good houses or good health for our people, their happiness will depend on something further, something that is spiritual added to the conditions among which they live. But these conditions have more than a merely material influence. Bad housing may make our days laborious and chilly; it also makes the development of family life difficult. The particular virtues of family love grow up with more difficulty when they have to struggle with dirt and discomfort. Undoubtedly, in some cases, they will develop in spite of all difficulties; but there are cases when the material difficulties will prove too strong, and what might have been a happy and virtuous home breaks up. In the same way health is the prerequisite for many kinds of happiness, and even for some types of virtue. The noble souls who can overcome disease and find happiness on a sick bed are few; for most of us health is necessary. Thus if we talk of good material conditions as playing a large part in happiness and virtue it is in this sense that we must be understood. Far more important than material aids to happiness are the psychological. We shall discuss later the conditions under which children develop happily and well. Again we must not be thought to mean that if the conditions are good happiness and virtue inevitably follow. Yet good psychological conditions have so great an effect that the training of children in virtue consists mainly in providing these conditions.

This suggestion that Happiness is the end, and virtue the means of achieving it, while in accordance with Greek thought, runs counter to the type of belief, common at all ages in connection with religious ethics, that suffering and self-sacrifice have deep moral value. If we try to understand why such unpleasant experiences are accorded a place of such high value, we can see two reasons. The suffering of the one can, in certain circumstances, produce the happiness of the many. This is expressed in its clearest form in the religious doctrine of Christ's sacrifice on the Cross, by which He bought the Salvation of the World. Where self-sacrifice is urged in a social setting it is for its value to others. There is also a value to the sufferer. Man's nature is so organized that a sacrifice for those we love brings a happiness of its own.

Suffering, apart from any good that accrues to others, has sometimes been regarded as meritorious in the sufferer, and entitling him to the reward of a heavenly bliss which may be quite materialistically, and even sensually, conceived—the "eternity of drunkenness" for which Plato jeered at the Orphics. The claim is baldly put by Tennyson's St. Simeon Stylites, in a poem which is an interesting psychological study:

> Who may be made a saint if I fail here?
> Shew me the man hath suffered more than I.

Such asceticism is often based on a mystical doctrine of the soul's need to be freed from the body if she is to attain her goal of the vision of divine mysteries in the other world, and to catch in ecstasy some fleeting glimpses of it while still in the flesh. Greek mysticism is generally mild and humane, but some of the Christian mystics practised incredible self-tortures. Henry Suso, a fourteenth-century mystic, has left us his autobiography (written in the third person) in which, among other details, he tells us:

"Then he devised something else: two leather gloves: and he caused a brazier to fit them all over with sharp pointed brass tacks and he used to put them on at night in order that if he should try, while asleep, to throw off the hair undergarment, or to relieve himself from the gnawings of the vile insects, the tacks might then stick into his body."[1]

It is unnecessary to discuss such excesses: the whole theory is clearly religious; it transcends and absorbs ethics, the moral life forming in orthodox Christian mysticism only the first, or purgative stage, in the mystic life.

But suffering and self-sacrifice in our earlier sense there must be. Sometimes this sacrifice is of something comparatively unimportant in order to achieve an end that involves much deeper interests, as when a parent foregoes a personal gratification in order to pay his

[1] William James, *The Varieties of Religious Experience.*

child's holiday expenses. Sometimes the sufferer abandons life itself for the interest of the group, as when a prisoner under torture refuses to betray his comrades. For self-sacrifice to be reasonable there must always be a compensatory gain to another. The Puritan tradition, caught up in the sentimentality of the Victorian age, made of self-sacrifice an end in itself, and thus deprived it of dignity and reason.

If we remember Aristotle's dictum that happiness must be predicated of a life as a whole, we can see at once that a happy life may include within it a considerable amount of austerity and even suffering. The words of Romola are often quoted, "We can only have the highest happiness, such as goes with being a great man, by having wide thoughts and much feeling for the rest of the world as well as ourselves; and this sort of happiness often brings so much pain with it, that we can only tell it from pain by its being what we would choose before everything else, because our souls see that it is good".[1] Hardship may be deliberately sought, either as the reaction of a noble nature against ease and luxury, or else as a necessary discipline for the training of character, as the young Marcus Aurelius, in his twelfth year, slept on the bare ground and only at his mother's request took to a pallet covered with skins. But there is apt to be a slight touch of priggishness about such self-sought hardship. The suffering that comes unsought, that is accepted and faced gladly for the love of another, is the kind that calls out our fullest admiration and answering love.

There is in Haering's *Ethics of the Christian Life* a very interesting definition of Love. Haering is trying to explain his conception of the Kingdom of God, the attainment of which he holds to be the supreme end of the moral life. It is difficult to understand exactly what he imagines the Kingdom to be, because, every time he comes near a definition, he grows dithyrambic; but his account of the Love that forms its basis is clear:

"Love is the endeavour of a society of sentient beings to realize from good will and benevolence, by surrender on one part and appropriation on the other, some common ends. In its final ground it is benevolence and surrender, altruism and self-renunciation, for pleasure without benevolence would be selfishness, and benevolence without pleasure would be the cold fulfilment of duty."

Explained thus, and correlated with an end beyond itself, self-sacrifice takes its place in an ordered scheme in which ultimate Happiness plays the largest part.

If the theoretic study of ethics has proved somewhat unprofitable

[1] George Eliot, *Romola.*

the practical aspect has recently became of overwhelming importance. This for two reasons. Of recent years we have learned much more of the workings of the human mind, and we are at last beginning to understand what kind of mechanisms come into play in the production of virtue and vice. Our study of children has taught us that most delinquents are the victims of their environments, and we have also learned, at least to a very large extent, which types of environment are favourable and which are not. Thus if a child is taken in time, most kinds of undesirable behaviour can be brought to an end. This is an extraordinary power and historically speaking a new one. The theories on which the whole modern practice of child guidance is based belong to this century. This century, too, has seen the comparatively general application of psychology to methods of education, with changes in the schools that only those in close touch with them can realize. In this century, also, the treatment of children in their own homes has undergone a great change. In many places through this book we shall give examples of Victorian child management. This is not done to raise horror, or to dishonour a great age. It is simply done to show that times *have* changed; that when we talk of the reforms brought about by psychological knowledge we mean something real and far-reaching, something that can be rightly claimed as having a profound moral effect on the nation.

Another change has recently come into our thought. In the past the individual—the subject-matter of ethics—has been imagined to exist, spot-lighted, against the background of the civilization of the author's own place and time. The study of anthropology has shattered this vision. We know that ethics is really a group matter. And with the study of anthropology we have learnt to know different types of culture, and can see how dependent the behaviour of the individual is on the environment in which he has grown up and lives. With this study of other cultures we have become able to criticize our own. It no longer looms, unanalysed and eternal, dwarfing our petty thoughts, and almost making it impious to dissect. We can see so many different patterns of life that it is becoming possible to say what features in our own way of life make for virtue and which do not. To take an example, the property structure that seemed to the Victorians to exist by Divine Right is now known to be a human construction, and alternative methods of economic life have been imagined and put into practice in other countries. We can thus judge whether, as compared with the others, ours is the best that could be devised; and we can discuss how it should be amended. We have then the realization that even individual virtue is a social matter, and we should try to produce virtue not only by appeals to the individual, but by attempts to rearrange the practices of the community to which he belongs.

The third reason why a study of practical ethics is so important to-day is that, for the first time for many centuries, states in their corporate capacity, and armed with the new psychological knowledge, have taken to large-scale training. Unfortunately for mankind, what the states, who have done this on the largest scale, have taught is not virtue but wickedness. In Germany we have seen a whole nation brutalized and corrupted. We have seen the mass production on the one hand of the S.S. men and concentration-camp officials, and on the other of citizens so inert and careless that they "never noticed" the chimneys of Maideneck smoking a mile or so away across the fields. This dreadful example of the power of a government to influence the minds of its subjects is not merely horrifying; it is also inspiring. If man can be made bad to order, it is at least worth trying to see if he cannot also be made good. It may be argued that the human heart is prone to sin, and that the downward path is always easy; but that we do not believe. It is part of the thesis of this book that man's nature is on the whole good, and to state this is to raise one of the most fundamental problems in ethical theory. If man is by nature bad, as priests in some ages, pessimists in others, have maintained, then the training in virtue must always be an uphill struggle and a matter of coercion. If man's nature is indeterminate, waiting to be directed to good or ill, then we can go forward with a new confidence. We can give the good room to develop, we can study the conditions that make the best of man's impulses. Human nature is a very composite thing. There are impulses that conflict, and there is infinite room for adjustment and arrangement. The same innate impulse may appear either as a virtue or a vice according to the way it is trained and the circumstances in which it is used. Our particular civilization is not an easy one. There are others which allow man's nature to develop much more favourably. Our culture is based on competitive industry and private property, and a proportion of the population has always in the past lived below the line of decency and comfort. The fear of poverty, cold, hunger, disease, is always present. A large part of the population has been denied full intellectual and cultural development. It is possible that, in these circumstances, the drag to evil is stronger than the impulse to good. But the order of civilization has changed in the past, and may change in the future. As we shall show later, the will to good in even a small portion of the population can produce social changes of the utmost importance. Were this will more widely spread the changes would be quicker and more far-reaching. Karl Mannheim remarks in one of his books: "I would venture to say that it costs a social organization at least as much deliberately to build up warlike attitudes as peaceful ones." This may be applied equally well to virtue and vice; except that it is definitely more expensive in money, thought and care to make a child virtuous than to leave him dirty, ignorant and frustrated, the breeding-ground for all the meaner vices.

In this century a great change has come over the teaching of virtue: the state, in its own person, has entered the field as educator. There have, for many centuries, been certain people and institutions specially charged with the duty of moral training. The foremost of these in Europe is the Church, and throughout the history of European civilization the Church, whether it was Catholic or Protestant, considered itself the guardian of the moral tradition. The tradition guarded by the Church was seldom the only one current at the time. There were others, and it is the measure of the richness of our European inheritance that there was always this diversity. The nobles had their own code, the craftsmen theirs. Even when the Church was at the height of its power it could only influence, never destroy, the courtly or chivalric morality. When schools grew up independently of the Church they claimed a share in moral education. There are schools which have for long prided themselves on their moral influence. Others, out-growing the stage of merely providing information, now see themselves as training-grounds for virtue. There are also professional associations, which, in the limited fields of their members' professional activities, take charge of morals. We shall discuss all these later. Even more important are the quite unorganized forms of moral training. There is the home in which each child grows up and receives a good or bad training according to the nature of his parents. There is the vast field of literature, the cinema, and the newspaper, that is more or less nobody's business, at least in England. Above all there is the inescapable pressure of the whole social system. Our standards of value, our objects of desire, our attitudes to our fellow men, our own development are all bound up with the type of state we inhabit. The Russian reared under Communism is different in countless ways from the American child of unrestricted, capitalist competition. The backwoods man and the factory hand have innumerable differences. Each has developed a system of morality and a belief in values corresponding to his way of life; his psychology is different, his approach to life. All this has now been consciously realized. When a government determines to produce a certain type of citizen it seizes, as far as possible, all these methods of influence. The Church can be supported, silenced or won over; schools can be specially staffed. The great mass of parents cannot be controlled, but children can be kept occupied in so many ways that they are hardly ever at home, or they can be taught to spy on their parents so that the elders dare not express their views. Above all, the state can take control of the vast field of unorganized teaching. Books of the right kind can be written, radio programmes can disseminate facts and theories, the cinema can give vividness to ideas. Lastly, and most fundamental, the social pattern can be varied. The worker on a communal farm is a different person, mentally and morally, from a peasant fighting for his living on his own individual patch of soil. The worker who in the factory receives good treatment, good pay and prospects of training and promotion is entirely unlike the driven and depressed slave who has no hopes and no vision of life. Or to take a less extreme example, an academic life develops a very different character and outlook on life from those of the "high-powered business executive".

It was part of the *laissez-faire* policy of the later nineteenth century to leave moral teaching and ethical standards as independent of government control as was commerce. So long as a man did not break the law he could develop in his own way and lead his children in the path he thought fit. The idea that government, as such, should intervene and deliberately mould the moral views of the nation is still repugnant to many people. They do not trust government. Yet during and since the war, even in England, where this hostility to control is strong, government in its full capacity has taken the task of moulding public opinion in what are essentially moral ways. Perhaps the most conspicuous example was the campaign for the maintenance of strict rationing immediately after the end of the war so that more food should be available for distribution to the hungry lands overseas. The papers were full of articles on the needs of different countries, and the beginnings of popular clamour for increased food at home were stifled by the representation of the needs of others. The very schoolchildren were banded together to send two ounces of their chocolate ration to Europe. This is an example of governmental intervention in morals in one particular matter. It was quite successful. No one would find it hard to think of many other matters which could be dealt with on the same level.

The nation to-day is bound together so closely by such things as the radio, the cinema, the popular papers and the posters on the walls that ideas spread very quickly. The higher level of education that is gradually coming makes it easier for ideas to be apprehended. We have learnt how to manage ration books, how to cook vegetables, and we could doubtless learn much about the moral conduct of life if it were resolutely taught by the same methods.

If a government merely teaches its own scheme of life, that is hardly objectionable. When it closes the ears of the nation to all voices but its own then there is a very different situation. We do not wish to live in a closed society with only one set of ideas and the stimulating effect of variety withdrawn. We do not feel that we yet know the best way to appoint our leaders and the moulders of our opinion. Even supposing we had confidence in our leaders we should still object to a society in which criticism was impossible. It is one of the hopeful aspects of social life that there is a progressive moralization of its different elements. We do not want any form of control that will prevent these changes, and ideas from without are as important as those that arise from within in bringing about this beneficent change.

When we say that ethics is an affair of the group we mean far more than that customs vary. In ancient Greece infanticide was moral, as it is in certain tribes in India to-day. For us it is a crime. Examples of this kind are endless, and they are too well known to need discussion. We mean here two things. The nature of the group modifies the individual, and without the group the individual can hardly maintain his moral life. We have already said how closely a man's psychology and morals are bound up with the life that society thrusts upon him, and that we could never find a common measure by which to compare much of the behaviour of the factory worker and the crofter maintaining a solitary and varied existence among his mountains and lakes. The Marxist doctrine of the overriding power of economic factors also embraces morality, and we must see men not as free souls but as bodies to be fed.

The other question of the decay of morality apart from the group has been commented on so many times that it is only worth giving one or two examples. The break-up of customary observations and sanctions has the effect of loosening all restraining bonds. The Romans, who were astute governors and knew the arts of control, were careful to preserve the forms that they found established. Gibbon describes their methods:

"The Roman magistrates under Trajan and the Antonines when dealing with subject peoples encouraged the public festivals which humanize the manners of the people. They managed the arts of divination as convenient instruments of policy; and they respected, as the firmest bond of society, the useful persuasion that, either in this or a future life, the crime of perjury is most assuredly punished by the avenging gods. But whilst they acknowledged the general advantages of religion, they were convinced that the various modes of worship contributed alike to the same salutary purposes; and that, in every country the form of superstition which had received the sanction of time and experience, was the best adapted to the climate and the inhabitants."

Later governors have been less prudent and too often, in their desire to introduce their own customs and beliefs, have destroyed what they could not replace. In the case of the Maori of New Zealand the European teaching destroyed the idea of tapu (the divine quality in things) and thus the power of the chiefs and the confidence of a man in his surroundings. In consequence, the characteristic qualities of the Maori were lost. He became lazy, dispirited, and the fertility of the women rapidly decreased. The traditional art has gone and been replaced by feeble imitation of Western styles. A similar example comes from South Africa. There the detribalized negro, brought to work in Johannesburg, loses the virtues that he had among his own people. The following comes from an interesting book by a missionary and deals with the effect of introducing the idea of civil marriage to men of this type.[1]

"European civil marriage—a welcome and legal way of evading his duties for the native. He will not say so because the 'boss' must not be angered, but his opinion of European social laws is not far from contempt. But the younger generations consider this liberty, this freedom from any restraining laws, as the expression of European civilization. They have been taught that anything native is sin and evil. They have accepted this judgment because it frees them from obligations and duties.

"They understand our civilization (and our attitude in the country confirms it) as a society where man has rights only, and is entitled by law to enforce them. Native life was of duties enforced by traditional custom; and between the two the choice of the younger generation is obvious. It is certain that women and children are much more strictly protected by native custom than by European law. This is quite logical since collectivism, which is the basis of native society (in opposition to our individualism) makes it a necessity to protect, very strictly, the woman and child through whom only society can live on, not to speak of the family or clan. When the native is torn from his group he loses his interest in its survival, and not having been made a member of the European group he has no

[1] D. T. W. Shropshire, *Primitive Marriage and European Law*, p. 73.

incentive to behaviour that will benefit others. A man closely bound to his group will obey its morality and strive to be at one with its members. When customs perish, as when the European tries to destroy types of behaviour he dislikes, the man is lost and unable to make a new adjustment. In place of his own customs that were unquestioningly followed and produced a society well integrated and stable, new ordinances come that he can amuse himself in evading. The customary authority is gone, the bonds of behaviour in which he felt himself safe, and he must now act without this security and guidance. He also loses, all too often, the forms of life and amusement he enjoyed. The Puritan attitude of many Christian missionaries forbids his dances, his games, his songs. He finds life dull, incomprehensible and hopeless. In certain cases race suicide, or at least race decadence, sets in."

It is probably this fear, this feeling of insecurity that makes a complicated civilization develop such acute psychosis when its system breaks down. The industrialized European is no better at adaptation then the primitive islander. When the security of regular employment is gone, when political changes have left the citizen without confidence, then he is apt to regress to a state of unreason and purely instinctive action from which his ordered state had brought him. In this condition he can be influenced to perform actions of all kinds. The violence of revolution, the passions of excited and frightened crowds, brigandage, the mass suggestibility and the belief that there is a good time round the corner if they can only destroy the Jews or some other enemies—all these are the results of the break-up of morality that follows the wreck of a social system.

Thus morality is bound up with the state. Aristotle pointed out that it was much easier to live well if a man inhabited a well-ordered state, and this is true. Virtue is impressed on him from outside till it becomes a habit, or at least a settled way of thinking. This is external morality, and will last as long as the fabric of the state is unimpaired. A deeper morality is produced by the process of individual education. If education is thought of seriously, it is not imparting certain facts (what they shall be is purely traditional), nor in teaching certain habits, but in producing a certain attitude of mind. The man who has been educated to virtue has some conception of the group and his place in it; some wider vision of conduct beyond that of his own immediate group. He has thought enough to be able to formulate standards of behaviour which at times differ in some degree from those immediately surrounding him. He also believes in his standards and is ready to follow them through difficulties.

Such an education cannot, perhaps, be given to everyone. Many have not intellectual abilities equal to it; but it can be given to a far larger number than receive it at present. In many cases it is given in the home, in others an attempt is made to give it in school. But if schools are to give it, a child must stay in them till he is old enough to understand principles, and this few can do till they are seventeen or eighteen. The amount of time that each adult gives to a child must be increased, because it is only by fairly close association between the adult and the child that traditions and ideas can be passed on fully. Such formative teaching can be done in class only with difficulty, and needs the employment of means other than those of traditional class teaching. Contacts of a more intimate kind are needed. Thus in the third section of this book we discuss the part that formal education has in training morals.

From all this it should be clear what relation we imagine to exist between ethics and religion. Religion roughly deals with the relationship of God to man, ethics with man's dealings with his fellow men. Some religions are not ethical at all—the ancient Roman was, for example, almost entirely ceremonial. Others contain a large ethical element, and devote much attention to man's behaviour. To this group belong the Jewish and the Christian. In these faiths the whole of a man's dealings with his fellow men is a part of his religious life, so that there is no ceremonial or ritual approach to God which has any validity if divorced from justice and mercy. Even the distinction between the "religious" and the secular life which arose in Christianity could be conceived in a form which recognized the latter as being just as fully the service of God. This is illustrated by the story of St. Richard, the Abbot of Verdun, at the beginning of the eleventh century, when the Emperor Henry I offered himself for membership at his Abbey. "Wilt thou then follow the rule" the Abbot asked, "and the example of our Lord, and be obedient in all things? Why then, I take thee for a monk. I will charge myself with the care of thy soul; and so I bid thee go back to govern wisely that empire which God has entrusted to thy hands".[1] No doubt they often lapsed back to an outlook which is essentially pagan in separating the parts of religion which prescribe man's worship of God from those which govern his behaviour to man. The anchorite in the desert was concerned solely with the relation between his soul and God, his dealings with men were reduced to a minimum and were regarded as hindering him in his attempts to gain sanctity. This attitude, though generally in a less extreme form, persisted for many hundreds of

[1] Kenneth Kirk, *The Vision of God*, p. 307.

years, and still is one of the elements that influence certain religious recluses.

In ethical conduct the sanction governing action is normally the desire for the earthly happiness and prosperity of the group. But this has often been reinforced by sanctions drawn from beliefs about life after death. The Christian religion looks beyond earth, and has often thought of this life mainly as a preparation for death. The sanction of the good life was then the fortune of the soul in its eternal existence. Mrs. Sherwood, with her faculty for putting a point better than any one else, sets out this view and the religious sanctions on which it is based. The following is part of the dialogue between the Fairchild children and Miss Augusta Noble. Augusta expresses wonder when she sees Emily and the others say grace before dinner.

"Oh, I forgot", said Miss Augusta, "your Mama is religious, and makes you do all these things. Don't you say your prayers four times every day?"

"Sometimes oftener", said Emily.

"Dear! How tiresome it must be to be so religious", said Miss Augusta, "and where's the use of it?"

"Why, don't you know", said Lucy, "that if we do not serve God we shall go to hell when we die, and if we do serve him, we shall go to Heaven".

This religion led in Emily's case to good moral behaviour, for she was a polite child and shut the door when the governess asked her. But the ultimate spring of conduct was not social. We must not of course fall into the error of making Mrs. Sherwood our final authority either on the upbringing of children or on the Christian religion. Her emphasis is no doubt a common one, but is far from representing the religious life of many sincere and orthodox Christians. "The central fact of religion is not survival but God," wrote a Catholic scholar. "To know God here is something and to have union with him here through our Lord, that would be enough without immortality."[1]

By contrast we may take Aristotle's conception of morality as the bond of the state, and the state as the means through which individuals are enabled to reach the highest way of life possible to man. Thus morality is sanctioned not by its effect in the hereafter but by its power to raise the standard of life here and now.

This does not mean that, in our scheme of practical ethics, religion plays no part. On the contrary, the teaching of such a religion as Christianity, with its very strong ethical element, is most important in the development of virtue. To some people ethics without religion

[1] Von Hügel. *Letters to a Niece*.

is unthinkable; they cannot imagine the two divorced, or see what authority ethical precepts would have without the support of religious ideas. With this we should not quite agree. Ethical principles have their own validity which arises from the very nature of man because man must live socially. But, for many people, ethical ideas are most easily apprehended through religion, and the two are often most conveniently taught together. Thus, later in the book, we have discussed religious teaching in the community and its part in formal education.

For most of us, free will is an essential element in religion and ethics. Religion stresses the responsibility of the individual soul and, without the free will to choose between good and evil, there can be no responsibility, no just punishments, no merited reward. In ethics, so far as there is a similar social responsibility, free will is necessary. In this book, when we suggest the cultivation of virtue by the proper treatment of children in youth, by the arrangement of society and by formal education, how far are we denying free will and setting up a system of determinism? We must, it seems, distinguish different elements in virtue. There is a part that seems to arise almost entirely from environment; just as the greater part of health can be produced by good food, fresh air and a sensible way of living. So a suitable environment can produce much of virtue. The stable emotional system, the mind free from hostility and anger, are results of an environment that has certain psychological characteristics. So, too, certain sets of ideas and beliefs must, almost inevitably, be accepted from the society in which a person lives. But just as the healthy need not be good athletes, or as bad acquired habits may ruin a good constitution, so there is much more involved in virtue than this habituation can explain. We find that in approximately the same environment individuals develop differently. What seems to produce virtue in one leads to vice in another. Among the most degraded surroundings truth, kindness and love will flourish; while in an otherwise virtuous family, one falls from grace. We can say that it is a happy temperament that keeps the one safe, and some weakness of character that destroys the other, but these are mere words. It would be as consonant with our knowledge to claim that divine care watched over one and that the other deliberately chose a path of sin. The facts are not in dispute, they lend themselves equally well to either interpretation. What matters is that in our hearts we *feel* free. It appears to us that we have choice, and that when we foresee our conduct we may adopt one course or the other with the experience of being our own masters.

In all moral action a large part is played by the ideals or ideas that we adopt. Our complex civilization offers so many patterns to us that each individual can select for himself those that he cares to take. A boy can choose to become a financier, with his mind set on riches, or a disinterested university chemist or scholar. These choices, which involve moral elements, seem as clearly a matter of free will as the choice between being a soldier or an engine-driver.

It is possible to argue that in all cases of apparent choice there are elements in the mind of the individual which, coming into contact with some elements in the environment, precipitate certain behaviour as by an almost chemical action. The appearance of choice is an illusion, and under the circumstances no alternative was possible. This psychological determinism can be neither proved nor disapproved: our knowledge is insufficient. All we can say is that it does not *feel* like that. We must thus in this book recognize the existence of two rather diverse things. The methods by which the basis of virtue can be produced and, secondly, the element of freedom and choice which the individual exercises in dealing with the environment. But since this choice is by definition free, we cannot set out rules by which it may be controlled. We shall therefore devote most of our space to the consideration of those conditions which seem to have a predictable effect. The book, then, falls into three sections, the psychological nature of man and his natural impulses; secondly, the elements in society that have most moral importance; and the last section will deal with formal education. In all sections we shall show that it is not mere facts, but what men make of them, that is important. We impress a pattern on the world, and the inventions of our own minds acquire the power to influence a whole civilization. Kant showed that we impose space and time on phenomena, science has demonstrated that things are never what they seem; modern psychology has taught us that we make our universe to a larger extent than Kant ever dreamed. If we ourselves are loving and orderly the world is the work of a benevolent God, if there is passion and disorder within, the principle of the universe is a devil in chaos. The myth, the pattern of emotional belief that we put upon the world, is a most important product of our moral life, and in its turn draws our moral life to its likeness.

It is hard not to feel that in this book we have taken all knowledge for our province. The subject is far too wide for one book, especially such a book as this, which attempts a general survey of the field in the simplest way. All sections are inevitably defective, and the deficiency is probably greatest in the second section, where the field is widest. Our excuse for writing this book is that, so far as we know, there is no other book on ethics of this kind; and when so many people are trying earnestly to understand how to reshape the world, we thought we might offer this small contribution to the solution of the problem.

Books referred to in this introduction:

Haering, *Ethics of the Christian Life.*
Brunner, *Divine Imperative.*
Isaac Watts, *Poems.*
Aristotle, *Ethics* (Welldon).
Whewell, *The Elements of Morality.*
Jeremy Bentham, *Principles of Morals and Legislation.*
W. McDougall, *Outlines of Psychology.*
Karl Mannheim, *Man and Society.*
D. T. W. Shropshire, *Primitive Marriage and European Law.*
Kenneth Kirk, *The Vision of God.*
Baron Friedrich von Hügel, *Letters to a Niece.*
Tennyson, *St. Simeon Stylites.*
William James, *The Varieties of Religious Experience.*
George Eliot, *Romola.*
Mrs. Sherwood, *The Fairchild Family.*

PART I

THE PSYCHOLOGICAL BASIS OF ETHICS

INTRODUCTION

THE first section of a book on practical ethics must inevitably be a psychological one. Till we have decided whether man's nature is inevitably wicked or whether it is capable of improvement, we can go no farther. If man is hopelessly and irremediably wicked there is nothing more we can do, except build up around him such a barrier of restraints and punishments that fear, if nothing else, keeps him on the right path. Man, on this view, is condemned either to break out into riot or to peer, tidy and helpless, from his confining walls. On the other hand, it may be that man's nature only needs training to become, if not perfect, at least a great deal better; and that we can safely allow him an ever-increasing measure of freedom in which to practise his virtues.

The first important point to realize in the study of man's nature is that it is very complex. This complexity is the cause of his moral struggles, and it is the reason why life presents so many problems to those who try seriously to understand it. It has, on the other hand, the advantage of enabling him to act effectively in a great number of different circumstances, and has given him his peculiar adaptability. It has enriched his life. For example, man shares his quality of gregariousness with many other animals, but man's gregariousness seems to include all animal types, from the mere desire of co-presence that we see in cows, to the co-operative hunting of wolves or the sociable chattering of monkeys. Nor is man simply gregarious. Many individuals have the quality only slightly, and are nearer the cat than the dog in their reaction to the world. Besides, important as gregariousness is in man, there are other impulses equally powerful. A bee, we must suppose, does not have to struggle against temptations which lead it astray from its duty to the hive: man must continually balance his desire for the welfare of himself and his family against the demands of the community.

The second point is that man's nature is very plastic; far more plastic than the nature of any other animal. Not only will men learn to live very different material lives, they will also learn to live very different spiritual ones. The insects represent a low level of adaptability. Bees will not modify their way of life, nor, presumably, their natures. A dog will learn to live either in a town or in the country, as a pet or as a working dog; and its character can be considerably modified by the treatment it receives. Man is still more variable, and the gap which separates one type of life from another, the best from the worst in any type, is far larger.

When all the variables are considered it is not strange that men are so different. The only limiting factor is the pattern of a man's culture. Any one of man's impulses may predominate, and lead to action that may be good or bad, and vary farther with the other powers a man possesses. Thus a man acting egoistically may be a black-market racketeer, or a virtuous politician who strives to satisfy his craving for power by a benevolent domination of others. These are two of the choices offered to an egoist by our culture. There are a great many others. In a different culture there might be a diverse choice, and a man might elect to achieve distinction as a magician or a head-hunter.

In consequence it is very difficult to say dogmatically what man's nature is. It is even harder to say if it is good or bad. All we are trying to do in this section is to point out, in very general terms, what are the main impulses that motivate man's behaviour, and to show how they are affected by his early experiences. It is these early experiences that first set the mark of good or ill on a man's character. They determine to a large extent whether he will be benevolent or the reverse, well disposed to society or its enemy. Through them he learns the particular forms of behaviour that are customary in his society, and from his experience he becomes either a conforming member of his group, or a rebel.

Chapter 1

BENEVOLENCE AND EGOISM

IF we consider the content of the moral mind we may distinguish three elements. There is a simple, unanalysable character that we can call benevolence. This consists partly of an expectation of good from our fellows, partly of an impulse to do good in our turn. These expectations and intentions are not formulated in the mind of the person, and are not matters of deliberate thought. Benevolence can be observed in others and felt in ourselves. It exists in animals. They show it to each other, and those that are domesticated show it also to man. In mankind it is one of the most easily distinguished characteristics in those with whom we have dealings. Without this characteristic, as we shall see later, other forms of morality are apt to turn to tyranny and oppression. The second element is the will to be good. This is a matter of reason and conscious thought. It arises at a much later stage than simple benevolence, but, for all that, it can be seen in the nursery. It appears as a voluntary conformity with the customs of the group in which the child finds himself, and it develops, as the child grows, into a desire to observe regulations as well as definitely to perform such acts as may be considered "good". The third element is the truly intellectual one, that concerned with the laws and customs of the particular society in which the man lives. The man knows what is done and what is enjoined. Some of this knowledge is acquired early, but much is learnt only as he grows to manhood, and moves through ever-widening fields of social life.

From the psychological standpoint of these early chapters it is the first two elements that are important, and it is in regard to them that modern understanding of children has made us so much more powerful than our ancestors. In general we now know how to produce in a child both benevolence and a will to conformity: and we can do it painlessly, if not positively pleasantly. At the same time we know how to prevent his developing the opposite vices: hostility, aggression, destructiveness, revolution.

Although we know this, we do not, of course, as yet always manage to use our knowledge. The child is not always in expert hands. Many parents have not learned the technique, either because no one has tried to teach them, or because, having been warped in their youth, they cannot now learn. In schools many teachers, working under bad conditions, fail to do for the children what they might; but, above all, society is so organized that frustration is inflicted in countless ways, and standards of value are taught that are contrary to the best elements in human nature.

Of simple benevolence we can say that it is merely the reflection in the child's mind of the environment. It is a characteristic producible in all animals that can be influenced by man. Cats, dogs, horses, even the larger carnivores, if caught young, can be made friendly and kindly. This is done simply by kind treatment. A cat brought up from a kitten without maltreatment does not scratch or bite, except by accident or under extreme provocation. It will even allow itself to be medically treated in quite a painful way without attacking the doctor. In the same way a dog, well treated from youth, is full of general benevolence, and will take active steps to please those with whom it comes in contact. An experienced breeder of racehorses will never allow a colt to be hit in any way, and the pretty shy creatures become docile and easy to train. In exactly the same way children take their general attitude to life from the treatment they receive. They could not learn it in any other way.

So easy is it to produce this characteristic that a child's experience need not be wholly good. If the good treatment extends over only one part of a child's life it will have its results in that part. A school that is well run and in which the children are consistently well treated has good, well-behaved children, whatever they may be at home. A school differently conducted may have quite unmanageable pupils. A change of method will, inside six months, change the one into the other. A child will even change in a few minutes on a change of environment, and from being a little devil at home become a pattern of virtue when he goes to tea with his grandmother three hundred yards away.

This apparently simple and obvious fact is really a modern discovery, or at least a discovery in so far as it is now consciously understood. Perhaps it would be truer to say a rediscovery, because there were educators of the later Middle Ages and Renaissance who practised such methods at least on the children of the aristocracy. It was the intervening ages of Puritanism and Calvinism that brought such misery to the child. When Shakespeare wrote of the primrose path that led to the everlasting bonfire he was expressing the belief that virtue was a thing of pain; and this belief showed itself in the management of life. The treatment of children was, particularly among the more religious parents, so harsh and unyielding that general benevolence had no chance to develop. The " good " man who had taken his character from his early training tended to be hard. He committed no sins himself and would pardon none in others. He would cast his erring daughter from the house, and burn her letters for the rest of his life. This hardness but reflected a religious conviction held up to horror by Robert Burns:

> Oh Thou, wha in the heavens dost dwell,
> Wha, as it pleases best thysel',
> Sends ane to Heaven and ten to Hell,
> A' for thy glory,
> And no for ony guid or ill
> They've done afore thee. . . .[1]

If God was like that, clearly man could be as capricious and unyielding. Quite simple kindliness would come as a surprise to a child reared in the harsh methods of a consciously virtuous household. In *The Way of all Flesh*, that brilliant indictment of Victorian morality, the point is well made. The narrator was out with Theobald and the family buying eggs from a cottage woman.

"A little boy, a lad much about Ernest's age, trod upon one of the eggs that was wrapped in paper and broke it.

" 'There now, Jack', said his mother, 'see what you've done. You've broken a nice egg and cost me a penny. Here, Emma', calling to her daughter, 'take the child away, there's a dear'.

"Emma came at once, and walked off with the youngster, taking him out of harm's way.

" 'Papa', said Ernest after we had left the house, 'Why didn't Mrs. Heaton whip Jack when he trod on the egg?'

"I was spiteful enough to give Theobald a grim smile which said as plainly as words could have done that I thought Ernest had hit him rather hard.

"Theobald coloured and looked angry. 'I dare say', he said quickly, 'that his mother will whip him now we are gone'.

"I was not going to have that, and said I did not believe it. And so the matter dropped, but Theobald did not forget it."

It was because so many people were cruel on principle, as part of their religious and social duty, that Tennyson could contrast the relative merits of kind hearts and coronets as if possession of the latter advantage completely precluded the former.

In countries where there is no tradition of unkindness and where children live under good psychological conditions, as in West Africa,

[1] "Holy Willie's Prayer."

and are hardly ever punished, the greatest friendliness prevails, and the adults are one of the most charming human types. Benevolence is, in essence, natural. It is not so much produced by training, it is allowed to develop by the absence of those conditions which make its development impossible; wherever it is permitted it will spring up. But for its fullest growth it needs positive affection and a sense of love and care.

This sense of being loved is, it seems, necessary for all satisfactory development, even physical. Charlotte Buhler records an experiment in which comparable groups of young children were treated with different degrees of affection, though the physical conditions were the same. The development of the two groups was measured at the end of six months and the petted group were found to be superior in every way.[1]

There has been a school of thought that has depreciated the showing of affection to children. This attitude is a not unnatural reaction from the excessive and hampering affection to which children were often subjected; and which by interfering with their actions and unduly stimulating their emotions had a very bad effect on their psychic make-up. The patients of many a psychological clinic have clearly suffered in this way in youth. But because there may be an excess, there need not necessarily be a defect in a mode of treatment; and for the development of true benevolence the child needs positive love, a little petting and the firm confidence that he will receive help in achieving his purposes, and will not have to fight for himself against a world that is in essence hostile.

All this means that, in the writers' view, man is benevolent if he is allowed to be. The more pessimistic and cynical writers do not share this view. Anatole France writes of the strongest forces "in human nature, pride and cupidity", and many another suggests that unbridled egoism, with all the attendant vices of greed and cruelty, is only held in check in each of us by fear and the pressure of a punitive system. We wish, therefore, in the rest of this section to show that man *need* not be bad, and that in fact he is generally *made* bad by mistreatment, either by his home environment or by the general arrangement of society. The first step in practical ethics is to understand man's nature and how it is acted upon by the conditions of life.

The first thing to realize about man's nature from this point of view is its composite character. There are roughly three sets of impulses that give the pattern of man's mind, and they are not consistent

[1] *Birth to Maturity*, p. 65.

one with the other. It is this inconsistency that makes it possible to produce such very diverse results by different kinds of training.

First, and earliest in development, are the egoistic impulses, those directed to securing the survival of the individual. Next appear the impulses of the group; and thirdly those connected with sex and parenthood that ensure the continuance of the race. It can be seen at once that egoism may easily be at variance with both group impulses and parental feeling, and if we are to believe the statement that "fathers of families are capable of anything", group and family loyalties may also conflict. When men are accused of being entirely egoistic, of being actuated solely by considerations of gain or suchlike, the speaker has forgotten the facts of human nature, or is speaking of a class or individual that has been vitiated by bad training. If we begin by considering egoism and its development we can see something of how this happens.

The tiny baby is, before all else, an egoist. He wishes, if he is a good baby, to live and thrive, and he is eager for food and anxious to avoid harm. His one weapon is the cry, and his powers are so limited that he must obtain all he wishes from others. He thus of necessity tries to control his environment, using the only means at his disposal. If he is hungry, wet, cold, tired, uncomfortable, he screams, and the success of his cries in bringing relief strengthens his impulse to use such means. Thus, unless the child is brought up on a regular system, he will begin to control his environment while still a baby, and will continue trying to do so as he grows older.

Adler would interpret this as a will to power based on a craving for security. It may be this, but it is also something simpler, the attempt, characteristic of all living creatures, to satisfy needs. In the child, as we have said, his helplessness makes it necessary for him also to control others if his wants are to be supplied.

This egoism of the baby is not absolute when once he has attained a certain age. From a few months old he makes contacts of quite a different kind with those about him, contacts that are essentially social and not egoistic in the narrower sense. A baby is sensitive to signs of emotion in those about him, and when well fed and comfortable likes to exchange smiles and inarticulate sounds of satisfaction with another. The curious toothless smile that the young baby gives in return for the adult greeting is very charming, and has no relation to any expected food or benefit.

Incipe, parve puer, risu cognoscere matrem

bids the poet, and most babies fulfil this first filial duty.

This emotional responsiveness appears in the child's imitation of the intonations of speech and his eagerness, when he is a year or so old, to include others in his activities and interests. A child sitting on its mother's lap and given a lump of sugar will offer it all round the table for every one to make movements of sharing before eating it himself. Or another, having listened to a ticking watch, will hold it up to his mother so that she too may enjoy the strange sound. Two-year-old babies put together in a cot will initiate friendly contacts, and show clearly that they are conscious of each other, and to some extent of each other's emotions.

Charlotte Buhler in her studies of the very young makes these points clearly.[1]

It is of the greatest psychological significance that the infant's first reaction to other human beings is a definitely positive one. Throughout the first year of life positive social behaviour predominates unless the environment is very unfavourable.

Differentiation of these positive social responses begins at a very early age. At five months the child becomes socially active, i.e. seeks contacts spontaneously with those who approach him, both by making sounds (babbling) and by physical contact (grasping and touching). At six and seven months the infant tries eagerly to include any one who is present in its play. Objects are given and received from others and the child enjoys especially those games that include a partner, peep-bo, exchange of toys. The partner's gestures are observed at the latest at the fifth month, and at eight months there is an astonishing capacity for interpreting and understanding them. Children of this age have been observed trying to comfort a frightened or crying child.

We have thus almost from the beginning of mental life two strands even in egoism, the claim to what he wants, and the consciousness of others and a desire for community of experience. This dualism continues to develop as the child grows older.

At first when a child attains more power and can walk and take what he wants his cruder egoism may seem to increase. He may pursue his desires without regard for others; the toy he wants, the attention of the adult, may be claimed and taken from another without apparent regard for the feelings of the attacked. This is true, but yet it is only a part of the child's behaviour. In what appears to have been, on the whole, a rather disorderly group with comparatively little adult help in social development, Anna Freud records instances of altruism and feeling for others.

[1] *From Birth to Maturity*, p. 55.

"Nurse Jean fetched Bridget (2 years) from the shelter dormitory. Since the dressing room upstairs was already full of children she only took that one child. When passing the other beds Bridget heard Jeffrey (2 years) cry. She stopped and said, 'Jeffrey crying, Jean'. The nurse explained that Jeffrey would have to wait a little, and proceeded to take Bridget upstairs. Suddenly on the middle of the staircase Bridget turned round and said, 'I go to Jeffrey', and went back. The nurse waited for her to return, and then followed to see what had happened. She found Bridget had opened the net of Jeffrey's bed so that he could get out, and had pushed the step-ladder to Bill's (2 years 9 months) bed to let him get out. She was just about to push the steps to Dan's (2 years 8 months) bed. She was holding his hand and saying, 'Not fall down'."

This is very advanced social conduct for that age. More common are the passing moments of shared emotion when one child comforts another's tears.

"Rose (21 months) watched Edith (22 months) petting Sam (22 months) who was crying. She went to Sam and petted him too, then went to Edith and Freda (22 months) and petted them, and finally she stroked her own hair and cheek, and, with a radiant smile, made an affectionate noise to herself." The happy end of social benevolence.

It is at this stage that social pressure first becomes consciously effective. The very small child was, of course, influenced by it. Regular hours for feeding, regular times when he was petted and talked to, conditions so stable that his whole organism became attuned to them, made him a conforming member of society instead of a tyrant. But at two and three years old the child can understand speech, can consciously share other people's feelings, and can be directly influenced, instead of being merely "conditioned".

When man is accused of egoism it is to teaching given at about this period that we must look for the decisive influence. He is taught partly by precept and partly by example. In the chapter on Socialization we shall discuss how the adult teaches the child social behaviour. It is enough here to suggest how he, perhaps unwittingly, provides the child with an egoistic outlook. The essence of egoistic behaviour is that the person demands preferential treatment for himself, and is insensitive to the emotions and needs of others. This preferential treatment may consist in having a greater share of goods than others, or receiving more service or affection or notice, or being excused the performance of certain tasks. In any case the egoist believes himself in some way different and to be preferred to others. This attitude is taught. "My little boy is not allowed to do this or that," because he is precious and superior. "My child must have this or that toy." As the mother of a far from satisfactory young man once remarked, "Of course, we never refused him *anything*".

At the same time while the adult claims much for her child, she is apt to claim even more for herself. It is frequently on the grounds of health that the mother claims special exemption from work or effort. Her "head" or her "back" or some other member is the reason why the child cannot make a noise or must forego some natural activity. Thus the child has both the pattern of egoism to copy, and has to develop an egoism as great or greater than his parent's if he does not want to be altogether submerged. There is the added danger for the egoist, that, as a spoilt child, he has an underlying sense of essential inferiority. He has things made easy for him, he is waited on, he is deferred to, and he very much doubts if he could compete with others on an equal footing. As Bernard Shaw's Duke remarks in *On the Rocks:*

"I am so accustomed, as a Duke, to be treated with the utmost deference, that I simply do not know how to assert myself and bully people. I'm so horribly hard up for pocket money, without knowing how to do without it, that I have lost all my self-respect."

It is this feeling of helplessness, quite as much as any other motive, that makes the privileged cling so closely to their privileges, and experience such acute terror at the thought of losing them.

This egoism is most generally bound up with the social institution of property. Children in our civilization are early taught to expect to own things individually and exclusively. They are taught that ownership confers the right to have a thing whenever they want it, and, if necessary, to take it away from another. They are given things of *their own* with suitable ceremony, and their parents have *their own* possessions that the children are supposed to respect. One of the common ways of making a child respect the possessions of others is to let him have things of his own, and ask him how he would feel if *those* were taken and destroyed. We thus build up in the child's mind a very strong system of ideas which is added to by the social prestige that attaches to property. From their parents children learn to attach prestige value to possessions. It may be that certain objects carry this in a higher degree than others: a silver teapot, a grand piano, a Persian rug, or merely, at a lower income level, a hat with a feather, or trousers with no holes in them. Children take over these ideas from their parents, and invent for themselves prestige-carrying possessions such as collections of cigarette cards, tram tickets, or postage stamps.

For most adults the prestige value of an article depends on its expensiveness; much less often on its intrinsic beauty. Things may be expensive because of their rarity, or because of the quantity of work expended in making them. Of the first class are radium or large flawless jewel stones; of the second, the elaborate articles characteristic of Chinese art: carved jade or huge jars of cloisonné enamel. The production of such articles as these, demanding an almost incredible amount of human labour, indicates power in the possessor, the power to command such labour; just as the possession of fine jewels indicates the power to obtain what is rare.

To some extent old, almost magical ideas enter into such appreciation. Gold is a substance with a magic quality; a quality which platinum, in spite of its higher price, has never possessed. The sparkle of diamonds has a prestige value greater than that of many other stones, and fur has a traditional value in excess of either its beauty or utility. In the main, however, the principle holds, and the association between wealth, prestige and privilege is established. The wealthy can develop a form of egoism that is far beyond anything possible without the particular social order. It is enough to be able to pay, and all else will be added. There is a little sketch by Arnold Zweig called "Kong at the Seaside"[1] which describes how a rich little girl tried to buy a dog which had spoilt her sand-castle so that she might shoot it; and how baffled she was by the small boy's refusal to sell his pet. The story is slight enough, but the state of mind of the very rich, who have come to believe that they have the *right* to the gratification of every whim, is a social satire of great importance. It is interesting that this story is by a German. It could never have been written in an English setting. Our more definitely ordered society, which for all the jibes flung at it is less of a plutocracy than many others, hardly admits the development of just that type.

Yet even amongst us the prestige of possessions is sufficient to make their ownership a normal part of our life and desires. To have them makes us feel more confident, more at home in the world; we realize our personality through them. To some extent we all become competitors in a field where it is scarcity that makes the interest of the game.

It is often easier to see our own customs if we compare them with those of simpler societies. In Margaret Mead's account of New Guinea we can see how an acute sense of property and all its accompanying egoism can be taught—even at the late age of some seventeen or eighteen years. The adult people of New Guinea are acutely property-conscious. The whole amusement of their social system lies in the elaborate system of exchange and gifts that is continually taking

[1] In *Playthings of Time*.

place. They teach their children very early never to touch anything, and to show the most exaggerated respect for property; but they give the children nothing of their own, apart from the little canoes and play-spears that they need for learning skills. Apart from these they have only the toys they make of shells and flowers, and they are completely excluded from the adult pre-occupation with property. It is only when they have grown up that the young men and women become part of the property system and have to adopt its standards.

"All about him the young man sees two types of older man, those who have mastered the economic system, become independent of their financial backers, gone into the gift exchange themselves, and those who have slumped and are still dependent nonentities, tyrannized over by their brothers, forced to fish nightly to keep their families in food. Those who have succeeded have done so by hard dealing, close-fisted methods, stinginess, saving, ruthlessness. If he would be like them, he must give up the good-natured ways of his boyhood. Sharing with one's friends does not go with being a financial success. So the independence of his youth goes down before the shame of poverty, the generous habits of his youth are repressed in order that his independence may one day be regained. . . ."[1]

The importance of this extract lies in the contrast it draws between the unpropertied generous child and the close-fisted contriving adult, and the extent to which the later characteristics are taught—though doubtless some are more apt pupils than others.

To make this point even more clear we can take an instance where the property standard does not exist in certain directions. Freya Stark comments on the dignity and independence of the nomad Arabs she knew so well.

"I have often wondered what gives the nomad Arab that superb and arrogant poise. . . . It is, I believe, his feeling of equality. Not American equality which depends on equal justice, equal opportunity, equal motor cars, frigidaires, permanent waves, equal comforts and movies. What he feels is something independent of possessions or even of rights: it is an innate quality of human beings, implicit in their subservience to God. If he wants a radio set, or a pen, or an education, he wants them for themselves, and not with the idea of being somebody else's equal: that he is already by the unalienable right of manhood. This gives him self-consciousness, his dignity, his easy courtesy and careless walk, whatever his contrast in rags may be with others in kingly garments".[2]

[1] *Growing up in New Guinea.*

[2] *East is West,* p. 16.

Here is the adult who has never come to regard the possession of property as his claim to social prestige. He would not understand the shame that a man, in a property-determined society, feels in being dressed in rags; or in having, as the law says, "no visible means of support". He has not learnt that particular set of ideas.

In certain Jewish agricultural settlements in Palestine there is a strict communism and no private property. The communism may be so extreme that clothing coming back from the wash is redistributed each week, and no one owns any particular garment. There is no family life as we understand it. The husband and wife share a room, but the children are brought up from infancy in the Children's House where they are cared for together and visited by their parents on holidays or after the day's work. The children, like their parents, have no individual possessions, and are taught to work for the community, not for any individual aim. It is very noticeable how easily the children accept these ideas, how exactly they observe communism even in circumstances where a breach of it might seem natural, e.g. when a child receives an individual gift of sweets, and how attached they seem to be to the system when they grow up. In most settlements it is customary to send the young people away for a year at eighteen or nineteen to see how they like the individualistic capitalistic system that prevails round about them. They almost all return to the settlements and profess a vivid preference for the communal way of life.[1]

The purpose of this discussion is to show that the form of egoism that shows itself in the acquisition of property is the result of teaching by the group, rather than of any innate impulse, and that with suitable training it need not be part, at least not a large or dangerous part, of human nature.

If we consider the purposes served by property, apart from its prestige value, it is simply that we should be able to satisfy our needs, and have things to use when we want them. When working it is exasperating to have to wait for one's turn with a tool, and it is painful not to get enough to eat. Among the characteristics of Western civilization is this, that it has always been a deficiency civilization, in which a portion of the population have never had enough. In this it differs from some, e.g. South Sea cultures where there is always enough of the essentials for all. In our European civilization, and the version of it that has travelled to America, a number of people have always been hungry, under-clad, insufficiently housed, cold, uneducated. There has always been, therefore, a competition for the

[1] Infield, *Co-operative Living in Palestine.*

fundamentals of life, and the fierce egoism of need has been a permanent part of the social pattern. The inevitable struggle of the lowest affected those above them. Struggle and fear were abroad in society. We must work, we must save, if we are not careful we shall fall into the abyss. This attitude has never been stronger in England than during the early nineteenth century, when the tide of human misery reached, perhaps, its highest point. Sections of the community which have suffered most from this fear, like the Jews, have developed the most resolute striving towards material success. In countries where sufficiency exists there is either comparatively little interest in personal property, or a fictitious standard of purely prestige values has been invented.

It is possible that we are approaching a time when even in Europe there will be enough for all, and it is possible that the urgency of egoistic competition will decrease. Even at present, groups exist in the community where such urgency is less than in the average of the population. Such a group is the academic one. University teachers are paid moderately, but in most cases sufficiently. Among them possessions have comparatively little prestige value, and are desired mainly for their real utility or beauty. There is not much internal competition, far less than, say, in the Civil Service, and none of the financial urgency that characterizes some other groups. At the same time the members are all highly skilled in their own way, and have no great fear of being helpless in any change of society. They are thus somewhat inclined to Labour and Communism in politics. Their children also grow up with a much less acute sense of property than others; and find no difficulty in adjusting themselves to the partial communism that often exists among familes, particularly in war-time, over such things as party frocks.

Under periods of acute strain and scarcity, as in some prison camps, egoism often developed in its most unpleasant forms; in many cases it did not. On the liberation of camps such as Belsen it was regularly noted that the children were in better condition than the adults, and this could only have been possible through adult self-sacrifice.

As so often happens with social arrangements this type of egoism has been given a philosophic and economic justification. Personal acquisition was, during the nineteenth century, made into one of the pillars of the State. Private property was national prosperity. The man who concentrated wealth in his own hands, had, inevitably, a large surplus above his needs, especially if he also believed the doctrines of prudence and economy that were so great a feature of the mercantile thought of the day. He thus had considerable capital to be invested, and the rapid expansion of industry and the installation of machinery provided continual employment for his capital. Thus, unless the State had men of concentrated wealth, how could industrialization take place and Britain grow great? Had this money been frittered away in giving each of the workmen enough to live on comfortably, there would have been no progress, and where would the poor have been then? Poverty on the one hand, and large personal acquisitions on the other, were the basic conditions of national well-being.

If this was the doctrine of the theorist and the practical man alike, it naturally affected the minds of even the best men, people who were really anxious to do their duty. To-day men of this type are influenced by a different theory of economics and have come to place individual acquisition less high in the scale of virtues. Moreover the enormous scale of modern industrialization renders individual accumulation less important. We may be approaching an age when purely economic considerations will bring about a great change in the egoistic attitudes of mankind.

A very different type of egoism arises from the natural inequality of man. It is impossible that those who are stronger and cleverer than their fellows should not know their superiority and put it to use. It is indeed necessary that they should. Those who despair of man's nature point to this inequality and declare that the strong will always despoil the weak, the clever the stupid, and suggest that inevitably men behave worse than pigs at a trough. Such a view seems singularly lacking in reality. Any society, even such as we live in to-day, would be impossible if there was nothing but a scramble for the best, and if brutality and chicanery were restrained by nothing but fear. On the contrary, we see clearly every day, among children and adults, that the strong protect the weak, and the clever devise means by which the stupid may live securely and well.

The fact is that this particular human characteristic of dominance is one that is generally socialized, and so far from leading to the break-up of society, provides its strongest bond. Every group requires a leader, and this leader must inevitably come from among the better endowed. This superior endowment is often associated with the character of dominance which belongs not only to men but to all social animals of a certain type. In man leadership is a quality apparently innate in certain individuals, and must find expression as other qualities do. Charlotte Buhler has noticed it in an infant of ten months, and at two or three years old the potential leader can be selected with perfect ease. When this dominant nature is associated with natural superiority there is very little difficulty in socializing the impulse. Even quite little children give way to the natural leader who is clever and strong as well as dominant. The task of society is merely to induce him to lead in the right direction. For reasons we shall discuss later this is surprisingly easy—if the general conditions, physical and psychological, are suitable.

It is, moreover, essential for the success of the group that this socialization should take place. No society can exist in which the ablest leaders are not allowed to lead legitimately, but are either destroyed or driven into opposition. Thus one of the first duties of the educator, acting for society, is to find ways of obtaining the service of these leaders.

In this respect English arrangements are superior to those in some countries and are comparatively successful. We can see how far from being a disruptive force this dominance and natural superiority is when we consider our army officers, our politicians, our teachers, even our leaders of industry. We know, not perhaps perfectly but fairly well, how to turn these potentially dangerous elements into the servants of the state and their less well endowed brethren.

A more difficult, but much less important, problem arises with the dominants who are not well endowed physically or mentally. They can never hold in society the place they desire, and must be content with a small sphere. They cannot dominate, by their direct personal superiority, even their own children, and become naggers, incessantly concerned with trifles, for ever dragging out moral clichés to justify their demands. They are in general among the worst of parents, and think themselves the best. Rationing has been a godsend to them, and in the exact administration of ounces of margarine they have been able to control their families.

Not only need egoism not be an inevitable source of vice in man, it is possible that some forms of moral excellence have at least part of their roots in a kind of ideal egoism. The characteristic which is called self-respect arises from a strong ideal of the self and takes its outer trappings from possessions and position. As we have said, these can in some cases be unimportant. What is essential is our belief in our inner qualities. Whatever it is we pride ourselves upon—our skill, our patriotism, our charm—to that we look in our estimate of ourselves, and that we strive to preserve. There is the egoism of the artist who will not allow the needs of his family to interfere with his art, of the saint who will not attend to necessary business, or the patriot who will perish for his cause. Without reaching such heights the ordinary man respects his own individuality and will not allow it to be invaded. He is an egoist in this ideal field.

Egoism is thus not the root from which all vices inevitably spring. It may be good or bad according as it is developed and trained. Nor does it hold sway over man's nature. Benevolence is in most cases stronger, and with proper treatment, can so transform egoism that it is only left as a competition in well doing.

Books referred to in this chapter:

Samuel Butler, *The Way of all Flesh.*
Tennyson, *Lady Clara Vere de Vere.*
Charlotte Buhler, *From Birth to Maturity.*
Anatole France, *Penguin Island.*
Adolph Adler, *Understanding Human Nature.*
Anna Freud, *Children Without Families.*
G. B. Shaw, *On the Rocks.*
Arnold Zweig, *Playthings of Time.*
Freya Stark, *East is West.*
Infield, *Co-operative Living in Palestine.*

Chapter 2

GROUP AND SEX

IN the last chapter we discussed the production in men's minds of benevolence, and argued that egoism is not inevitably evil. The impulses to protect oneself, necessary if the individual is to survive, can be so moulded by proper treatment that they are harmless or even beneficial. In this chapter we shall show that in much of man's life his actions are directed to the welfare of others and are positively altruistic. The two great biological impulses that are directed outwards are those concerned with sex and parenthood, and that which leads to group feeling and activity. In both cases the impulses are intended to secure the welfare of those on whom they are directed. The maintenance of the group necessitates a certain type of behaviour to the other members, such behaviour that they remain members and do not break the association apart. The effect of sex emotion, in all but the very small group of the perverted, is to lead to the cherishing of the partner, and parenthood is directed to the well-being of the child.

Those who find man's nature essentially evil and talk as if egoism of the most unpleasant kind were the core of his being, should reflect that in times of crisis both these altruistic impulses take precedence over egoism. Horace assures us that it is sweet and fitting to die for one's country, and though after two wars no one really believes him, men still die and sacrifice themselves for their comrades. The devotion of husband and wife and parent and child stands for most men far beyond the claims of individual safety, and most parents forgo the minor indulgences that are a daily craving because they feel the money could be better spent on their children.

There is thus in man's nature a stronger impulse towards altruism than there is to egoistic behaviour. This is not to say that all men are the same. In each individual these innate impulses vary in strength, absolutely and relatively, and from these variations very different characters emerge. The predominantly egoistic, even when he has been well socialized, will never have the gentleness and altruistic thought of the essentially paternal type. He may use his power unsparingly for the public good, but the effect on those who come into contact with him will be very different. Combinations of types also give characteristic modes of behaviour. The egoistic paternal, with its attempt to secure its own glorification through its children, is perhaps the most unpleasant; while the "complete family man", uxorious and humbly devoted to his children, lends himself most easily to amiable satire. These different types find various ways of life pleasant. The gregarious like occupations that keep them in close touch with their fellow men. Other characteristics being suitable, they are happy in the army or on the staff of a school. The markedly deficient in gregariousness make independent enquirers, if they are sufficiently able, or good lighthouse-keepers if their powers are differently directed.

Sex has always attracted much attention, especially in our type of civilization. It is for most people the source of the keenest gratification, and contemplation of it, in one form or another, provides pleasant excitement. Each culture has its own definite sexual customs, and usually attaches the greatest importance to their observance, believing that the Deity is deeply concerned that they should be obeyed. Thus for a certain type of moralist "sin," without qualification, means failure to observe the sexual pattern approved in his community. More recently psycho-analysts, licking their lips, have hunted the aberrations of the libido down all the nights and days from birth to the grave. From the point of view of this book, these variations from the normal are of comparatively little importance. What we desire is to see what part sexual and parental feelings play in the moral life, and how social arrangements affect their moral character.

It is impossible to say what type of sexual behaviour is *natural* to man. In every community sexual behaviour is regulated and controlled by custom. We can only say that, as in vigorous stocks the number of females considerably exceeds the males, some form of polygamy suggests itself as the reasonable arrangement; and, further, that the highly emotional monogamous marriage characteristic of modern Western civilization is a fairly recent development, and still confined to a comparatively small section of the world's population. We can only, therefore, accept the custom of each group as the standard for its members, and we shall discuss the advantages and disadvantages of our own type.

In a later chapter on frustration we shall show some of the psychological difficulties that arise for spouse and child from our modern arrangements. It is a matter of the utmost importance because the family is the first group of which a child is a member, and his attitude to so many matters is coloured by the treatment he receives, and the ideas he is taught in this group. If the sex and parental instincts waited till maturity before they developed, the pattern of family life learned in childhood would not be so significant, but in fact these impulses appear in very small children, and affect their minds long before they appear in full-formed activity.

For instance, even before children begin to play with dolls they are interested in a baby brother or a kitten, and there is a considerable element of the parental in their behaviour. In the same way, very small children of eighteen months or so will be sexually excited by children of the same age but opposite sex, and will display this excitement in behaviour so characteristic that it might be modelled on a performance in a silent film. By five or so they possess adored boy or girl friends and treat them with loving concern. A child of this age caring for his friend or showing off his merits gives human nature a very pleasant aspect.

If parents choose to stimulate it, there is also sexually toned feeling towards adults. We learn most about it from the records of cases where disastrous consequences have followed; and in the greater number of these records it is clear that it is the parents or other adults who are to blame by forcing their disordered emotions on children who cannot safely endure them. The sensible parent receives a flirtatious glance or giggle from a two-year-old. For the rest he is an object of slightly detached friendliness which warms, with time, into a tie of respect and affection, tinged, as the years go on, with a paternal quality which is the reversal that age brings.

In animals we see a very clear type of family life, a cat with kittens, a bitch with her puppy or a flock of cygnets escorted fore and aft by their parents. If one may judge by appearances, animal parenthood, though strenuous, must be a source of calm pleasure and happy co-operation. A pair of swallows in trying to get their fledglings to fly are using encouragement, not punishment, and the perfect discipline that swans maintain appears to meet with no opposition from the cygnets. Among men things seem less easy. It is possible that some of man's difficulties are inevitable, being due to his greater variability and his more complicated environment, but many undoubtedly arise from the overcharge of emotion that is now customary, and also from the association that has grown up between property and the family. The overcharge of emotion arises, as we shall show later, from our type of romantic monogamy and from the institution of the small family. The idea of ownership is transferred from inanimate things to wives and children, and in the case of children it is sanctioned by law. Thus all the property-conscious possessive attitudes are directed to members of the family. Moreover the family becomes a vehicle for the display of property. Men like to see their wives wearing jewels as indications of wealth and power, and mothers assert their riches by the way they dress their children or the schools they send them to, and their own might by domineering over them and displaying their accomplishments. Thus the family relationship, instead of being as simple as that of a cat to her kittens, becomes affected by the whole social complex, and the essential love and benevolence is lost among the greed and urgency of private property. If children are taught these attitudes by the close contagion of the family group, they will carry them on into their adult behaviour.

But whatever the dangers, the family and its characteristic emotions are the pleasantest part of human life.

> We thank Thee, Lord, that Thou hast made
> Love to abound,
> So many gentle thoughts and deeds
> Circling us round,
> That in the darkest spot of earth
> Some joy is found.

For the child and for the adult the successful family means safety, care and the tender interchange of the pleasantest emotions. Without it there is a hostile world that must be continually contended with, and in which loving thoughts have little share. The importance of this "ingroup"—as the Americans now call it—cannot be over-estimated. It is probably the chief necessity for the development of benevolence. Benevolence can only grow up when a person feels safe and when he is surrounded by kindly emotion. Nowhere does this happen so fully as in a family. In those races which we shall mention presently where a particularly humane culture has developed, the family plays a large part. It is generally the extended family of aunts and cousins, and in it the child grows up surrounded by a forest of kind friends. So too in England, the large Victorian family, with relatives who were always coming to stay, and aunts and uncles who provided tips and treats, allowed the child to grow up with a great feeling of security and well-being. It is possible that the supreme horror of *King Lear* arises from the fact that the family has failed him, and that the basis of all trust has crumbled.

From the family, benevolence and kindliness radiate outwards. William McDougall has claimed that the parental instinct is the basis of all moral virtue. This is to ignore a great part of human nature, but it is certain that family affection and the sexual instinct provide the strongest elements in developed benevolence, and direct that benevolence into characteristic channels. Just as in animals parental impulses lead, say, a cat to mother the young of other animals, so in man the benevolence generated in one's own family spreads to other fields. Children in general, the small, the weak, all become the recipients of this extended kindliness. Women as a sex acquire some of the characteristics of a man's own wife. The history of social legislation shows how strong is this protective impulse.

The emotion accompanying both sexual and parental activity is called love. The use of this one word, though the emotions it denotes are so different, indicates that there is something common to them all. From the passionate embraces of youth to the placid affection of old age is a long journey; the parental emotions change also with the growth of the child. What unites them, as well as the love of David and Jonathan, is a quality of tenderness, of care, and a gladness in the heart.

From the family the individual child learns the pattern of life. He sees before him a path leading to marriage, some such marriage as his parents', to his own parenthood, to old age. He sees his grandparents, and learns about different forms of work from his relatives. If he sees the generations bound together in amity, if he learns how to behave and what to expect, he has received the most valuable moral training in the simplest and most natural way. These ideas are impressed on him by the strongest form of group contagion, and it is difficult for any child to resist the ideas that come to him from his immediate family. The moral character, therefore, of the family is transmitted to the children and has the greatest effect on development.

In the family we see the close manner in which the impulses of parenthood, sex and group co-operate. The child from a very early age feels the family as a group and wishes to remain a member of it in emotional sympathy with the others. He is afraid of isolation. He is also unwilling to inflict any harm on the other members. He desires that they should be well and happy. He values them absolutely for themselves, and he also shrinks from the sympathetic pain that their suffering would inflict on him. A little child will offer consolation to his mother if she hurts herself, just as he desires to receive it from her if he falls. Quarrelling between parents inflicts the cruellest suffering on the child, because he feels his group falling apart, and cannot achieve a synthesis.

The group instincts play a much more complicated part in moral development. They can, like all human potentialities, be put to good or bad use, and they are closely connected with the family, because it is in that group that a child is first moulded. Like the sexual impulses they appear very young, long before they are fully developed. There is no doubt that children feel towards the family as a *group*, and desire the experience of having an adult who is in charge of a definite number of children and whom they can feel is their own.

In Anna Freud's nursery the experiment was tried of putting a nurse in charge of a special group of children, and letting the children, aged two or three, regard the nurse as "our Jean". The children had, it is true, already inbibed the current belief that a child belongs especially to one person and that there is a special reciprocal relationship, but, even so, the eagerness with which the children appropriated their "own" nurses, declared that they were the best and resented other children outside the appointed group sharing in their ministrations, indicates how important this tie appeared to them.

Inside the family or pseudo-family group we find the characteristics of group emotion. There is in the first place community of emotion. The child experiences the emotions shown by those about him. He is eager to promote group happiness and to enjoy the pleasures of shared satisfaction. Closely following from this is the acceptance of the ideas of the group, and the identification of himself with its members, so that he is prepared to defend their ideas as his own and to derive his self-importance from the group. Quite a small child shows this, asserting that "Daddy says so", or that his father is rich or has other desirable characteristics. The child also becomes conscious of group purposes and likes to identify himself with them: "helping" to sweep or clean, scrub the floor or lay the table for dinner. He is anxious to be included in the life of the group and attaches great importance to his effective membership. All this takes place long before the child has developed the power to form a group of his own without adult direction.

The group impulses are so strong that severance from the group is, perhaps, one of the most serious punishments. This is one of the reasons why the individual seeks to please the group and why he is so extremely sensitive to its opinions. In that textbook of past practice, *The Fairchild Family*, we can read how an outraged parent fell back on this exclusion when other means of coercion failed. We have quoted in a later chapter the full story of little Henry's refusal to learn Latin and of his obstinacy under the whip. His determination collapsed when he was cut off from all family intercourse. The rigid mother of the last generation, who continued to "show displeasure" several hours after a fault had been committed, was carrying on the tradition.

If we consider primitive ways of life, or the tendencies of children, we see that group activities are really the normal. Work as a group, play as a group, are on the whole much preferred to individual labour or amusement. The child always seeks the companionship of his contemporaries, or, if he be younger, the vaguely protective eye of an adult, and then plays or works in this social setting. It is the same with most adults; solitary work is unpopular and puts far more strain on a man. The membership of a gang or group or organization that can be felt is one of the first considerations of a satisfactory occupation. The housewife, working alone in the house, longs to get out to the sociability of the café or the queue; and the complaint of the occupant of the ribbon-built suburb is that it offers so few opportunities of friendly intercourse. In the wider field of society the group is of the utmost importance. The vast group represented by the state is far too big and its purposes too incomprehensible. We need a variety of groups small enough for us to feel noticeable members of them, and simple enough for us to understand. The child passes from the home group to the school, and within the school smaller groups are organized, forms or teams or societies. The adult has his place of work, his church, his club, and once more his family. If the organization of these groups fails, the child or youth sets up his own groups, often of a somewhat anti-social nature. One American writer, Thrasher, has given these reasons for the existence of gangs:

"Gangs represent the spontaneous efforts of boys to create a society for themselves where conditions more adequate for their needs exist. What boys get out of such an association that they do not get otherwise, under the conditions that adult society imposes, is the thrill and zest of participation in common interests, more especially in corporate action, in hunting, capture, conflict, flight and escape. Conflict with other groups and the world about them furnishes the occasion for many of their exciting group activities.

"The failure of the normally directing and controlling customs and institutions to function effectively in the boys' experience is indicated by the disintegration of family life, inefficiency of schools, formality and externality of religion, corruption and indifference of local politics, low wages and monotony in occupational activities, unemployment and lack of opportunity for recreation.

"The gang functions with reference to these conditions in two ways. It offers a substitute for what society fails to give, and it provides a relief from repression and distasteful behaviour. It fills a gap and it affords an escape."[1]

A sound society must provide abundantly for the group impulses.

[1] *The Gang*, p. 37.

One might think, from the activities of children, that group activity was always play activity and often anti-social. On the contrary, the natural group has purposes co-extensive with life, and children are never happier than when they are meeting real needs in the group way. A party of adolescents camping derives vast satisfaction from cooking, and more still from hunting or fishing for the pot. They like fetching water and conducting the business of life. Even at home children who are taken through domestic duties as a group do not find them irksome. It is "fun" done all together.

Among primitive peoples any large enterprise beyond the power of an individual is accomplished by group activity. A new canoe or a new house is built by communal effort with singing and presents of food from the eventual owner. There is nothing that corresponds to our payment for work done. The workers are fed, and the chief craftsman may receive presents, but the essence of the activity is that the group freely undertakes the task of providing for the needs of its members.

In our money-ridden society such group activities are looked at very doubtfully, except in war-time. The "voluntary worker" has acquired a bad reputation for fussiness, incompetence and unreliability, because, in the ordinary way, any one possessing desirable powers prefers to sell them. War, calling for the maximum effort from the whole community, has reinstated the principle of voluntary group work, and in very many cases it has brought deep pleasure. A.R.P. and Home Guard, in spite of strain and the anxieties of war, were a source of much satisfaction to many, and their members disbanded with real regret. The W.V.S. seems to intend continuing its benevolent activities. Some religious bodies have always maintained a high standard of unpaid social work.

When the group is well established it claims from its members a degree of service and loyalty that is striking from the ethical point of view. Work that a man would never do for himself, he does for the group. Very few women will cook if they live alone, nearly all will do it with satisfaction if there is a group to feed. A teacher will work for his school, a secretary for his club, or a politican for his state far harder than they would do for their own interests or than the circumstances really demand. This is important when we consider later the motives that can be used to encourage work and effort. When we pass to circumstances of danger the list of V.C.s and G.C.s shows that some men are careless of their own lives and constant under torture for the sake of the group to which they belong.

How extremely varied the activities of a group can be, even under our conditions, is shown in the account of one gang that Thrasher gives in sone detail. The story has some extraordinary features from the English point of view, especially the successful intervention in school politics, but in a land of intrigue and pressure groups it is not really surprising. Moreover, the continuance of the group into adult life may seem strange, but American men and women maintain the gang all their lives in a way that is foreign to British custom.

"About a dozen boys in a three-block area who had played together since childhood took the name of the Tri-street Athletic Club. A clubroom in a barn was furnished with a stove, cots, pennants and other equipment. The usual fights took place with other gangs, but at worst we were friendly enemies and there was never bitter enmity.

"Our activities included football, basket ball, and baseball in season, and we scheduled many games. In the winter there were bobsled parties ending with oyster stew at the home of one of the members, girls were usually included on these occasions. The parents knew of the group activities and some of them took an active interest in us. The father of one boy always brought back a possum from his hunt and put on a big dinner for the gang. The gang got out a weekly school paper, censored by the principal of the grade school. Although published for the whole school, it bore the initials of the gang, T.A.C., which had now become secret symbols and were also worn as a monogram on our sweaters.

"The T.A.C. existed for about three years. When most of the gang entered high school a change in the organization soon took place. The new principal tried to shut out all activities and make the high school a routine. The opposition of the gang was intense, but they dared not oppose him openly."[1]

The gang thus became a secret society determined to "get something on" the principal. They were so successful that he had to resign.

"When the new principal arrived we took him into our confidence and came out into the open. The gang now became conventionalized as the 'Forum' and took up the role of debating club and school boosting organization. We backed all worth-while activities, sending cars to haul the teams, rubbing the men down in the gymnasium, getting out placards and so on. During this time we continued to function as a wire-pulling faction and the nucleus of a political party in the school. . . .

"The group is now a perpetual organization of from 30–35 members with an alumni council which attempts to keep up our moral

[1] *The Gang*, p. 69.

standard. Our men are pushed in school activities, in which they occupy places of prominence. Dances and banquets are frequent, and meetings are held in the boys' homes."

Without wishing to imitate the American pattern exactly one can see how an important a source of effort is lost if the gang is neglected.

This tendency to form gangs being so strong it is clearly the duty of the community to provide groups that are socially desirable. Gangs degenerate very easily. The criminal gang grows up without adult guidance in bad social conditions. The children have never been properly socialized at home, and they have no facilities provided for them for healthy enjoyment. They must play on patches of waste ground, and all the occupations that children really enjoy are impossible. They thus feel hostility and resentment against the environment and fall back on the lowest forms of activity, such as thieving and fighting. The gang lends itself especially to both of these. The group enjoys the excitement, the group gives a certain sense of security, and combat with the law or another gang enhances the feeling of solidarity and internal unity.

In the bad old days in England, fist fights on Sunday mornings were the regular amusement of the young men from a mining village. They were comparatively orderly fights and a village who employed a professional pugilist was felt to have transgressed the conventions. American toughs tend to rely on knives, and, though there are deaths, the activity continues and street fights are always occurring because they give activity and excitement and a sense of power. The attempt is always being made to turn these impulses to athletics, to boxing or some other approved activity. Often the children themselves try to run their gang as an athletic club. Circumstances are often too strong for the children, and when adults step in their presence may spoil things. The young do not like their gangs managed too obviously from above. One of the pleasures of the gang is the opportunity it offers for leadership; and if an adult annexes the position of leader the most aspiring children will go off to some organization of their own.

If children are to organize their own groups, they need practice in it. The power to maintain group activity does not arise till children have achieved a certain stage of development. Small children do not even play together for more than a few minutes, and till the age of five most children tend to play separately rather than in groups. If two children of six or seven are left alone to play together, they are for ever quarrelling, and an adult has to suggest a change of occupation every fifteen or twenty minutes to keep them at peace. By ten, children play in groups very happily, but they cannot yet manage to conduct an organized game, such as cricket.

There are a number of important psychological characteristics involved in successful group activity. The chief is purpose. The group exists to achieve some end, and till a child is old enough to imagine a purpose and the steps to its fulfilment he will never have any steady group consciousness. Furthermore, there must be constancy. The small child, who gives things up, will break away from a group, or fail to join one. This state of mind is very clear in a nursery school, where the teacher has the greatest difficulty in attracting the attention of the whole group, and has no sooner got it than she finds one child wandering out of the door. Next, inside a group there must be at least a rudimentary structure of leadership. This leadership need not be absolute in any way, and the leader may change fairly frequently; but at any moment someone must be in charge of the activity, and the others must accept his leadership. The satisfactory co-ordination of leader and led is an important part of organization. Most people are never *taught* how to deal with the problems of group activity. If they learn, it is by chance in their own spontaneous groupings. For successful moral development group impulses must be harnessed to socially desirable ends, they must also be trained. The earliest training, naturally, needs to be given in the family. With the child constituted as he is, it is not difficult to teach him consideration for the group, and an understanding of his place in it. The well brought up child knows what type of activity is for the public good, and also how he should treat each member of his group—from his grandmother to the cat. This learning can be very rapid. A child of five can be a good member of his family group. This training should be carried farther in school, but the majority of schools, while talking much of the "group" or "team spirit" and insisting on their own corporate existence, are so arranged that the true group spirit is either completely denied or seriously perverted.

Group activity, as we have said, demands a purpose pursued in common with differentiation of tasks and leadership. School, by insisting on identical individual work, by raising up a bugbear called "cheating" or "it isn't fair", and developing a system of competitive marks, renders group work impossible. If forty boys must all do the same set of sums, and all do them themselves without helping each other, no real team work is possible, even if the marks of every ten boys are lumped together and said to "count for their side". Moreover, in a class controlled continuously by the teacher, no child has the experience of leadership, and no one has the opportunity of being a voluntary follower.

In order to have any real education in group activities, a totally different conception of school-work is needed. The atomism which was so strong a feature of moral and educational systems in the nineteenth century must be forgotten, and children must be allowed to work in the natural way in groups. This will involve a rearrangment of the curriculum and the realization that it is not necessary for children all to do the *same* work, and that the teacher must abdicate much of her autocracy. In an experiment to be discussed in a later section of this book, it was found that children aged nine, if given the opportunity in school, would form small groups for such activities as photography, cooking or play-acting. They would make their own arrangements, and carry through their projects to a successful conclusion. In doing this they had to have an idea, foresight, a leader, a simple organization and sufficient constancy of purpose to achieve their ends. They must also establish control over their emotions, learn to bear slight frustrations and disappointments and to influence others in the way they desire.

This type of training is of the utmost importance for adult life. Much work is already co-operative, and in the future far more will be. The pattern of society towards which we are moving, when socialized industry will have largely taken the place of individualism, depends for its success on a strong group feeling leading to hard, conscientious work for the community. At present this feeling is largely lacking, and it is one of the saddest commonplaces of the day that a man only works really well for himself. We shall discuss later some of the ways in which this may be improved. But before we can have successful group activity men must acquire the moral characteristics that are its basis. There is much vague talk to-day in books on education about the education of the emotions. There is very seldom any statement about which emotions are to be educated, or how. The emotions connected with group activity are of first-class importance, and they can be educated quite easily by suitable experience given to children.

Schools not only fail to give this kind of training in group work, they base a training on competitive athletics that is often dangerous. The idea that the only purpose of a group is to defeat another group is morally harmful. English games are so constructed that they *must* be played competitively. There are only a few physical activities that can be done for their own sakes, as swimming, or done against a standard, as jumping. But in life, competition should not, and generally does not, play a very large part. The more civilized people become, the less competitive life is. The idea that national honour consists in defeating someone, or owning a bigger territory than someone else, is on a spiritual level with being the best football team in the school or having a larger stamp collection than anyone else. When the level of civilization represented by the arts is reached, competition is quite out of place. In music, particularly, the highest forms depend on the most intimate co-operation. The Arts Council, realizing this, has adopted as its motto, *To compete with none and to co-operate with all.* Nor with art only. There is no need to organize a competition for the best garden for men to enjoy growing vegetables and flowers. Competitions are popular because we have not developed any proper system of minor rewards for merit. We confer the O.M. on our most distinguished citizens; there should be similar distinctions for the suburban gardener, or the enthusiastic amateur choir-master. What a man wants, and what competition gives him in a crude form, is the consciousness that his merit is noticed and approved. The Russians have developed many methods of encouraging excellence, and many of them do not depend on competition. The child trained to think of group activity solely in terms of competition, whether at games or some other occupation, is given a wrong outlook on life which affects his whole conception of action and incentives.

Further, the schools fail to realize how great a part purpose plays in consolidating a group. They are always trying to make children group-conscious about entities that have no real purposive existence; the school, the house, the form. The purposes of these entities are so vague and general that for the child they have no dynamic existence. They are thus kept in existence solely by competition, something to fight someone else with. If this competitive interest were withdrawn, there is nothing left to engage a child's interest. A group formed for some concrete purpose, to cultivate the garden and supply the school with vegetables, or to produce a play, is quite different and, having a real reason for existence, requires no competition to keep it alive—though it will thrive all the better if its merits are noticed and applauded.

From this attempt to keep fictitious groups alive come many distortions of the group virtues. There is no doubt that almost all the virtues that do not spring from sex and parenthood arise from the group. The group is the true centre of moral life. In striving to maintain the group and achieve its purposes all kinds of devotion, consideration for others, loyalty, hard work, intelligence are involved. Where the group has no purpose, these virtues have no proper sphere of activity. There is no need for work, it does not matter if the group disintegrates; and loyalty, which in essence involves reference to purpose and may demand criticism and disobedience of a bad order, becomes slavish adulation. The curiously perverted morality found in so many schools arises largely from this.

There are certain groups that, because of their closeness, because every member has an obvious and responsible part to play, exercise a very strong effect on morals. The old recipe for dealing with a naughty boy, send him to sea, had much to recommend it. It was not merely that at sea he was out of his family's way, but because the group of the ship's company, even when sailors were a brutal lot, had a moral effect. He might be beaten, but he would find that he was important, that he was developing skill, and that the egoisms and self-indulgences of his previous life had here no place.[1] He was one with the purpose of the ship; to survive, to trade, to destroy the enemy. He was bound to the service of his fellows, as they were bound to him.

This consideration of the force of the group leads to important general considerations. One we raised before: why does a man *wish* to conform to the customs of his group? The answer undoubtedly is that he fears severance from it; and he is unwilling to risk that, or the experience of sympathetic pain if the group suffers. This adhesion to the group may easily prove stronger than other moral impulses. Benevolence issues in action intended to please the patient, but benevolence might well not lead to socially approved action. In fact, where the customs of the group are cruel, benevolence would tend to make a man dissent. There is thus a conflict between the demands of the group and a man's own benevolence, and in this opposition lies one of the bases of moral conflict. For conflict can be between two moral principles as well as between a moral and an immoral one. When the conflict is between the group impulses and individual judgment, the group principles generally prove the stronger at the time. Men are swept with the crowd, and learn their ideas from the behaviour of those about them; and if they repent afterwards they do it in solitude. The moral innovator, who looks beyond the customs of his day to principles of pure benevolence, must usually be equipped by nature to endure a degree of spiritual severance from his fellow-men. If he is not persecuted, he is at least regarded as queer and perhaps dangerous. The Englishman in Italy who spat at Shelley admirably typifies the truly gregarious, when confronted with the more individual type of moralist.

The second point is even more fundamental. It is characteristic of morality that it includes a sense of obligation; a sense of "I

[1] Kipling, *Captains Courageous*.

ought". It has been somewhat of a puzzle to moralists to find a satisfying origin for this imperative; and as long as it was looked for in the realm of metaphysics, or as a quality pertaining to the isolated individual, it was bound to be illusive. It is, in fact, a psychological matter, and can be understood only in that sphere. It is essentially the appreciation of means to an end. The acts that carry this sense of obligation are those which are bound up with some deeper purpose. "I ought to write an essay", says the student, and the force of the moral imperative depends on the vividness of his desire to pass his examination. "I ought to feed the baby." And again the imperative loses its force for the careless or degenerate who do not much care for the baby's survival. Most of our oughts are of a more indefinite obligation and therefore its nature has not been so clearly observed. It is bound up with the welfare and maintenance of the group. Without this underlying purpose the greater part of our sense of obligation would lose its authority. Honesty, truth, loyalty, only have meaning within the group setting. When Kant bids us "so to act that we would wish all men to do the same", the injunction loses force it we do not think of ourselves and the others as forming a community anxious for survival.

It is perhaps a slightly fanciful observation, but this sense of obligation appears to exist in the gregarious animals, but not in the solitary ones. The animals that can be trained to work are all gregarious (not all gregarious animals have been trained). The solitary ones, especially the cat, refuse to work, though they will catch mice, or perform other useful functions in their own time, like gentlemen. A dog shows clear signs of having a conscience and a sense of duty, the cat none. Among men there is a certain proportion observable between gregariousness and willingness to work—work here being distinguished from activity which is self-motivated, however arduous and useful that activity may be.

Among most primitive peoples the group ideas are very strong, and group obligations are readily accepted. In many cases, with a more advanced culture, individualistic tendencies have superseded the sense of group obligation. The change from the idea of group obligation to the individual atomistic conception of duty is nowhere more marked than in religion. In the Old Testament group responsibility is strongly marked, and the transgression of one man may involve the destruction of the whole group. The prophets Jeremiah and Ezekiel protest against this conception and in the New Testament man is an individual, personally responsible to God for his actions, and the penalty for his transgression falls on his head alone.

This change has not been in all ways advantageous. Pitt-Rivers, thinking, it is true, of magic rather than religion, emphasizes the greater social order that results from the earlier conception:

"When it is remembered that in communities that have no organized and paid police force, no monetary inducements or financial prizes, no prison system, the position and power of the chiefs, the minute observance of the codes and taboos and communal undertakings entailing arduous labour cheerfully and willingly borne, even when it taxes the limits of human endurance, are sustained by the *ultima vires* of magic, and when we remember how much more effectively and with how much less friction this factor operates than the forms relied on to preserve order in civilized countries, its position and social significance must surely be recognized."

This is part of a plea for the understanding of the place of the sorcerer in primitive society, but it seems doubtful if it is really the sorcerer who is responsible for the remarkable order and observance of custom. In societies where sorcery does not seem to exist, the same thing can be observed. It is far more likely that the difference lies in the extent to which the individual is merged, morally, in the group.

Our religion, based on the New Testament, has, if anything, increased this moral atomism; and in this it has been seconded by law and custom. The hymn writer who sang:

> My daily question, Lord, is
> Lovest Thou me?

was concerned with the relation of the individual soul to God, and did not feel that any one else was involved. The law makes each man stand his own trial, and it is held contrary to justice that the family and friends of a convicted man should be in any way punished for his offence. The German taking of hostages, and the destruction of villages for assault committed in or near them, was held in universal execration, and yet it was in accord with the custom and belief of many centuries. The eighteenth and nineteenth centuries, with their excessive individualism in industry, religion and education, have tended to sunder man from his neighbour and make him responsible for himself alone. Once a man has been thus isolated, what moral obligation does he feel? Not the obligation of the group. Too often society cuts him off even from the family group, and his home offers him nothing of security or assurance. He may have no work, no club, nothing to which he can give his devotion, and receive in return the pleasures of purpose and co-operation. As we shall show later, the first step in reforming the juvenile delinquent is to make him a member of a group.

Instead of establishing this group feeling and responsibility, society threatens the individual. If you do this, so and so will happen . . . Penalty for improper use, £5. . . . And how often have we wished we possessed, like the man in *Punch*, sufficient notes in our pocket to justify us in stopping the train. The penalty becomes a challenge, and the sanction behind it fear: fear of Hell, fear of prison, of the whip. If a man yield to fear he is dishonoured. If he so determine, he can brave the threats and decide that, come what may on earth or in Hell, he will have his fling. Besides, punishment is not quite certain. A death-bed repentance, the perfect crime, and he may get away with it. Thus begins the deplorable competition between the invention of punishments of ever-increasing severity and the hardihood of the wrong-doer. And all the time the criminal is quite free from any sense of moral obligation. The matter rests on him, and ends solely with himself.

Thus, as regards the control of action, the group impulses take first place. Through them a moral code is built up, and they supply the impulse to observe it. It is only the very exceptional who can made a code of their own, and hold to it against the voice of the group, be it the family or some outside body.

This becomes particularly clear in certain types of religion. The religious teacher is hard put to it to render sin sufficiently hateful to his flock. His difficulty arises from the fact that he is dealing with the individual soul, and must make his appeal to egoistic sentiments. Thus death and Hell play a large part in his morality. Death is the first stage on the mystic and fateful journey, and death must have its full terrors and be seen in its proper light. The episode in *The Fairchild Family*, when the children are taken to view a corpse, is very displeasing from the modern point of view, but a necessary stage in moral education when looked on with the eyes of the time.

Mr. Fairchild is told of the death of his gardener and at once decides to make full use of the event.

" 'Have you any desire to see the corpse, my dears?' asked Mr. Fairchild. 'You never saw a corpse, I think?'

" 'No, Papa', answered Lucy, 'but we have a great curiosity to see one'.

" 'I tell you beforehand, my dear children, that death is very terrible. A corpse is an awful sight'.

" 'I know that, Papa', said Lucy, 'but we should like to go'.

They set off all of them after dinner, the children skipping ahead.

But when they came to the house, where death was, they grew quiet and the children fell behind.

"When they came to the door leading to the staircase, they perceived a kind of disagreeable smell, such as they never had smelt before; this was the smell of the corpse.

"The body of the old man was laid out upon the bed in the upper room. The face of the corpse was quite yellow, there was no colour in the lips, the nose looked sharp and long, and the eyes were closed and sunk under the brow; the limbs of the corpse stretched out upon the bed, and, covered with a sheet, looked longer than is natural; and the whole appearance of the body was more ghastly and horrible than the children expected.

"They all three stood looking at the corpse for a long time, without speaking one word. At last, Mr. Fairchild said, 'My dear children, you now see what death is; this poor body is going fast to corruption. The soul is, I trust, with God; but such is the taint and corruption of the flesh, by reason of sin, that it must pass through the grave, and crumble to dust. And this shows the exceeding sinfulness of sin, and its horrible nature, that the soul that has sinned, must be born again, and the sinful body be dissolved, and fall to dust in the grave. Remember these things, my children, and pray to God to save you from sin'.

" 'Oh, Sir', said Mrs. Roberts, 'it comforts me to hear you talk'."

The corruption of the body having made clear the sinfulness of sin, it was necessary to bring home fully to the mind, especially to the mind of the young, the dreadfulness of Hell. In the middle of the century this was a matter of much concern with the divines of a certain type. "Little child", writes the Rev. J. Furniss, "if you go to Hell there will be a devil at your side to strike you. He will go on striking you every minute for ever and ever without ever stopping. The first stroke will make your body as bad as the body of Job, covered from head to foot with sores and ulcers. The second stroke will make your body twice as bad as the body of Job. The third stroke will make your body three times as bad as the body of Job. The fourth stroke will make your body four times as bad as the body of Job. How then will your body be after the devil has been striking it every moment for a hundred million of years without stopping?

"Perhaps at this moment, seven o'clock in the evening, a child is just going to Hell. To-morrow evening at seven o'clock go and knock at the gates of Hell and ask what the child is doing. The devils will go and look. They will come back again and say that the child is burning. Go in a week and ask what the child is doing; you will get the same answer, it is burning. Go in a year and ask, the same answer comes—it is burning. Go in a million of years and ask the same question; the answer is just the same—it is burning. So if you go for ever and ever, you will always get the same answer—it is burning in the fire."[1]

This delivered with a blazing eye and the tones of conviction must have been terrifying. If children still sinned, it can only have been due to incurable vice in the young or to a defect in the general nature of the method.

Bishop Jeremy Taylor, in rather more adult language, could raise a picture equally dreadful. "The slavery of the damned in Hell is such that all their senses and powers of soul and body are subject unto eternal pains and torments. With their touch they are to serve that burning and never-consuming fire; with their taste, hunger and thirst; with their smell, stink; with their sight those horrid and monstrous shapes, which the devils shall assume; with their hearing, scorns and affronts; with their imagination, horror; with their memory, despair; with their understanding, confusion with such a multitude of other punishments as they shall want eyes to weep for them".[2]

The failure of this type of moral control is now a matter of common experience, and it has been abandoned over the greater part of religious and social life. Its place must be taken by a development of the group impulses and their use in the control of action.

[1] Rev. J. Furniss, C.S.S.R., *The Sight of Hell.*
[2] Jeremy Taylor, *Death, Judgment, Heaven, Hell.*

Books referred to in this chapter:

William McDougall, *Social Psychology.*
Thrasher, *The Gang.*
Rudyard Kipling, *Captains Courageous.*
Rev. J. Furniss, *The Sight of Hell.*
Jeremy Taylor, *Death, Judgment, Heaven, Hell.*

Chapter 3

SOCIALIZATION

In the past two chapters we have spoken several times of socialization as the process by which man's innate impulses are made to fit in with the framework of society. From our point of view it is perhaps the most fundamental conception of practical ethics. All men are born with the same needs, all men, so far as we know, with approximately the same innate tendencies to the satisfaction of these needs. The manner in which these needs are satisfied, the behaviour to which the innate tendencies lead is the difference between one culture and another. Each child, as he develops, has imposed on him a way of life and thought by his environment, and this imposed culture goes far deeper than differences of dress, houses or even diet. It is something that penetrates into the very fabric of his mind. This conditioning of the child by his environment is generally so thorough that it is very difficult for each of us to see clearly the society of which we form a part. All the mental and physical conditions of life are so much a part of ourselves that they are unnoticed or accepted as laws of God or Nature. It is from this that the curious blindness of so many writers on ethics seems to arise. It is perhaps, then, a good thing to look for a moment at some other patterns of socialization before we consider our own.

There has recently appeared a very interesting article on the Japanese character.[1] It begins by setting out the principles on which any study of this kind should be based, and then analyses, most acutely, the elements in the Japanese child's upbringing which form the attitudes of mind that issue in characteristic social and political behaviour. In essence these principles give, in slightly more technical language, the thesis of this book. That man's behaviour is *learned*, the learning modifying the ways in which he satisfies the needs of his nature. This learning is imposed by the whole society, its ideals, its rewards and punishments, its institutions. In the first place the teaching is done by the parents, through their treatment of the child in infancy. The treatment they give a child impresses on his mind, in the most potent way, certain sets of values, certain attitudes towards the world. These rewards and punishments of infancy, largely

[1] Geoffrey Gorer, *Science News*, I.

because they are non-rational and non-verbal, produce in the mind beliefs that are almost immune from reason and argument. They appear as laws of nature and there is no means of altering them. When the child grows up, these acquired attitudes realize themselves easily in the social pattern that has predetermined them.

The Japanese baby has no physical freedom. He is always bound tightly to someone's back. He is indulged in the matter of food, but treated harshly in an attempt to make him acquire control of evacuation before his muscles are really strong enough. While yet a baby he must learn to sit still in uncomfortable positions, and move about the house with great care. When they leave babyhood, the girls must remain submissive to all; but the boys, while retaining their submission to all older and superior males, may bully and tyrannize over all girls and women.

The pattern here of socialization is very different from our own, and the resulting ideas of justice, individual worth, and the proper type of social behaviour quite impossible to compare with ours.

It would be possible to take countless examples of different types of culture and show in the same way how the full social pattern was reflected in the treatment of children, and how, in turn, attitudes impressed on the child by his early training appeared in his adult behaviour. It is sufficient to mention the treatment of children in England one hundred and fifty years ago. At that time the patriarchal family, dominated by an absolute father, with a mother, subservient indeed to the male, but still by her adult status supreme over the children, was in existence, and would continue to exist for many years. Mrs. Sherwood records that she never sat down in her mother's presence, and the physical attitude of reverence was not without its spiritual counterpart. The position of the child at home was reflected in an aristocratic government in the State, and an acute class consciousness that made one man incalculably superior to another. The child of to-day belongs to a democracy that is as marked in the home as in the State. In some homes, it is true, the patriarchal pattern perssits, but it is the exception. So is privilege, of class or sex. We are not completely free of it, as some countries claim to be, but it is so greatly reduced that its faint wraith hardly obscures our view of essential justice.

This concept of the pressure of society in moulding the mind of the individual runs all through this book. In this chapter we wish to discuss some of the ideas and ways of thought that are formed in children in our civilization to-day. We must remember that, in essence, these ways of thought and behaviour are now democratic, and thus they are different from many formed in the past, or in other societies. Not all the modes of behaviour that are taught to a child are of signal importance.

Socialization extends over a very wide field and covers action of very different degrees of moral worth. Through it, on the one hand, a child learns the common conception of justice—the natural justice that we shall talk about later. This is in essence the modification of man's egoistic impulses so that he can live as a member of a society. On the other hand, it includes modes of behaviour that might more aptly be described as manners. There is one characteristic of educated behaviour which possesses considerable social importance, but very little moral significance, and that has to be taught carefully to a child—not to make a noise in a public place. The children who rag in a bus, girls who giggle and scream at a dance, are not doing anything morally wrong; they are merely offending against the code of educated manners; and yet parents spend many years hushing the too-loud-voiced four-year-old, or rebuking the child who, at ten, is still too exuberant in public.

With such a wide field there are naturally many aspects of the process. Some are connected with the production of benevolence, and we have discussed them in part. Others concern the manner in which benevolence is shown. It is necessary for a child to learn the particular pattern of social adjustment accepted in his group, and the type of behaviour that is the necessary basis of group life. Then he must learn to adjust special powers, such as leadership, to the needs of the community, and lastly, he must learn those details of behaviour which we call manners. In addition, though we shall not discuss it in this chapter, he learns the ideas that lie behind the social structure, and he absorbs the beliefs that make up his society's myths. If we take the child's life we can see how various elements in his nature, various impulses to action, are socialized.

The little baby lives under a regular regime and this enables him to develop habits that help him to conform to his environment. If he is fed at regular intervals, he ceases to be hungry except at feeding-time, and thus, apart from some few minutes before the appointed hour, he is placid, expecting nothing and untroubled by pangs. He learns, too, when he may expect adult attention and to amuse himself for the long periods when he is left alone. The healthy, well-managed small baby lies quiet for hours in cot or pram contemplating his toes or a coloured ball, and making no demands on his environment. He eats when he is fed, and goes to sleep alone. He has ceased, already, his attempt to dominate the environment because in the first place the environment attends to his wants so that he has no anxiety, and in the second because his attempts end in failure and so give no ground for repetition. This happy condition is important because it lays the foundation of so much good in later development. It provides, in fact, the pattern of the social life. The group holds the individual in its kindly grasp and the individual makes no opposition. On the contrary, he develops benevolence, a desire for social contacts, and an expectation of pleasure given and returned. From this peaceful state of babyhood comes also a certain tranquillity of character that seems to be a permanent asset. The really healthy baby—quiet, well-managed, friendly—does not seem to develop neurotic characteristics. He may prove talented, exceptionally able, even excitable, but he retains a fundamental steadiness that helps him to make the best of his powers.

When the child passes to the next stage and learns to walk, he presents his parents with a new set of problems. His greater activity, his eagerness to take part in what is going on, his complete inexperience and ignorance, make it unsafe to leave him unattended and exceedingly difficult to control him. The child at this age needs a special environment—an environment which is safe, interesting and his own; where he will not be continually impeding the activities of adults. If this is given him, as in a nursery school or in a well-arranged home, the placid tide of his development is not checked. One can stand and watch a group of nursery-school children playing with their toys for an hour or so, and no one does anything to annoy anyone else. If there is space, open air and plenty of occupation of a suitable kind, there seems no impulse to interference, bad temper or unkindness. It is very different when the children are shut up in a room or are bored.

But for most children, especially those who are made to live in a largely adult environment, the age of two brings other problems in addition to those of activity and inexperience. Individuality begins to develop and the child shows the first real signs of independence. A child of this age, when called, will run away; just as a kitten does when at the same stage of development. He will contradict and have views of his own. One of the writers was once flatly contradicted by a three-year-old as to which platform a train arrived at, though the child had had no previous experience of the station. One small boy of this age when rebuked for being naughty, smilingly replied, "But I *like* being naughty". Three years later his philosophy was more fully developed, and, in reply to some of his mother's reproaches, he expounded it. "You see, Mother, you are one person, and I am another—and I won't be bossed". What the child is claiming in all these cases is the right to think for himself and the admission that his personality should be respected. The child has begun to feel that he and his environment are separate, and to see his own development largely in terms of escape from too close care. And yet it is during this age, between two and five, that a great part of the socialization of the child takes place. Under the age of two he is too young to learn, or indeed do, very much. After five, if fundamental attitudes are not formed, it needs much more labour to bring them into existence.

During these years the benevolence of babyhood is strengthened. The child continues to find the world kind and to respond to it kindly. The unfortunate child who is wrongly treated—slapped one moment and fondled the next—or who is kept from a natural development by excessive strictness or care, grows nervous and hostile or depressed. The basis for further teaching is lacking; for all moral training must be based on this underlying benevolence, and without it loses all meaning.

Where this benevolence exists, teaching follows two lines, though they are not at all separated in practice. There is the guidance given to the child in the methods by which this benevolence should be shown, and there is the help given in solving the problems that arise in group life.

There is a school of thought that believes that children should be left to find things out for themselves. They should struggle and fight till they learn that fighting is unsatisfactory. They should pull the cat's tail till they find she scratches, and they should refuse to learn to read until they are driven to their books by boredom and the disadvantages of ignorance. These beliefs are characteristic of the maladjusted adult, and seem generally to be associated with the further belief that children *wish* to swear and use bad language, that smashing windows is good for their souls, and if they haven't yet done it, it is time they began. It is assumed, and it might as well be stated as an axiom in the works of these writers, that any normally well-behaved child is suffering from parental tyranny and, unless it at once learns a number of dirty words and when to use them, will be a permanent neurotic wreck from repression. This belief in the tyranny under which a child lives appears even in such a book as Herbert Read's *Art and Education.* The whole premise is, of course, false. Well-managed children are good *and* free, and those who believe that virtue is always constrained have themselves been badly educated. There is no doubt, as we shall say in the next chapter, that parents can be too careful, too anxious to guide, too omnipresent; and the poor child in his search for independence revolts from the confining tie, and perhaps breaks a window as the first claim to autonomy. But that a thing can be carried to excess does not condemn it in all degrees. It is sheer cowardice in an adult to refuse to a child the help it needs. The adult who accepts the care of children should have sufficient self-confidence to believe that he can give what is wanted, but not too much; and he must be sufficiently analytically minded to understand what exactly it is that he wishes to give.

In organizing social life the chief problem is the accommodation of natural egoism to the needs of the group. It is, as we have said, perfectly possible to do this. Perhaps the first lessons in the control of egoism are concerned with food, the enjoyment of adult attention, and the use of possessions. Where children are adequately fed, it is very easy to teach them to take their food in a manner socially approved. In the nursery school little children serve each other, wait to begin to eat till all are ready, eat at a proper pace and talk amiably while they do so. Once adults have established the routine the children carry it out and accept it as their own with pleasure. If the same standards of behaviour were maintained all through the school at school dinners, one of the natural forms of egoism would largely disappear, and meals become, as they should be, a pleasing social function.

This communal feeding would also put an end to a particularly tiresome form of adult egoism: the insistence on having the particular food each man prefers served as he likes it best. It was an egoism particularly characteristic of the Victorian male. The Holy Grail had the characteristic that it gave such feasts to all in the hall when it appeared; the Victorian father and the working-class husband expect such consideration without the aid of heavenly intervention, and refuse to eat in their canteen, thus inflicting much unnecessary work on their families. If every child learned to eat what he was given, and his individual preferences were ranked as low as his individual impatience, life would be easier.

Closely connected with this are the first lessons in justice. Countless occasions arise when the wishes of individuals, held together in a social bond, conflict. It is then the duty of the adult or the ruling power to show how these conflicts can be resolved most satisfactorily. To the general body of these solutions we have given the name Justice or Law, and for the child, as the adult, individual cases are resolved in accordance with the general principles; and from the individual cases the child builds up, in his turn, the idea of the general principles. In the nursery conflicts arise over the division or use of desirable things or the right to monopolize an adult's attention. Where there are sweets, cakes, or anything else desirable, the uninstructed child tends to fight or snatch because there might not be sufficient for all to take to satiety. If children are allowed to fight, the victor gobbles, pursued by the disappointed cries of the vanquished, and the loser gets nothing. Here the adult has a real part to play. He suggests the solution of equal (or almost equal) division; and this way of dealing with the problem is so obviously the most satisfactory to both potential victor and vanquished that it is normally accepted, and becomes the basis of much future action. With a child in whom greed is strong, the adult may have to intervene more than once, but as soon as the idea has been clearly apprehended it takes possession of his mind, and generally determines his behaviour; even if now and then there is a lapse.

This method of division ensures that everybody gets something. Moreover, since emotion spreads quickly in a group, the victor of a fight can hardly enjoy his triumph if surrounded by hostile and aggrieved neighbours. Besides, he never knows when an attack may be made upon him, and he cannot eat in peace. All this is clear to the child, and the belief that Justice is somehow concerned with equal division and the quiet enjoyment of what is received becomes part of the mental furniture that man brings with him from the nursery.

In the nursery this principle of equal division is varied by special claims. The youngest gets the odd sweet, the eldest is taken out for a ride in the car, and so forth. This also is accepted from the environment. In the days when boys were far more important than girls, the inequality of treatment of the sexes did not usually produce jealousy; and variation in treatment, even between strictly comparable individuals, if elevated into a constant principle, was tolerated.

It is interesting to see how this principle has fared in certain crises of adult life. In the 1914–18 war rationing was introduced late and then was very laxly applied. It was roughly accepted that the rich had a *right* to more than the poor, and practically nothing was done to prevent their having it. In the second German war the rights of the rich had been successfully challenged. The nation as a whole accepted the nursery principle of equal division, and the exceptions were expectant mothers, the sick and other categories that took preference for certain reasons. In other countries where the black market flourishes the principles of 1917 still hold sway, and wealth, in itself, is accepted as a reason for preferential treatment.

The well-to-do did not, of course, resign their privileged position in England without a struggle. The indignant letters that appeared in the papers as each new commodity, e.g. eggs, was brought under control show that a principle so long established dies hard. But once the protests proved unavailing the relentless march of equality trampled down another privilege, and on the whole the rich accepted their loss quietly.

Closely connected with this is the adult solution for such problems as the use of toys, or other desirable possessions. "Take turns", says the adult, and gently enforces his command. Again, it may take two or three repetitions of the lesson before some children learn it, but, as in the case of sharing food, the advantages are so obvious that it is not long before the child has learnt to do it the adult way.

One of the things a child desires is adult attention. He does not want it all the time, and certainly not directed on him too closely, but the little child in particular likes to have an adult handy to attend to him and, if necessary, help him. He also likes to be included in the adult activity and group, and to take part in what is going on.

For the adult, as for the child, this constitutes a problem. The adult is busy, she wishes to get on with her work, or to talk to another adult without infantile interruption, or to help another child without disturbance. The socialization of the child depends largely on the adult's behaviour. A child who is pushed aside resents his treatment, and redoubles his importunities; a child who is told kindly that he will be attended to in a few moments, and that mother is at present busy on her own affairs, rapidly learns to accept this position; only, of course, the promise must be kept and attention given in due course. The basis for the child's consideration is the understanding, by sympathy, that other people have their own affairs, as he has his, and the confidence that in time he will get what he wants. If there is a strong bond of sympathy in the family group the realization of others' wishes comes easily, and if there is confidence it is unnecessary to make too many claims. The children who find it impossible to learn this adjustment to others' needs are those who have been pushed aside, or who have never had the claims of others reasonably represented to them.

The second part of the adult's task is to teach a child the proper methods of showing his benevolence. It is not always easy for a child to know how to do what he wants to do. The elder child who feels affection for his younger brother does not know the right way to approach the infant. It would be perhaps natural to imitate mother and try to hold him in his arms, or, in excited interest, to poke a finger into his eyes or mouth. A mother shows a child what he can do: how to touch, how to amuse, how to admire the baby. The same thing applies to the other creatures in the environment. The cat, an aunt, grandparents—all require treating in a certain way, and, when so treated, display satisfaction. The benevolent child is hurt and disappointed if his advances are not well received, gratified if they are liked. Thus he needs adult help, if his good intentions are to be strengthened by success.

This type of help and training can be given without imposing on a child a burden of formal manners, and should be represented to him in its true light, as a means of increasing public happiness rather than as an empty formula that must be gone through. Grandparents, like ducks, wish to sleep after lunch; cats object to being grasped round the middle and hauled up; aunts will play the gramophone, and fathers enjoy being "bears" in the garden. The child wishes well to all these creatures, and is ready to fall in with all their desires, if he understands them and if they do not run too contrary to his own purposes. In return, of course, he expects his own wishes to be respected, and he thus learns to build up a system of mutual tolerance and reciprocal care.

It is interesting to see how far a child of five has advanced in socialization. If he has been well looked after he is an amazingly social being. The cruder forms of egoism have been largely overcome. He is not greedy of food or attention. He shares his toys, discussing amicably with his little friends who shall have which. He is polite in his simple way and friendly; and, in addition, he has made some progress in understanding the more abstract standards of his society.

This social teaching, based on benevolence, is reasonable and can be explained to a child. It relies for its force on impulses existing in his own nature, and is not felt by the child as something imposed arbitrarily from without. This is very important. Where the standards imposed are felt to be purely external, they may be resented and resisted. Where they are founded on his very nature itself, where the force comes from within, they will be welcomed as helping the child to resolve satisfactorily the contradictions that exist in himself. Man's nature, as we have said before, is not simple and all of a piece. To be happy, some harmony must be achieved among the diverse parts of man's make-up. The younger a child can learn to bring his egoism into accord with his sociability, the happier he will be and the brighter will be his prospect of a successful life. The effect of this type of training is totally different from that based on fear, which will be discussed later.

In the training given to a child there is also a large part that is conventional and concerns manners, and, being conventional, can only be partially explained. In most cases all one can say is that such and such an action is "customary", or that "people like it". With the strongly gregarious child, or the very benevolent, these reasons may be sufficient; where more is needed to secure obedience a display of emotion is the natural way of ensuring conformity. Margaret Mead has given a good example of this process in her account of the teaching of modesty to Manus's children.

"Children must learn privacy in excretion almost by the time they can walk; must get by heart the conventional attitudes of shame and embarrassment. This is communicated to them not by sternness and occasional chastisement, but through the emotions of their parents. The parents' horror, physical shrinking and repugnance are communicated to the careless child. This adult attitude is so strong that it is as easy to impregnate a child with it as it is to communicate panic."

In this it is to be noted that the parents really feel the emotions they display—there is no hypocrisy of the kind that a child readily detects. The emotions are merely shown, and no punishment is inflicted on the child to make him conform. Since within the group there is so strong a contagion of emotion, it is clear why this display of parental emotion is so effective in influencing the child's mind.

A European child learns many items of behaviour in this same emotional way. These are actions that cannot be justified rationally, and for which no appeal can be made direct to a child's nature. For example, our standards in regard to excretion and nudity. In a family where children and often parents go naked, the child is unashamed; where there are different standards, the five-year-old is already learning them. In the same way the child has learnt, more or less, what his group considers fit to eat, and he does not experiment with grubs and caterpillars—which some races find very nutritious.

Two very important sets of ideas that are taught in this manner are those connected with sex and status. As we have said, man's "natural" sexual behaviour can hardly be ascertained, and each group has very strong conventionalized ideas as to what is proper. A child grows up in an atmosphere in which sexual behaviour is the subject of strong emotional feeling. He hears praise of those who conform; and hushed voices, dark hints or outspoken condemnation come the way of those who transgress. When this emotional expression is genuine, and not just adopted for the edification of youth, it makes a deep impression on the minds of the young. The dreadful phrase of the past, "gone wrong", covered a multitude of horrid thoughts and suspicions, and its emotional effect must have been considerable on those who still kept the right track. It is through this unreasoning, often unformulated emotional teaching, much more than through any direct discussion or prohibition, that the young person of seventeen or eighteen comes to have an idea of what is expected or condemned. If parents do not express honest emotions, or if the relationships within the family are such that emotions are not accepted, then this kind of teaching is ineffective, and the child has no body of established prejudice to guide his actions. Most parents in the past were, many in the present are content to leave the whole matter to this emotional contagion, in the hope that children would conform without their being put to the trouble of special instruction. The child was left to its emotions, and ignorance cast the shadow of its wing to darken the gulf of horror.

The idea of status is quite foreign to a child's nature. Children left to play together will distinguish each other by strength or skill but not by riches or race or any other of the marks that adults use in assessing status. The belief in superiority for reasons of a purely social nature is always taught, and is one of the most difficult elements in the organization of society from a moral standpoint. The idea of status once established arouses the most violent emotions, and they are very difficult to overcome by reason. The problem arises in its most acute form in countries where two races with differently coloured skins mix, e.g. the negro problem in America,[1] the racial discrimination in South Africa, or the relations between British and Indians in India. Even where there is only one race, the sense of status may be extreme. In many American studies of New England the existence of a superior element in the population is emphasized. Generally it is the older-established families who feel themselves superior to the newer immigrants. In other countries it is birth or wealth, or the school one was sent to that establishes superior status. We shall later discuss some of the general consequences of these emotions; it is enough now to quote part of a letter published in *Picture Post* to show the violence of feelings to which these ideas lead. The letter concerns an article on pre-natal care that was published in the paper.

"I was positively incensed at your article and pictures of the birth of the Lewington child. We have heard much in recent years about the working classes having the same as the other classes, but when it comes to having *much more*, where is it going to end?. . .

"My mother did not deprive the poor people, who to-day are known as 'non-priority', of their milk, because she did *not have any*, only her meals and the milk she always had during normal life. . . . This milk has not been diverted to the working classes because they

[1] Cf. *Black Boy*, Richard Wright.

need it, but to starve the better classes, to whom it has always been a great item of food, thus the evil, common man hopes to rid the country of the better classes in time."

This attitude, and others like it that are rather more restrained in expression, is taught by parents who are always keeping their children away from others whom they regard as nasty and dirty, who find it impossible to eat a meal at the same table with one they regard as of "lower social position", and imply that the world is divided into "pukka sahibs" and others. In a country where there are different races, children by the age of two or three have already recognized their position as members of a superior race, even though they continue to play with coloured children of their own age. Once this attitude is established it is extremely difficult for an individual to free himself from it, and in most cases it remains with him for life.

The force and persistence of such emotional attitudes and the ease with which they are established makes parental behaviour of great importance. A silly mother who gives a display of emotion over a mouse may quite easily be making a lasting impression on the child's mind. On the other hand, genuine enthusiasms, genuinely expressed, will also have their influence. If the parents care for art, music, learning, if they are strictly honest, generous and genuinely kind, these attitudes are also learned emotionally. The careless, stupid parent, while he fails to rebuke vice, also fails to show enthusiasm for virtue.

One of the attitudes learnt in this way is willingness to work. We shall later discuss incentives to work, but it is important to remember that, quite apart from anything else, a person is predisposed to work by the emotional ideas he learns when quite young. The industrious parent who himself works willingly, and takes a pride in his skill and efficiency, produces in his child a firmly held conviction that work is necessary, honourable, and largely pleasant. On the other hand, the idle, grumbling, the unskilled and worst of all, the snob who thinks work "beneath him", establish in their children a state of mind that will never let them do an honest job. At school a child begins to show the effect of parental example, and the idle scholar generally has behind him a mother who, by example and precept, maintains that work is a curse.

A parent must therefore be careful about his emotional behaviour before a child. He may dislike a thing very much, but before he expresses his dislike he must consider if he wishes his child to go through life attaching great importance to the matter. The busy housewife who holds up her work while she decides whether the hem of a sheet should go this way or the other has not much to thank her mother for. An officer, whose peace of mind was entirely ruined all through the Italian campaign because he could not acquire a certain sort of face-flannel, was suffering from a too careful upbringing. A woman who was taught by her nurse never to use the same towel for her face and body, found the emotionally engrained habit a hindrance all through a wandering life. It requires considerable self-control on the part of the adult to restrain the expression of an emotion, but it is often necessary.

There also takes place during these early years a teaching in the actual customs and modes of behaviour in the particular community. A child learns how to eat in the customary way, how to address his elders and friends. The modern child who grins "Hullo" to an adult as he passes on his cycle, has learnt a way of behaviour that is accepted to-day, but would not have been in the past. The child in his home learns how to set a table or make a bed, how to turn on the bath taps and what a toothbrush is for. In fact, by five, the well-instructed child is well on his way to knowing how things are done, and will often astonish his elders by his knowledge of ration-books or the rule of the road. All this knowledge makes a child part of the group, and he pursues it eagerly.

Nowhere does the essentially democratic nature of the pattern we impose on children show more clearly than in our treatment of the superior child. Of the various ways in which a child may excel, leadership is the most socially important. There are two completely different conceptions of leadership, the despotic and the democratic. The despot rules without regard for the feelings or desires of anyone else. He may hold his position as part of a social order that has given him his place by birth or custom, or he may win it, as did many rulers, by his own prowess. But, however acquired, he enjoys his rule and spares not a thought for the feeling of slaves or women. The democratic ruler is the servant of the group he leads, and even when they are bound to follow, as by military discipline, he must still serve them and put their welfare before his own.

This latter conception of leadership is early impressed on children. In the first place, they are themselves treated by their elders with consideration, and have no pattern of despotism impressed on their mind. Next they are taught that superiority, as in size or strength, gives them no right to tyrannize over their younger friends. Groups are fluid, and a child who cannot make himself accepted as a leader is rapidly left alone. In the scheme of thought that we encourage no term of abuse is worse than to call a child a bully. It is an accusation that he has failed in one of the most important aspects of life. The position of the leader is emphasized in myth, religious and social. The Son of God, the divine leader, sacrificed himself for his people; the motto of the Prince of Wales is *Ich dien*; the captain of a vessel must see to the safety of all his crew, including the ship's cat, before he tries to save himself. Thus the growing child has a clearly defined pattern of ideas before him while he is learning the art of leadership.

Leadership is a characteristic that shows early in many children. In some it is clear at the age of three, in others it appears a few years later. The natural leader has apparently no difficulty in establishing himself in command of his group. He is fertile of ideas, ready, amiable, and is always surrounded by contented followers. Other children have more difficulty. They find it hard to acquire the manner that becomes a democratic leader and find their authority flouted. They then fall back on violence to establish their position. In these cases the influence of the adult is used against the would-be leader, and the child is told that he must be better behaved or more persuasive. It is interesting to watch the stages by which one of these children arrives at the power to control others.

While still quite young, children are made conscious of the obligations of leadership. They are expected to take care of the younger members of the group, to help them and not require them to do things beyond their strength. But, naturally, a child is not confronted with the deeper problems of leadership till he is older. As groups become larger, as they pursue purposes of more complication and difficulty, the leader becomes more and more clearly the servant of the group. On him falls the planning and organization; he must see that all the group are well looked after, and he must take the responsibility and devise remedies if anything goes wrong. In the public schools the older boys are taught both aspects of leadership through the prefect system. They must learn how to give orders in such a way that they are obeyed, and through their duties as games captains they learn the kind of organization that is necessary to make a project a success. The training is carried further by such bodies as the Junior Training Corps. There the newly appointed corporal finds that one of his chief duties is to inspect his men's feet after a march, and prescribe suitably for the blisters and corns that may have arisen. After a little of this sort of thing a boy understands clearly that leadership is not all glamour. The potential leader develops a sense of responsibility to the group, and only those with ability and a real desire to lead are prepared to accept the duties. It gives the group in return a sense of security and a confidence that its needs will be understood and respected.

Parallel with the training of leaders goes the training of followers. In a successful group the follower is of great importance, and on his behaviour the stability of the group depends. He must be obedient, but within the bounds of intelligence and general moral standards. In certain circumstances, to obey an order may be as serious a fault as to fail to obey one in another. He must understand to a point what is going on; make intelligent suggestions, and endure to have them criticized or refused. All this children learn by trial and error, though they can sometimes be helped by the kindly criticism of an adult.

It is interesting to consider at what age a child is more or less fully socialized. Many children by the age of ten or twelve have ceased, as their parents say, "to be any trouble". They have learnt the ways of their group, learnt to accommodate themselves to others, learnt to control their emotions, and how to please those with whom they come in contact. Other children develop more slowly. They may not have had as careful and intelligent help, or they may have had to struggle against some of the difficulties we shall discuss in the next chapter. These unfortunates often seem to grow worse as they grow older. At seventeen they are more unmanageable than they were at seven, and have acquired strength without learning sense. Some improve when they leave home and discover by experience just what the environment will and will not stand. This is what the more fortunate child learnt before he was five, and when this is learnt late it is rarely learnt so well, or learnt without leaving considerable resentment behind. This resentment is always ready to break out in the repression of others as soon as it feels safe to assert oneself. Thus defective socialization is the most frequent source of troubles in later conduct.

Books referred to in this chapter:
Herbert Read, *Art and Education.*
Margaret Mead, *Growing up in New Guinea.*
Richard Wright, *Black Boy.*

Chapter 4

FRUSTRATION

In the last three chapters we have given, on the whole, a pleasant account of man's nature. He is a social animal, anxious for the welfare of his group: a kindly creature filled with benevolence, loving his family and friends, and even, if he has been fortunate, extending his care to the world outside. He speaks gently to stray cats, and pushes bits of broken glass into the gutter. This picture of social virtue may fit some inhabitants of the world, but it can hardly be held to apply to all. We know differently, and we do not need the pessimists to tell us that man, clearly, is much less virtuous. He is manifestly bad on many occasions, and vile on some. The organization of society is based, in some respects, on greed and self-interest; and after the happenings of the last ten years it is hard not to believe the doctrine of original sin, and agree that man, given the chance, sinks at once to the lowest level. If we mean to make good our thesis that the achievement of virtue is largely in the power of the community, supposing it sets about it properly, we must explain why man falls so far short of what we claim he could achieve.

In the optimistic days of revolutionary thought, Godwin and others preached the infinite perfectibility of man. Treat him right, they said, and he would develop all the virtues. Unfortunately, in spite of Liberty, Equality and Fraternity, he did not. The Revolution, instead of starting the millennium, let loose on the affrighted nations a tornado of bloodshed and anarchy. This sad fact was felt to be explained when psychologists discovered that man had instincts that predetermined his behaviour, and they could only agree with the theologians that man's nature was radically evil—and alas, not really susceptible of a new birth. This discovery of the instincts allowed the pessimists to convince the world that wars, sin and cruelty were inevitable, and that man would always engage in the scramble for egoistic advantage. It came to be said, quite without thinking or proof, that "pugnacity" made wars inevitable, just as "egoism" rendered it certain that the strong would always plunder the weak. In addition, we had the psycho-analysts with their depressing picture of man's sexual nature, and their hobgoblin host of perversions which rendered a pleasant emotional life almost impossible. Thus we must show that instincts do not make vice inevitable, as well as *why*, in certain cases, they issue in bad actions.

On the first point we have already said a good deal. We can repeat that man's instincts, though in their essence fixed, result in actions that vary with the teaching a child receives, and the kind of society into which he grows up. We can thus accept the theory of instincts, at least as first formulated by McDougall,[1] without committing ourselves to pessimism. If the instincts lead to undesirable modes of behaviour, at least part of the fault lies in the environment and training the child receives. The good man and the criminal may not differ greatly in natural endowment; and most of the neurotics could have developed normally if they had grown up in different surroundings. In fact, in so far as man is bad, we must look for this cause outside in his environment quite as much as inside in his nature. We can thus hope to find what are the conditions of that goodness the perfectionists believed in, but had not enough knowledge to attain.

If we look at the external conditions of life that have a psychological effect, the unfavourable circumstances fall roughly into two groups. There is the damage done to the man's nature by the mismanagement in youth, and there is the wider influence due to the structure of society and the ideas which are current in the community. When these are bad, man's behaviour must necessarily be adversely affected. We shall take the first of these groups in this chapter. The structure of society, as it affects ethics, will be the subject of the next section.

The earliest training a man receives comes necessarily from the family. It is thus a very personal thing dependent on the quality of the persons who make up the group. Thus it is very difficult for the practical moralist to control that part of life which is in many ways the most influential. Thousands of mothers bringing up thousands of babies are a very difficult subject for anyone to influence; and thus we cannot hope for a sudden change in the moral character of a nation. Clever and stupid, careless and careful, must alike be instructed and controlled. So long as the structure of family life remains the same, the moral training that each receives must vary greatly. Moreover, the ideas that must be learnt are not easy for some people to grasp. During the last twenty years or so, pre-natal and baby clinics have taught a large part of the nation's mothers how to care for a small baby. They have learnt, practically, how to supply his physical needs, and the result is a decrease in infant mortality, and a surprising increase in the health and beauty of the babies one meets in the streets. But when

[1] *Social Psychology.*

a child grows older the skill needed to deal with him psychologically is much harder to acquire, and no widely spread organization has started to teach it. For one thing children present many different types of character, and, while fresh air and regular feeding hours are good for all babies, it is harder to lay down precise rules for dealing with the behaviour problems of childhood. Moreover, to deal with a child physically only demands a rather simple intelligence and rudimentary self-control; to guide a child through his growing years demands a far higher level of ability and a far more developed self-control. Thus if the nation set itself to the task of giving psychological education to all parents, only a proportion would be able to profit from it fully. Still the difficulty is no adequate reason for the complete neglect of this most important element in the moral education of the nation, and until it is remedied we shall never advance as far as we might along the path of virtue.

In this chapter we are discussing the part that the home plays in the child's moral development, or rather the harm that mismanagement in the home can do. We shall divide the topic into two parts. In Europe we have developed a special type of family unit, and we have exported it to America and other countries that have come under the influence of our religious and social ideas. This special type of family produces problems that are very fundamental, and deeply influence the development of each child. A discussion of this we shall keep to the end of the chapter. In the first half we shall discuss matters less fundamental but still of the greatest importance for behaviour. We have taken the topics in this order because we do not want to suggest that the European type of family *need* have bad results, or that the ills that undoubtedly occur are directly attributable, in most cases, to this type of family structure. They are due to mismanagement in matters which are far less fundamental and which can be put right without in any way disturbing the family.

We can group the troubles of a child roughly under the heading of frustration. Frustration is the state of mind experienced when a purpose or desire is thwarted—when we fail to get what we want. It may be an ice-cream we have set our hearts on, or doing well at school, or marrying the girl of our choice; but in any case, if we do not get it, we experience an unpleasant emotion which varies in violence according to the importance of our desire. The more fundamental our desire, the more violent our reaction: unless indeed the desire is so fundamental that its frustration produces despair and lethargy. This emotion that accompanies frustration is hostile to someone, or some thing; it is anger, at least in part, and desire to harm the cause of our suffering. We are not, of course, always very clear who *is* to blame for our unsuccess; and our anger will often lead us to attack some quite innocent person, because we do not know, or cannot easily get at the real cause. Thus children who have been thwarted by one person will often attack a sympathizer, or a child who is tired will grow angry with anyone he meets. On the political level the victims of economic depression will lynch a negro, or develop anti-semitism. This anger and aggression arising from frustration is the very antithesis of benevolence; and that, we have already said, is the basis of moral behaviour.

For a long time this sense of frustration and its expression in aggression seems to have escaped the notice of psychologists. They made no protests against the theories of discipline which were in essence based on the frustration of all natural desires. Now suddenly it has become a truism that the two are linked, and we have pointed out to us[1] that frustration is the root of all vice and that if we wish people to be good we must make them happy and give them what they want. This does not, of course, mean that no child must ever be thwarted. In that way we should produce a race of monsters—of Neros or Loebs and Leopolds, the sons of American millionaires who killed a school-fellow for the pleasure of committing the "perfect crime". It is part of the process of socialization that each gives way to others, but it is necessary that this surrender should be in accordance with nature. It is also necessary, on occasion, that the individual should accept absolute frustration; as when he fails to secure a coveted appointment, or when it rains on the day of a picnic. But the frustrations of which the psychologist writes are somewhat different. If, over a considerable period, a child or an adult is placed in a position in which his natural impulses and desires are continually thwarted, in which he feels the victim of injustice or maltreatment, then anger grows and his attitude to the world becomes one of aggression and hostility. He cannot develop benevolence, and he tries to revenge himself for his unhappiness on any one who comes in his way.

If we take the main desires of the baby, they are love and food; of the growing child, independence, security, self-esteem, with love and food as important as ever. The youth needs increased independence, a place in the community, the right to use his powers and think well of himself. If these desires are seriously thwarted, then deep damage occurs, and it is hard in later life to repair the harm that has been done. After the first years, other influences besides the home affect the child. Perhaps the chief of these is school, and too often in the

[1] *Frustration and Aggression*, Dollard.

past, and even sometimes in the present, school has been the scene of the cruellest frustrations. The child has been ridiculed, punished, deprived of all the legitimate pleasures of self-esteem, with the result that he has learnt little in school, and forgotten most of it as soon as he has escaped. But, compared with the home, school has a lesser effect. A bad school may make a child hate learning; it is not likely to make him a criminal, as a bad home may. Because a school is less intimate and fundamental than a home, it has a more localized effect.

For the tiny baby the two things that matter are food and love; love carrying with it the assurance of safety and care; and it is the lack of these two that causes most damage. It is not only insufficient food that affects a child, but also irregularity of feeding-time, so that he grows hungry and frightened. The modern baby, as we have said, who lives by rule is saved from distress. He gets what he expects, the rhythms of his body adjust themselves, and there is no frustration. The other frustration that can affect the small baby is lack of love. There is no doubt that in some way, not very clear, affection penetrates a child's mind and affects his whole being. The unloved baby does not really thrive, though it may receive physical care. The stupid mother shows her lack of love by careless handling, a slap now and then, and the absence of the tenderness that a baby early begins to respond to. If one reads the accounts that psycho-analysts give of their patients one can only assume that many of them were unloved babies, or that they had mothers so stupid and egoistic that they could not show such affection as they felt.

The psycho-analyst Melane Klein and her follower, Susan Isaacs, maintain that children as young as a few months have already developed psychological difficulties. It is very hard to accept the account of the babe's mental processes as given by these writers, but even without believing what seem to be rather unpleasant fairy tales—stories on the level of Grimm's ogres—it is possible that they are right in general, and that mismanagement has already, at this early age, had an effect, and is preventing the normal physical and psychological development of the child. On the other hand, the number of children so affected is probably not large, at least in England. Modern methods of baby management and the teaching given to young mothers lead to healthy, friendly babies who seem very free from psychic trouble.

When the child passes from early infancy, the matter is much more difficult. The child is a little nearer to the adult—not much, but a little—and parents begin to apply adult standards to him. This is liable to produce a double frustration. On the one hand, the parent is pushing him forward and demanding actions from him that are in advance of his power. When he fails he is blamed, or punished, and he suffers in his self-esteem. On the other hand, in order that he may not give trouble, his independence is curtailed. He is not allowed to do things for himself or in his own way. He is prevented from engaging in activities that seem natural and charming to him, like making mud-pies. He may, further, not even be sufficiently controlled, so that he lacks the sense of security, and feels continual, if undefined anxiety. It is impossible to discuss all the ways in which a child can suffer;[1] we must confine ourselves to a few examples. Perhaps the most conspicuous example of an attempt to push a child beyond his own stage, and one which according to the psycho-analysts causes most trouble, is the training of a child in cleanliness. If a child does not learn control of excretion many mothers become anxious. They punish the child for being "naughty" and attach a vast amount of emotion to the performance of necessary functions. If children are left alone till their muscular development is adequate, they become clean quite suddenly—as a puppy does—and suffer from none of the inhibitions and fixations that seem to trouble less intelligently treated children. It has been argued that all kinds of psychological abnormalities, such as miserliness, arise from troubles in evacuation in childhood.[2] Psycho-analysts have never applied the canons of formal logic to the presentation of their theories, so that their assertions do not carry great conviction; but there is no doubt that a child is damaged both by the attachment of emotion to the act—the anxious mother imparting her own urgency and uneasiness to the child—and also by the sense of failure and inferiority when he fails to do what is required of him.

Similar difficulties, though perhaps not so important, arise from the demand that a small child keep his clothes clean, that he "eat nicely", that he does not "make a mess", or that he "keeps quiet". Failure to comply with these demands is usually punished with a smack, or, if a blow is not given, it is threatened. The child then finds himself up against a wall of adult demands that he is unable to satisfy, and his failure brings a continual atmosphere of hostile emotion. The child is unhappy, and tends to give up activity if all he does involves him in trouble; or he may, if of a very resolute disposition, decide that as he is wrong anyhow he had better indulge his natural tendencies to the full, and give up trying to please.

In his relation to adult demands the modern child has certain advantages over his predecessors. His clothes are so simple that they are easy to wash, and his mother is so busy shopping or doing the

[1] *Poil de Carotte*, Jules Renard.

[2] *Psychology of Social Movements*, Pryn Hopkins.

housework that she must leave him to his own devices for much of the time. She may even send him to a nursery school, where he will have an environment specially suited to his needs. On the other hand, the child of an earlier day had a nursery where he was lord. He did not have to live continually in the company and among the furnishings of adults, where he cannot set out his trains without getting in the way, or model in clay without dirtying the table.

There is a third cause of difficulty, less now than in the past. Parents are anxious to "get their child on". They want the glory of an unusually accomplished child. In Victorian days it was not uncommon to teach a child to read at three and to start Latin at five; and paragons of precosity, like the young John Stuart Mill, sometimes resulted. But with the ordinary child things were very different. Unnatural demands had to be made on the child's power of attention, and the result was irritation for the parent and misery for the child. Samuel Butler[1] has described with admirable cruelty this type of education and its results.

"Before Ernest could well crawl he was taught to kneel: before he could well speak he was taught to lisp the Lord's Prayer and the general confession. How was it possible that these things could be taught so early? If his attention flagged or his memory failed him, here was an ill weed that would grow apace, unless it were plucked out immediately, and the only way to pluck it out was to whip him, or shut him up in a cupboard, or dock him of some of the small pleasures of childhood. Before he was three years old he could read and after a fashion write. Before he was four he was learning Latin, and could do rule-of-three sums."

Or again more shortly.

"When Ernest was in his second year, Theobald, as I have already said, began to teach him to read. He began to whip him two days after he had begun to teach him. 'It was painful', as he said to Christina, "but it was the only thing to do, and it was done'."

The poor child, exposed to this kind of teaching, could never acquire proper confidence or self-respect. At best he could become a prig, and console himself for his own sense of deficiency by thinking himself superior to the others. When the child went to school, even up to a few years ago, he was faced with the same unnatural demands. He must sit by the hour on hard benches and con his A, B, C, or he must control his attention through the weary stretches of an oral lesson on the *Tails of Animals*. If he failed he was beaten, for the teacher taught with the book in one hand and the cane in the other.

[1] *The Way of all Flesh.*

To-day, in infant schools at least, methods have been devised which by fitting the child's nature, have rendered success possible, and saved his self-respect. Moreover the part that maturation plays in learning has been realized. It is of no use teaching a child a thing before he is ready; but once he has developed sufficiently his progress will be swift, and he may easily overtake one who has started earlier with a less complete endowment. Thus even those most anxious for a child's progress may be content to wait.

If a child was pushed on too fast in one direction, he was all too frequently restrained in another. The child's attempt to develop individuality and independence was resisted for a variety of reasons. Adult convenience played a large part, so did a too possessive maternal love, fear of what independence might bring, and even religious conviction. A Protestant theologian of the end of the last century discusses family life as part of Christian ethics.[1]

"It is clear that it is rightly said that love in relation to education shapes itself in the form of authority and respect; and the sole, all-comprehending virtue of the child is obedience based on gratitude and trustful reverence."

To a child things look very different. He may love his parents, as a child can love, but already by two his great urge is for his own development, and this means independence in which to develop. It is not absolute independence that he craves—he would be utterly lost without the authoritarian background to his life—but an independence that grows with his powers, and that is never frightening. It is, of course, an independence of small things. He wishes to do for himself the things he feels within his powers, or almost within them. The great cry of the enterprising baby of this age is, "I can do it", and he resists help from adults in such matters as putting on his clothes, feeding himself or washing. Frequently it is more convenient for a parent to do these things herself, but the effect on the child is very bad. The thoughtless, busy or domineering mother or nurse will not wait while a child fumbles with buttons or solves the problem of getting his foot into a sock. The matter is one in which the child's honour is involved. If he is picked up, dressed, and then set down again, he may have suffered severe psychic damage. It is possible to render a child a nervous wreck at six simply by doing everything for him; and it may take years in a more suitable environment before he recovers normal self-confidence. Mme Montessori was the first to point out in regard to school work how important it is that a child do everything possible for himself. She would have toddlers trained on

[1] Haering, *Ethics of the Christian Life.*

special apparatus so that they can learn to undo buttons and fasten shoelaces. There should be chairs and tables they can move, low cupboards and free access to them. Babies trained in this way became self-reliant, competent, quiet and orderly. On the other hand, a child thwarted in his efforts to help himself, who has thus had his self-confidence destroyed, develops characteristic vices. He is unenterprising and timid, or else violent and noisy. He refuses to try and do things or does them deliberately badly. He needs to be reassured, and so he boasts and lies. He attempts to control his environment by his incapacities rather than his efficiency. He is peevish and exacting.

Closely akin in psychological demoralization is the spoilt child. He too has had everything done for him, and, in addition, he has been taught to believe that he is of supreme importance, talented and wonderful beyond others. To the defects already mentioned there is added a weak pride that will not let him accept his proper position, but also will not let him really make the effort to gain a better one. He is the student who stages a nervous breakdown on the second day of his examination, or the brilliant *flâneur* who goes through life, like Waring, with "No work done, but great works undone", living in a fantastic world of splendid projects and no fulfilment.

At first this desire for independence has little philosophical basis. A child wants to try, to do things, to experiment. But by six a child is conscious of himself as an individual, and experiences the need to assert himself against a too enveloping environment. He wishes to question or transgress the conventions of his group.

In this he is claiming the right to individual judgment and intelligent thought. This independence is greatly helped by going to school. There, escaped from home, a child makes his own friendships, has his own experiences, and learns to manage his own social contacts.

It is hard for a fond mother to realize the urgency of a child's desire for independence or the success with which he assumes it. At a communal dining-room the authors saw a boy of five fetch his own plate of food choosing which dish he preferred. Then, having got it, he refused to sit with his own family and walked to another table where he entered into amicable conversation with three strangers. His behaviour was completely correct and adult. When, two years later, he was sent to boarding-school he accepted the new life without the slightest difficulty. It is hard for a mother who believes in the sacred and peculiar rights of parents to find that, so soon, her child can do without her, or at least appears to do so. In fact, the child who claims, and achieves, a great deal of personal independence relies on the background sense of security that home gives; but even this can be dispensed with. A girl who had left Germany with her younger brother as a refugee when she was about twelve was telling of her experiences. She was asked: "Were you frightened?" "Oh, no", she said, "not so long as we had enough to eat". And she managed quite happily in the strange country for two or three years till her parents were able to join her.

The child who refuses to be parted from his mother, who cries distractedly when left alone and dreads independence, is the product of the mother's own egoism. She has clung to him till he has come to believe in the dangers of the world and his own helplessness. He has had no experience, acquired no skills in dealing with the world. He will remain incompetent through life, unless, before it is too late, he is forced to face an independent existence. One conscript gifted with certain powers of self-observation wrote to his mother after some two months' service: "I have learnt a lot of things in the army, I can change my clothes, clean my boots and be on parade in five minutes; I can punch another fellow on the nose; and I have discovered that if I don't look after myself no one else will look after me." It is this last discovery that comes easily and naturally to the child allowed independence.

In his play *The Family Reunion*, T. S. Eliot deals with the infinite damage a dominating, disappointed mother can inflict on her sons. She took from them completely the sense of freedom and discovery. Harry and his cousin Mary are discussing their memories of childhood.

MARY: Well it all seemed to be imposed upon us;
Even the nice things were laid out ready,
And the treats were always so carefully prepared;
There was never any time to invent our own enjoyments.
But perhaps it was all designed for you, not for us.

HARRY: No, it didn't seem like that. I was part of the design
As well as you. But what was the design?
It never came off. But do you remember

MARY: The hollow tree in what we called the wilderness

HARRY: Down near the river. That was the stockade
From which we fought the Indians, Arthur and John.

MARY: They never found the secret.

HARRY: Not then. But later, coming back from school
For the holidays, after the formal reception
And the family festivities, I made my escape
As soon as I could, and slipped down to the river

To find the old hiding-place. The wilderness was gone,
The tree had been felled, and a neat summer-house
Had been erected, "to please the children".
It's absurd that one's only memory of freedom
Should be a hollow tree in a wood by the river.

Even when a child is not unduly restrained, the period round about seven is often one of great difficulty. In these years the child often suffers acutely from nightmares. Perhaps the most characteristic form is the dream in which the sleeper, with clogged feet, runs away from something that will catch and swallow him. The emotion that accompanies these dreams shows that the problem they represent is not solved, and their cessation generally coincides with the child's further acquisition of autonomy, often through the power to read more freely. R. L. Stevenson, in his fable, *The House of Eld*, was probably trying to express, perhaps for an older rebel, the felt necessity to escape from the too close grip of family life.

A similar crisis occurs in adolescence. Again the child feels that he has outgrown the frame in which he is held, and longs to achieve yet more fully his own individuality. Hence to a large extent the bad behaviour of the adolescent who is anxious to achieve what his environment seems determined to deny. He wants his own life, and therefore his own money. He wants the status and self-respect of a job and the plaudits of his friends for some achievement, legal or illegal. He wants a place in the adult group, and often it seems long before he is accepted.

This doctrine of the importance of independence is, of course, completely opposed to the belief, held explicitly in the past, and too often implicitly in the present, that rigid, unremitting discipline is the best preparation for life. The child who was always under the eye of his parents at home, or sitting quite still at school—when he wasn't standing in line, equally still—was assumed to be getting a training that would fit him for the trials of his future life. The girl educated in a convent, where she was never alone for a moment, and where her whole life was patterned, down to the smallest detail, was supposed to be acquiring habits that would later keep her safe from temptation. In that mine of information about the beliefs and practices of an earlier day, *The Fairchild Family*, there occurs a very good account of this kind of training, and also the incompetence and untrustworthiness to which it leads. That this kind of thing is only just passing into history is clear from the life-story of an eminent general, still under sixty, who was brought up for a large part of his youth under rather similar conditions. His exceptional abilities enabled him to succeed later, but his path was undoubtedly made much more difficult by the effect on his character of this training. In *The Fairchild Family* the chapter is ominously headed: "Story of the constant bent of man's heart towards sin", and tells the sad tale of how the children fell from grace on a day when their parents had to leave them at home under no other charge than John, the faithful man of all work. Their normal guarded life is first described.

"I will tell you exactly how they lived and spent their time: Emily and Lucy slept together in a little closet on one side of their mamma and papa's room, and Henry had a little room on the other side where he slept. As soon as the children got up they used to go into their papa and mamma's room to prayers. After which Henry went with papa into the garden while Lucy and Emily made their beds and rubbed the furniture: afterwards they all met at breakfast, dressed neatly but very plain. At breakfast the children ate what their mamma gave them, and seldom spoke till they were spoken to. After breakfast Betty and John were called in and all went to prayers. Then Henry went to his papa's study, to his lessons, and Lucy and Emily stayed with their mamma working and reading till 12 o'clock when they used to go out for a walk all together; sometimes they went to the schools and sometimes they went to see a poor person. When they came in dinner was ready. After dinner the little girls and their mamma worked, whilst Henry read to them till teatime; and after tea Lucy and Emily played with their doll and worked for it; and Henry busied himself in making some little things of wood, which his father showed him how to do. And so they spent their time till Betty and John came in to evening prayers. Then the children had each of them a baked apple, and went to bed.

"Now all this time the little ones were in the presence of their papa and mamma and kept carefully from breaking into open sin by the watchful eyes of their dear parents."

How artificial this life was is shown by the disasters that descended on the children the moment the parental eye was removed. Business compelled the parents to be away for a day and John was left in charge. Things went wrong at once. They got up late, failed to make their beds or wash their faces, ate too large a breakfast of toast and butter, chased a pig out of the garden, down the lane and through the mire. Wet and dirty they visited a family they had been strictly forbidden to speak to, as "these people were not such as lived in the fear of God, neither did they bring up their children well".

In spite of this ungodliness, they treated the children kindly, but gave them a drink of cider, which made them so sleepy that they sat down on the bank on the way home and fell asleep. When they got back to poor harassed John he could think of nothing better to do with them than to *lock* them in the play-room, where they sat and thought about their sins. He later *fastened* them in the barn where they fell out of a swing, and at last he *tied* two of them to the table in the kitchen with his blue pocket handkerchief. It is clear from all this that they could not be trusted alone, though the youngest was over seven, and they were so unskilful that even swinging was unsafe.

At the end of this disastrous day, when they hear the sound of the horse bringing back their parents they are terrified.

"'Oh, John, John! What shall we do? What shall we say?' said Lucy.

"'The truth, the truth, and all the truth', said John. 'It is the best you can do now'."

Then the parents enter:

"'Oh, Mamma, Mamma, Papa, Papa', said Lucy, coming forward, and falling on her knees before them, 'we have been very wicked children to-day. We are not fit to come near you'.

"'What have you done, Lucy?' said Mrs. Fairchild. 'Tell the whole truth and pray to God to forgive you for His dear Son's sake. These are the only things that children can do, when they have been naughty, to make their peace with God and their parents'."

It is worth noticing in passing, how different these children are from the efficient, reliable and independent children who appear to-day in books by Arthur Ransome.

If a child is exposed to the frustrations that we have so briefly surveyed, he cannot develop as well as he might. The schools all know the surly dunce who is always trying to make trouble and can only be won over, if he can be won at all, by being given encouragement, consideration, and even respect. So, too, the child in the family, if deprived of love, or care, independence or security becomes hostile, and, instead of benevolent, anxious to inflict on others the obscure unhappiness from which he suffers. The child who remains consistently naughty after about nine or ten years of age has almost certainly been mismanaged. The adolescent who is rude, dishonest, disobedient or destructive has certainly not received a proper upbringing. The figures of juvenile delinquency show such a high proportion from homes where the conditions can be assumed to have been bad, that it is clear that, in most cases, the criminal is made by incorrect treatment in his early years. If the mismanaged child is not criminal, he is apt in adult life to be domineering, unkind, unreliable. They may be successful, and make bad masters; or they may fail, and make bad workmen. When they exist in large numbers, the State cannot be healthy, and their presence makes possible many dangerous forms of political activity.

We have given only a few of the ways in which a defective home life can disturb the moral development of children. In the second part of the chapter we wish to deal with problems on a different level. Neither writers on ethics nor psycho-analysts seem to have realized that the European family, as we have developed it, is a special type of institution which has the deepest effect on all its members. Within the family a number of our strongest instincts have their field of activity; it holds and moulds us during our most impressionable years; it continues the centre of our material and emotional life to the grave. For our particular type of family, we claim religious sanction, and many people regard it as almost impious even to suggest that other forms exist and give satisfaction in other cultures. To criticize our own is wicked, and yet, when the European family fails, it fails so badly that hardly anything could be worse. Perhaps this is due to its being a very difficult form of association requiring very high qualities in the partners; and where there is egoism, silly or ungovernable emotions, these have the fullest opportunity for making everyone miserable. The marriage that is more or less unsuccessful is known to everyone, and the partners live out their lives as best they may. The books of psycho-analysts are full of accounts of the emotional disturbances of family life. Sons are jealous of their fathers and love their mothers to excess; daughters adore their fathers and cannot bring themselves to marry; elder and younger are at strife or in an orgy of self-sacrifice. All these disturbances, if we look at them carefully, have their cause in the special emotional atmosphere that grows up only too easily in our type of family. They have no natural place in many other types; and when anthropologists are looking for them, they express surprise at their absence. To say this, is not to condemn our type of family. At its best, it is an institution leading to the noblest association among its members. It is only to say that, in many cases, it proves too difficult for those who are involved; and in those cases grave moral and psychological damage results. The practical moralist must try to understand the special difficulties and seek ways to overcome them.

There are several elements in the theoretic structure of the European family that are important. It is monogamous and patriarchal. The spouses are bound by a unique emotional tie, and a similar unique tie connects the children to the parents. Lastly, the family is an individual property-owning unit. In the last fifty years or so, it has become cut off from the clan, and the characteristic attitude is: my house, my husband, my child, my silver tea-pot; and the best-natured mothers will say such things as "Of course, grandparents are *a menace*", while the idea that a mother can be friends with her son's wife is slightly shocking. In return, the clan ceases to function as a support to the individual family, which thus shrinks to one pair of adults and one or two children, alone, without aid, left to manage as best it can. All these characteristics have important effects both on the spouses and the children, and it is perhaps worth considering them separately.

Monogamy, except in the parts of the earth brought under the influence of European Christianity, is rare as a social institution. In the European type of family it is, nowadays, connected with the special emotional tie between spouses that will be discussed later. This emotional tie, the religious sentiment and the property element in the family have established it firmly. It is indeed a necessity if these ideas prevail; but over the mass of mankind they do not, and polygamy is practised for its greater economic convenience. The one-man-one-woman group is, practically, very unsatisfactory. Every woman after the war years knows it. Endless cooking, no one to take charge of the baby when you want to go out at night. A recent play[1] demonstrated the hopeless parallelism of the lives of two neighbours. Both up early, both making the fire and preparing the coffee, both hurrying the children off to school. Neither with a moment's free time. In the past, among the well-to-do, the inconvenience was avoided by importing other women of inferior status—domestic servants—to complete the group. These maids, excluded from the privileges of wives or even concubines, held, at least before 1914, a semi-servile position, and were expected to wear a definite uniform as a badge of status. They could not claim the dignity of an economic share in the family, and their relation to the children was special and peculiar. So long as these servants were available, the difficulties of the monogamous grouping were not felt by the richer classes and the leaders of opinion; while the poor took their ideas of what was suitable from their betters. When servants are not available, polygamy springs up naturally, and is in no way necessarily connected with a low position of women. Among the Mormons colonizing Utah, the custom of "living plural" grew up not through lust but convenience, and the woman's position was carefully guarded. When Utah joined the Union, and these polygamous wives, now technically "living in sin", were offered asylum by a benevolent federal government, they showed

[1] *Our Town*, Thornton Wilder.

no desire to accept the proffered change. The West African women, who are farmers and traders, are horrified at the fate of the European woman for ever tied to her cooking-pots and children. They have well-developed arrangements by which each does the work of the family in turn, and the others are free for their own business. So far from this form of polygamy making for the subjection of women, there is no surer way of depressing the sex than binding them to perpetual housework and baby-tending, and thus making any profession or outside work practically impossible.

A great deal of the unhappiness and frustration of family life has come from the position of woman as continual houseworker. Clever and ambitious women have felt their gifts unused and have turned them to domestic tyranny. The comparative solitude, and the endless repetition of jobs, which are only done to be done again, bring on many women nervous ailments that render them bad wives and mothers. While maintaining the monogamous structure of the family, it is essential for happiness to get rid of the chief disadvantages and set women free for a wider and more interesting life. In order to achieve this the absolute individuality of the family must be broken into. There must be some advance towards communal living. The two fields in which this is most important are food and the care of children. The simplest devices of all for communal living are canteen and school dinners, which set the individual housewife free from cooking for some hours of the day. The other is the crêche or nursery school that keeps the child in a satisfactory environment for the hours that a woman needs to transact the business of life. Some groups go farther and devise a life which, while preserving a sufficient element of privacy, allows housework and child-minding to be done communally, thus setting free some of the women every day. Settlements of Jewish agriculturists have evolved communistic experiments that go much farther. The essential elements are communal life and work and communal care of the children, who are placed in a Children's House and not left to the individual parents.

Such experiments grow slowly, even under the stress of war. There is a myth connected with the family which stresses very strongly the necessity of individualism, the myth that any larger grouping *must* be bad. This is often a male idea, because it is mainly the men who get what advantages there are to be had from the separateness, and wish to perpetuate it. The girl who is taught that "in *my* house I mean to have things done *my* way", is as much a victim of the myth as the man who says "I want *my* wife at *home*".

Almost the same sort of position exists in regard to children.

Monogamy means, in general, a family of not more than four or five children. It may mean one or none. On these children the parents lavish a possessive affection, often far in excess of what a child can bear. Moreover, the child may be cut off for many years from the company of children just his own age. The polygamous family has many children, several probably of just the same age; and the West African child, running in a group with his siblings, would not exchange places with the solitary Britisher. For the child over five the answer is, of course, provided: the child goes to school. Enterprising and determined children often get there earlier, forcing their parents to let them join the wider world which they know provides amusements not to be found at home. The number of schools for the younger children is gradually growing, and the more progressive housing schemes are incorporating nursery schools in the buildings. But there is still a failure to understand what children need. In Vienna, the great blocks of flats always contained a children's room where the children of different families could meet and play. This is important both for the children and for the parents. The small modern house has no place for fidgety, restless children along with grown-ups, and unless there is a place for children to go, there is frustration for both older and younger. In England we have not realized this. The idea of the segregated home still maintains itself, and the child, unless it goes to a club, has nowhere to go where it can find company, escape too close adult control, and have space for its own activities.

Among the causes that have delayed the realization of the importance and desirability of at least partial communal living is the theory of the dominance of the father and the ownership of the children. The feeling that a man must be able to come home and find his slippers put to warm, his wife waiting to serve him, and that for the rest of the day she must sit one side of the fire and he the other, ignores the difference in their day's activities. Not only must the wife be there to give him her exclusive attention, but the children must be under control so that they shall not acquire too much independence or learn ideas that are not in accord with their father's wishes. This patriarchal domination is gradually dying out, but it has been almost incredibly strong in the past; strongest, because of its connection with Jewish thought, in those who were the most religious. "I stand to you in the place of God", says Mr. Fairchild to little Henry reluctant to learn Latin, and as God he chastised his rebellious subject. This attitude has received the support of law and economic custom. Till the passing of the Married Women's Property Act the husband was in supreme financial control, and the wife as dependent on his bounty as the children. So long as the woman is tied to the house and the children, she can do little, even to-day, to achieve any economic independence. In the past, when all girls and many boys were brought up ignorant of any means of earning their living, the father's grip of the family purse made him lord of their fates. Papa might be a genial tyrant or a benevolent despot, he was none the less tyrant and despot, eager in many cases to keep his flock in absolute subjection. The picture of family life in such a household as the *Barretts of Wimpole Street* raises a number of problems to-day. One of the chief is why the young men of the family remained under their father's roof and there submitted to his domination. There is no doubt that the economic motive was strong, but it seems hard to believe that they could not have asserted themselves to some extent. The probable answer is that their father had beaten them into submission in early life, and that they had never recovered from the dread he then inspired. To quote again from *The Way of all Flesh*, we can see how fathers then instilled in their children the absolute fear that a more liberal age finds surprising.

It was Sunday evening and the children were singing hymns. . . .

"Ernest was, however, very late in being able to sound a hard 'c' or 'k', and instead of saying 'come', he said 'tum, tum, tum'.

" 'Ernest', said Theobald from the armchair in front of the fire where he was sitting, hands folded before him, 'don't you think it would be very nice if you were to say *come* like other people instead of *tum*'?

" 'I do say *tum*', said Ernest, meaning that he said *come*.

"Theobald was always in a bad temper on Sunday evenings. He noticed the fact that he was being contradicted in a moment. He got up from the armchair and went to the piano.

" 'No, Ernest, you don't', he said. 'You say nothing of the kind. You say *tum* not *come*. Now say *come* after me as I do.'

" '*Tum*', said Ernest at once. 'Is that better?'

"I have no doubt he thought it was, but it was not."

The miserable dialogue continues for some minutes. Then it approaches its foreordained end.

" 'Now, Ernest, I will give you one more chance, and if you don't say *come* I shall know that you are self-willed and naughty.'

"He looked very angry and a shade came over Ernest's face, like that which comes over the face of a puppy when it is being scolded without understanding why. The child saw well what was coming now, was frightened and, of course, said *tum* once more.

" 'Very well, Ernest', said his father, catching him angrily by the shoulder. 'I have done my best to save you, but if you will have it so, you will', and he lugged the little wretch, crying in anticipation, out of the room. A few minutes more and we could hear screams coming from the dining-room across the hall, and knew that poor Ernest was being beaten.

" 'I have sent him to bed', said Theobald as he returned to the drawing-room. 'And now, Christina, I think we will have the servants in to prayers.' And he rang the bell for them, red-handed as he was."

It would be rare to find a father to-day who claimed such a position of absolutism, but, in so far as there remain people who were brought up under the system, we are still suffering from it. There are, moreover, parents who still claim greater rights over their children's lives than freedom sanctions, and in so far as these exist there must be emotional strains in the family that can only injure the members. The unhappy disagreements that some people seem to regard as an inevitable part of the relations of parents and children take their origin largely in this claim to absolutism.

The law still supports the idea that up to a certain age the child is *owned* by its parents. It is illegal to take a child away from the custody of the parents, unless they are guilty of cruelty or gross neglect, and, at a later age, the parents have a ground of action if they are deprived of the child's services. The father also has the right to decide what kind of religious education a child shall have, though he has lost the right to deprive him altogether of any education. Again, in law, the father's claim is superior to the mother's, and, should disputes arise, the father has the superior right.

The British child is thus in a very different position from the child, for instance, in New Guinea. There no-one has any rights over him. He goes where he will, and if he does not choose to come home he can stay with any of his relatives (and the whole village is intermarried) till he is tired and desires a new roof. When we consider our own theory of the parents' property in the child, and the fact that he has no rights in any other home, it is clear that a child in an unhappy family environment has no escape. Under certain circumstances this may become a matter of extreme importance to the child. Even where he is not treated with cruelty something in the environment may not suit him. A home that appears quite good to the casual observer may contain elements that make a child acutely unhappy. This unhappiness shows itself in all kinds of ways. The least severe type of strain may appear in nail-biting, the more in playing truant and lying. The child of well-to-do parents may agitate till he is sent to boarding-school, where he will probably improve rapidly. The child from a poorer home has no refuge, and may have his character permanently warped. If the conception of the parents' ownership of the children was slightly relaxed, and the provision of public boarding-schools increased, many children would have a far better chance of growing up good and happy.

All this is complicated by the belief that children must necessarily love their parents. It was for a long time a literary convention that parent and child, when meeting, unknown, after years of separation, experienced a peculiar attraction for each other that could not be gainsaid; and while they lived together any other emotion than love was wicked. Aunts and more distant relatives were included in this obligatory affection, though at a lower level as became their inferior status. Where the parent's conduct was worthy of love, he received it—as did anyone else who ministered to the child's life—but where experience showed the parents harsh or unreliable, too restrictive or possessive, the child's emotions were different. There arose, therefore, in many children a state of mind to which psycho-analysts have drawn special attention. The child both hates and loves his parents. The love is the natural reaction to such care and friendliness as he has received; the hate, to harshness, excessive control or hurtful emotion. There are few people who have engaged our emotions at all strongly who are not to some extent the recipients of this ambivalent emotion.

Odi et amo. Quare id faciam fortasse requiris.
Nescio. Sed fieri sentio, et excrucior.

The danger for the child arises when he is taught that any other emotion than love for his parents is wicked, and then this "love" is made the basis for excessive control and emotion. The mother demands service in the name of love, and "loving mother" becomes the keynote of life, the reason why he must forgo all natural and desirable amusements. If he plays with his friends or has his own ideas, mother is "hurt", and to hurt her is a *sin*. Thus the child builds up a complex state of mind in which love and hate conflict, and a sense of guilt poisons his actions.

Possibly the most important element in our type of marriage is the tie that is assumed to exist between the spouses. Among most races to-day, and in Europe up to a comparatively recent date, the relationship between the parties was supposed to be fairly simple. They were agreed on sexual relations, they intended to do their duty by each other as was customary in their culture, but there was not supposed to be the intense emotion that is considered proper to-day. It was not even thought amiss to start with a "little aversion". Moreover, when the marriage bond was essentially economic, a matter arranged by the clan for settling property, the inclination of the parties was little considered. Emotional relationships, when they occurred, were generally extra-marital, the lover was the natural correlate of the *mariage de convenance*. When free choice became usual the emotional excitement was expected to increase. It became the common belief that, suddenly, one man felt an overmastering love for one woman, and she responded with a similar passion. "From all the girls in all the world, I chose thee", sings the sentimental tenor, and the truly infatuated is prepared to endure all things for this overmastering attachment.

"I know", says the heroine, speaking of the man of her choice, "that he will sometimes be brutal and often drunk, he will beat me, he will take me by the hair and drag me through the gutter, but he is my man, and I will follow him to the ends of the earth".

Where this sudden passion comes from no one says. In the Middle Ages it was recognized that Idleness was the gatekeeper of the Garden of Love and it is to sexual diversion, as we shall say later, that the unoccupied turn. A much admired recent film suggests that this passion, undesired, unexpected, may fall on anyone, and tear our hearts, if we have not the strength to give up all and follow where it leads. The existence of this belief about *Love*, fostered by the poets from Shakespeare's day to ours, and now strengthened incalculably by the cinema, to some extent obscures the real facts. A large proportion of actual marriages are entered into with much more reason, and there is a real appreciation of worth of character; but so long as the theory holds its position in the minds of the young, it is possible for a girl to accept a proposal of marriage from a man on one day's acquaintance, only to find out a week or so later that he is a murderer.

Bernard Shaw points out that, at a certain age, almost any man will do for almost any girl. This is, of course, not strictly true. But there are probably thousands that would be quite suitable to each other, if they chanced to meet when they were in the right mood. In the days when families arranged the matches of their children, suitability of character was a serious consideration, and there is no doubt that many happy unions resulted from this care. As the moral character of the spouses is the basis of success in the European type of marriage, the greatest change for the better must come when there is a return to a more thoughtful method of choice—one which is based on considerations of character rather than a moment of sexual passion.

A successful marriage is probably one of the highest types of human relationship, and within the bond love and virtue grow. On the other hand, if it fails there can be nothing worse. Moreover, without sinking to the lowest depths, much serious psychological damage can happen in marriage precisely because there is this myth of the unique emotional tie. In most marriages, in fact, this highly emotional relationship changes, in the course of a few years, into a constant affection; and this endures through life. When both parties are agreed emotionally, all is well; but one partner may crave excitements and raptures long after the other has tired of them.

So we'll go no more a-roving
So late into the night,
Though the heart be still as loving,
And the moon be still as bright.

For the sword outwears its sheath,
And the soul wears out the breast,
And the heart must pause to breathe,
And love itself have rest.[1]

When one partner fails to recognize this necessity there may easily be misunderstandings and recriminations; and disappointments and frustration follow. This was made all the easier by the convention, so strong in the past, that kept the woman comparatively unoccupied, if she was rich enough to have servants, and so gave her no other outlet for her energies and thoughts. When this lack of occupation reaches its highest point, as among the very rich or women in the tropics, quite characteristic results follow.

In theory the European marriage demands chastity before marriage and faithfulness after, thus ensuring that all the sexual energy shall flow into one channel. Milton, who was anything but a libertine, writes bitterly of the disadvantages of this custom. A man who has been quite chaste in his youth has not the experience to choose wisely, and, when passion does assail him, it is so violent, so irresistible that he can hardly bring his reason to bear on it. Milton's marriage, entered on in a month of roses, brought unhappiness to both. To Milton, who could never like poor Mary nor overcome his sexual passion for her, it brought the deepest humiliation and shame. To Mary it brought a life of loneliness, insult, and an early death. Nothing could show more clearly the failure of this theory of "*Love*" as a basis for married happiness. In a vast number of other cases the parties are not faithful to each other, and the passion that one person no longer inspires is renewed with another. It must almost always be sad for one spouse when this happens, but convention has added

[1] Byron.

further suffering; the "wronged party" feels disgraced as well as deserted, and the "guilty", if he has any conscience, oppressed by a sense of wrong-doing. These acute emotional disturbances are dependent on the theory of marriage, and do not arise in anything like the same form where the theory is different. In extreme cases the strains put upon the individual become intolerable. The law has recognized this clearly. When one partner to a marriage murders the other no further "motive" is looked for. It is understood that provocation within the marriage bond is quite sufficient to account for any act of violence.

All this is recognized by thoughtful people. The Committee on Procedure in Matrimonial Cases recognizes both the fundamental causes of failure in so many marriages, and also the difficulty of those who wish to preserve the marriage. They write in their Final Report: "We have been much impressed by the evidence of experienced workers in this field that the basic causes of marriage failures are to be found in false ideas and unsound emotional attitudes developed before marriage, in youth, or even in childhood . . . . The difficulty of the marriage relationship is attested, not only by the number of divorces, but by the growing demand for guidance by those who genuinely wish to preserve their association, but do not know how to do it."

When parents suffer emotionally from any of these causes the strain is passed on to the children. In the close group of the family the child is bound to be conscious of his parents' emotional condition. Even when the parents do not deliberately invite the child to take sides he is made part of the atmosphere of strife and unhappiness. Worse still, he is almost always directly involved. One parent or the other makes him an ally, and tries to establish with him a relationship only fitting between adults. The father claims his daughter as his support, or the unsatisfied mother tries to get from her son the emotional response her husband refuses. Such a situation almost always affects a child most unfavourably. If the child responds he naturally tries to possess one parent in an exclusive way, and there is jealousy and suffering. It is fairly clear that a child, while needing love, shrinks from passion; and the attempt to impose on him an adult degree of emotion, or even forcing him to be a continual witness of it is definitely harmful. Passion is an affair of maturity, it cramps and warps a child's growth. Thus the highly emotional family, especially where the emotion is partially frustrated, produces nervy children with all kinds of psychological defects who are unable to develop in their turn the calm affection that leads to virtue.

In one of his plays, T. S. Eliot, with the poet's insight, describes the effect on children of this kind of family situation.

Amy, the mother, having failed to keep her husband's affection—he felt so strongly about her he thought of murdering her—tried to get from her sons a compensatory adoration. This is Harry's account of the resulting education.

HARRY: . . . . What about my mother?
Everything has always been referred back to mother.
When we were children, before we went to school,
The rule of conduct was simply pleasing mother;
Misconduct was simply being unkind to mother;
What was wrong was whatever made her suffer,
And whatever made her happy was what was virtuous—
Though never very happy, I remember. That was why
We all felt like failures, before we had begun.
When we came back, for the school holidays,
They were not holidays, but simply a time
In which we were supposed to make up to mother
For all the weeks during which she had not seen us—
Except at half-term, and seeing us then
Only seemed to make her more unhappy, and made us
Feel more guilty, and so we misbehaved
Next day at school, in order to be punished,
For punishment made us feel less guilty. Mother
Never punished us, but made us feel guilty.
I think that the things that are taken for granted
At home, make a deeper impression upon children
Than what they are told.

It is perhaps not surprising, after all this, that Harry, in his turn, wanted to murder his wife.

There is also the isolation of the individual family. In its extreme form this is a comparatively recent growth. Even at the end of the last century aunts and uncles, cousins to many degrees of remoteness, played a large part in life. There were inter-family visits that lasted for months. Weddings, christenings, funerals were all occasions for meetings, and the family council decided important matters. The clan gave a sense of security to the individual and his immediate family. They knew they would not be abandoned or allowed to suffer too much.[1] Clan charity migh tnot be very pleasant, but it was an insurance against the worst. Moreover, there was always help in trouble. A maiden aunt was part of the equipment of every family,

[1] For a pleasant picture of such a Victorian family see Guy McCrone, *Wax Fruit*.

and in all crises she was telegraphed for, and came, uncomplaining and helpful, from one end of the British Isles to the other. Family councils now rarely meet to discuss what is to be done about poor Alfred's boy; and maiden aunts have to hold jobs and earn their living. At the same time the excessive individualism of modern families might well be decreased. Schemes of partial communal living help both parents and children. A less possessive attitude by mothers would encourage grandparents and uncles to take more interest in the children of the clan. Once the family comes to think of itself as an integral part of the larger unit a new spirit could easily grow up. The advantages would be many. A greater sense of security would make the growth of benevolence easier. Most of all the children would gain by having alternative homes. It is an immense advantage to a child to have another house to which he can go as a right. He may be bored or cross in one; he goes to the other and finds amusement and a different set of ideas. No home is perfect. By combining the best in more than one, he can form a picture of how life should be lived. Lastly, if the family were a more integral part of the clan, the aunts and uncles, especially if unmarried, would be able to satisfy their parental impulses by aiding the development of the children. Dr. Schiller[1] has emphasized the importance of the clan spirit from the eugenical point of view. It can also be beneficial from the financial. Often a young couple feel unable to face the expense of educating a large family though they want children; and yet in the clan there are members who could easily contribute to the charge, and enable them to have more children happily. Among people of other cultures the extended, clan-type family is generally associated with a benevolent and pleasant way of life. In England the tendency to separateness in the family is so strong that a really "united family" is something to wonder at, and almost deride.

When we come to look at the dreary catalogue of spoilt lives that psycho-analysis discloses, it is fairly clear that they are mostly due to those characteristics in the family structure that we have discussed. Perhaps they may be roughly grouped as arising from over-emotional emphasis in the relations of the spouses, and in their relation to the children. The belief that *love*, in the sense of sudden sexual attraction, is sufficient basis for marriage is partly responsible for one group of difficulties, and the further theory that this *love* remains constant through life for another. While children need the sense of care and security that a family gives they do not need or like any appeals to their passions. It is perhaps because of this that in many ages the

[1] *Social Decay and Eugenical Reform.*

children of the European élite have been educated away from home. The Spartans took a child from his mother at about the age of seven, the children of the age of chivalry were sent to be pages in the household of another; for centuries the British upper classes have sent their children away from home to be educated, even in places of such brutality and degradation as the Eton of Keate, believing apparently that it was better for a child to suffer even this than be left in his family during his formative years. To-day those who speak most movingly of the sanctity of family life usually send their children to boarding-school at seven. Why exactly they do it would be hard to say. It seems too often to be a combination of custom and laziness. Yet, except in an unusually loving and intelligent home, a child runs more risks of having his character warped by living with his parents than by the impersonal gregariousness of the school.

Some recent writer has suggested that what is taking place to-day is the democratization of the family. The period of patriarchal or even parental domination is passing away, and children are being asked to become partners in their own upbringing. This is not a bad summary of the changes that have been advocated in this chapter as far as children are concerned. Families seem to need federation. The independent, entirely autonomous family must be integrated in a wider grouping, either of the clan or the neighbourhood; and life readjusted, emotionally, as well as practically, to allow of this partial loss of sovereignty. If all this happens the decay of family life that is so deplored may cease, but a family of a rather different structure will emerge, one that is perhaps better fitted to give a life free from frustration and the causes of unhappiness and vice.

Books referred to in this chapter:

William Godwin, *Enquiry concerning Political Justice.*
Dollard, *Frustration and Aggression.*
Jules Renard, *Poil de Carotte.*
Melane Klein, *The Psycho-analysis of Children.*
Susan Isaacs, *Social Development of Young Children.*
Pryn Hopkins, *Psychology of Social Movements.*
Guy McCrone, *Wax Fruit.*
M. Montessori: *The Secret of Childhood. The Montessori Method.*
J. C. Flugel, *Psychoanalytic Study of the Family.*
T. S. Eliot, *The Family Reunion.*
R. L. Stevenson, *Fables.*
Thornton Wilder, *Our Town.*
Report of Committee on Procedure in Matrimonial Cases, H. M. Stationery Office.

PART II

THE ORGANIZATION OF SOCIETY IN RELATION TO MORALS

Introduction

In the first part of this book we have tried to present three ideas, mainly psychological, that man's inherited nature does not necessarily predispose him to evil, rather to good; secondly, that his potentially evil traits can be so modified by training that they may become highly beneficial—as when egoism is realized through socially helpful work. Lastly, we have suggested how the evil and malice in man's nature is produced by frustration and certain of the arrangements of society. It is now necessary to go further and consider society itself; how it attempts to organize virtue, or, too often, how its arrangements are directly hostile to the development of the better qualities in man. We shall, of course, mainly discuss our own type of society, but we shall remember that there are other types and that their effects on the moral behaviour of individuals may be different.

The actual arrangements of society are important, but the beliefs of the members of the group in regard to these arrangements are even more important. It is not possible to say exactly where the effect of the actual arrangements in creating ideas ends, and the effect of the ideas in creating arrangements begins. But it is true that a revolutionary idea will translate itself, on some occasions, into fact, and thus transform society; while just as truly, under other circumstances, established institutions will prevent the development of revolutionary ideas. The more successful the institutions are, the fewer ideas will grow up opposed to them, and the less change will take place.

There are some arrangements frequently found in states that are clearly morally bad. One that we shall discuss in the next chapter is the existence of great inequalities of wealth and power; another, closely akin to it, is the coexistence in one group of complete idleness and excessive work. The successful community must avoid such arrangements as these. It must also attain control of the ideas that correspond to these arrangements. Thus it must manage to devise a way of life that gets rid of ideas that have long possessed our minds, e.g. the essential superiority of certain persons because of birth or some other characteristic, and the corresponding inferiority of others.

Important from this point of view are the theories at the base of capitalism. No one doubted for many centuries that a man might do as he liked with his own; or that he had the right to possess exclusively things, such as land, essential to the life of the community. The assumption on which the owner of mining royalties bases his claim to be supported by the labour of others involves ultimately a whole theory of the nature of man and his relation to the world.

These ideas are directly bound up with the economic arrangements of the state. It is also necessary to consider ideas of a different type. These are concerned more with moral patterns of behaviour.

The Germans have used the word myth for an idea that has enough power and emotion associated with it to determine action. For them the supreme myth is the superiority of the German nation and its inevitable dominance of the world. But there are many thousands of other myths; some, like the myth of chivalry, affecting European thought for centuries; others, like the myth of the Victorian woman, having a place but for a season; but all potent by their control of conduct while they lasted.

It is necessary, if we are to understand morality in the state, to have some idea of how these myths arise and who produces them; we must also consider what definite steps the community takes to teach its myths or moral ideas. Many are forced on the individual by the economic organization of society, such as those of private property; others are gathered haphazard from house or street; and some are deliberately taught by the Church, the school or some professional body. We shall suggest that if there were more intelligent and deliberate teaching, and if the teachers had a clearer idea of what they wanted, social morality could be greatly improved.

Not only must myths be taught, they must be changed, and, in a healthy society, the mechanisms for changing ideas must be as well developed and as conscious as those for teaching them. As an example, let us take a myth which grew up under certain circumstances, to be discussed later, and which we now think bad: the belief that wealth and idleness are the supreme ends of human endeavour. For a satisfactory state this must be got rid of, and replaced by a belief that work for the public good—whether as scavenger or Member of Parliament—is a noble and desirable thing. If this idea is to supersede the other there must be an alteration in the beliefs of certain groups of people. When a state sets out to effect an alteration in ideas we call the process propaganda, especially if we disapprove of it, or education, if we think the change desirable. No state that takes its moral duties seriously can afford to let the control of ideas slip altogether from its hands. In the last section of this book we shall deal with these conscious agencies of instruction.

Before we start a consideration of the nature of our state it is important to consider the meaning of a fundamental concept. The word used to cover the sum of moral ideas in the state, and in many social relations, is justice. Justice has always proved very difficult to define, and the arrangements considered just have varied conspicuously from age to age. And yet men have felt that they knew what justice was, and have been prepared to fight for justice and the 'rights' that they felt justice conferred. Possibly the best way of thinking of justice is that suggested by Brunner in his *Justice and the Social Order*. In his view, justice is the working out in practice of a social pattern that has been generally accepted.

"Thus in virtue of the idea of Justice mankind is placed in an order. He is part of a structure, fits in a certain place in that structure, and it is a structure that orders the whole of life, the relationship of every man to his fellow-men and at the same time the relationship of every man to the natural constants of life. By justice every man is fitted in, and hence in a way disposed of."[1]

For Brunner the order to which justice makes us conform is a divine one, immutable, unchallengeable. There may be this ultimate order; but even if we do not go as far as this, the idea is most valuable. If justice is the practical application of a pattern which has been accepted, that explains its force at any time, and its variations in different times and places. With this as a guide it is even possible to understand such apparent contradictions as Hitler's idea of Nazi 'justice'.

Justice in this case meant the acceptance of a completely new pattern: the right of the stronger to do as he thought fit in his own interest, and an ordering of the community into degrees of strength and importance on the new valuation. Once this pattern was accepted—and it was by many—justice quite naturally and logically became those actions which were in accord with it. A failure to distinguish the exact use of the word justice when on Nazi lips made the statement of an idea appear needless hypocrisy.

For most men this idea of justice is expressed as 'having a right'. We claim 'rights' which are in accordance with the pattern we have accepted. Thus individuals in different, but closely similar circumstances, have very different rights. A man engaged on one kind

[1] P. 24. This is virtually the same as Plato's definition that Justice means 'minding one's own business.'

of contract has a right to a week's notice, a man on another to a month's. In many civil suits it is good law to plead "the custom of the trade". Thus the serf under a tyranny does not claim freedom. He suffers, but neither law nor custom suggests any escape from his sufferings. When Dingaan's defeated impis marched over a cliff to their deaths they were remaining true to a pattern of government that they had accepted. There must be the repudiation in these cases of the whole pattern before men can escape the consequences of parts of it. Few people can imagine such a complete repudiation, and thus great revolutionaries are rare; at most we can criticize details of our pattern and suggest minor modifications.

When we ask on what grounds we criticize any pattern under which we live there are two answers. We have perhaps some ultimate and divine pattern, such as Brunner believed in, and this vague overriding scheme can be used to criticize our actual arrangements, as when Paine wrote of the *Rights of Man* or the American constitution claimed for all of us "Life, Liberty and the pursuit of Happiness". At other times all we do is to bring one part of our pattern into conflict with another, for no pattern is completely consistent. What we admit in one case we refuse in another. The suffragettes invoked the admitted principle, no taxation without representation, and claimed that it should be applied to them. Their opponents felt that the part of the pattern which excluded women from public life was the more fundamental principle. There are countless conflicts of this kind in any complicated social pattern, and the reformer who claims that what one man has all should enjoy can ask for almost any change he wishes in our constitution.

In many cases when new circumstances arise in society conditions of great hardship or theoretic injustice occur. The Industrial Revolution saw in England the growth both of the pauperized urban proletariat and the great industrial middle class. The fight for the economic rights of the one and the political rights of the other dominated the nineteenth century. Another example of the same kind was the loss by the Church of England of most of its hold on the religious life of Wales. Church endowments then seemed an injustice, and the disestablishment of the Church admitted the claims of the Nonconformists to equal treatment.

The conceding of a 'right', therefore, particularly an important one, is a serious business. It is the first step in an alteration of our pattern of living. It also lends vigour to the protestants. Once a right is admitted the claimants have something to fight for, a rallying cry, to which they can gather supporters. We may illustrate this with a curious case, that was brought during the agitation for the reform of the franchise, against certain citizens of Birmingham who were claiming the right' to have a Member of Parliament to represent them. At a perfectly orderly meeting they had elected one of their number to go to Westminster and petition to be allowed to take his seat. The leaders of the meeting were prosecuted and received sentences of up to eighteen months' imprisonment, and one a fine of £100. The Judge took the case very seriously, since, as he pointed out, to claim a 'right' which does not exist is to lay the way open for all sorts of disorders and disturbances.

"Now that a man may have many rights and that he may not have the right of choosing and specifically concurring by his own vote in the election of Members of Parliament, no man alive can doubt. Can any man doubt that those speeches were calculated to raise in the minds of those persons by whom they were heard, dissatisfaction and belief that they were deprived of certain rights, and to excite contempt and dislike to the Commons House of Parliament. History, however, will not, as far as I can judge, prove the existence of such a right, and if he had taken some little advice from those persons in his own station of life upon that particular subject, it might be that the delusion would to a certain degree have been removed and he would not have given his weighty sanction to a meeting of this description, which meeting might have brought very dreadful consequences on many of the persons who were present at it, and concurred in the proceedings that occurred at the meeting. Suppose any of the members of the meeting had believed that they had the right to elect their own members—suppose that they had insisted upon attempting to carry into effect by force that which they thought was their right—What would have been their offence?"

The learned judge was not without justification for his view. The citizens of Birmingham were in fact demanding a complete revision of the franchise of the country, and a radical change in the nature of the governing class. To have conceded them their 'right' would have been to admit that this reorganization was necessary. No judge could countenance such a claim. Once admit it, and there must either be an outbreak of revolution and violence, in the attempt to turn a right into fact, or there must be prompt action to grant the change.

It was eleven years after the unfortunate meeting at Birmingham before the right claimed by Mr. Edmonds and others was conceded; and, before it was granted, part of the dreadful consequences foreseen by the judge had taken place, and windows had been broken. Other agitations have had to be long continued before the wall of conservatism crumbles. The suffragettes, conscious of their rights, attacked policemen and went on hunger-strike, and only got the vote after the nation had undergone the dissolving experience of a war. More recently the right to better education has been conceded piecemeal; and again, only after a war.

It is not only necessary that a right should be conceded by the bulk of the nation—there must also be a political organization such that the judgment of the majority shall not be held up by a minority in a key position. This has happened in the past. The House of Lords for just one hundred years prevented the passage of the bill prohibiting the use of boys as chimney-sweeps. At the end of that century popular clamour forced the noble lords to agree to a measure almost identical to that originally proposed. The modern impotence of the Upper House, first established in 1906 when Lloyd George's budget was forced through, has taken away from the English constitution this particular danger. A country which wrecked its political structure by not being able to act in accordance with the general policy was Poland, where the *liberum veto* allowed one man to block any change.

It is of the utmost ethical importance, therefore, that not only should there be people to have ideas and suggest new rights when changes of circumstance make them desirable, but that the order of society should be such that these rights can be conceded without breaking the social order in pieces, as happened to France and Russia when the State had reached such a degree of rigid inefficiency that it could not be reformed, only destroyed.

Books referred to in this introduction:

Plato, *Republic*. (The best translations are those by Lindsay and Cornford.)

H. E. Brunner, *Justice and the Social Order*.

Thomas Paine, *Rights of Man*.

Howell, *State Trials*. (Thirty-four vols. 1-10 compiled by Wm. Cobbett; 11-21 compiled by T. B. Howell; 22-23 compiled by T. J. Howell; 34 [=index] by D. Jardine.

J. L. Hammond, *Life of Lord Shaftesbury*.

Chapter 1

SOCIAL PATTERN

In a previous chapter we discussed some of the frustration that an individual suffers in our society. In this chapter we want to go further and discuss the moral pattern of society as a whole. Our own society, and the whole of the Western European civilization of which it is a part, has certain characteristics, and these must react on the moral type of the members.

It is not fanciful to find the type of society and the moral type closely connected. If we look to other types of society we see that there are great differences in the presupposition on which society is built, and also differences in the moral type of the inhabitants. If it is true that variations in family structure greatly affect the psychological stresses to which the child is exposed during the early part of his life, and thus determine to a large extent his basic moral character, it is also true that differences in the larger social structure affect him during the rest of his life and appear in his adult behaviour.

Certain types of social organizations seem to result in, or at least to be associated with, certain types of behaviour of the kind we call moral. If a society is organized in a certain way the people are amiable and friendly, honest and generous; if in another, hostile, thieving and savage. Once this characteristic is established, it is transmitted to the children by example as well as by the structure of society, so that we rapidly get a very strong characteristic tone in the group. This tone cannot be removed merely by the teaching of the individual. This may have some effect if the child is separated from the immediate example of his parents, but so long as society remains unchanged the strains and stresses will be produced in his mind, and be liable at any moment to issue in conduct characteristic of the group. In the book *The Individual and his Society*, Abram Kardiner takes some contrasting types of society and shows how individual conduct appears to be affected by them.

The Zuni, who live along the Zuni river between the desert and the Sierras, are an agricultural people with a few sheep. The climate is harsh, hot or cold, rain is irregular and scanty, but for two thousand years they have made a living, and continue to do so. Occasionally they fight their neighbours, but within the group they are completely peaceful and amiable. They live in large matrilinear families calling all the female relatives mother, and the fathers' brothers father. There is little sex repression and though marriages are monogamous they are easily broken. Property is in common. There is no system of dignity and honour, and the framework of government, imitated from the Spaniards, has no reality. There are no punishments. Children are very kindly treated and the function of religion is to produce rain—there is no fear of the dead, no suicide and no murder.[1]

At the other end of the scale are the Chuckchee, a sub-arctic people of Siberia. They are a people dependent on half-wild reindeer, and where their beasts go, they go also. They are nomads, the deer are private property, and thus with the accidents of fate there are great differences in wealth. They are dogged, ruthless, hostile. "The basic weakness in their society lies in the fact that neither the economy nor the social organization fosters any high degree of trust. Mistrust is the rule, and the inability to trust in any one must lead to a deep-rooted suspicion, eagerness to exploit, fear of being exploited; grandioseness, fear of degradation, brutal methods of ensuring security at the expense of others, and a lack of a sense of responsibility to any one."

The children are made to work at ten and if lazy are sent away to fend for themselves, probably to die. The old are often killed as useless. Murder and suicide are common and the religion is one of dread and attempts to subdue dangerous spirits.

Midway between these two examples comes the society of the Kwakiutt who live on Vancouver Island. They live by hunting and fishing and this gives them a secure livelihood, but the major part of their lives is concerned with prestige and privilege. There are nobles and commoners. Eldest sons are superior to the later born. There is a currency of blankets and coppers. There is a conspicuous waste at feasts, when what is not eaten is burnt. The use of wealth and prerogatives is to shame rivals, and the religion is violent with a tendency to cannibalism.

If we look at these three examples it is clear that safety and confidence are one of the bases of a peaceful mind and quiet ways. This is given among the Zuni by the large family, and by the absence of punishment and of the mechanics of government. The communal ownership of property ensures enough for all, and as no one has any authority over another, there is no aim for ambition. The Chuckchee suffers the maximum of insecurity. His reindeer are difficult to manage, and liable to disease or death. He has not even the stability of a

[1] For further details, see Ruth Benedict, *Patterns of Culture.*

settled home. He has no security in youth and none in age; and the emphasis on private property, wealth and status, prevent any comfortable merging of the individual in the group.

The Kwakiutt are an example of the disturbing influence of a prestige system based on birth and wealth. Among the Zuni such distinctions as there are lie open to everybody. No one is frustrated or excluded. Among the Kwakiutt success is by triumph over another, and a large part of the community starts handicapped in its desire to succeed.

These sketches suggest that we must look to the pattern of our own society for some of the fundamental characteristics of our morality. By analogy with the types we have quoted, two elements in society seem important: the degree of equality and the degree of security. Until comparatively recently we have had no conscious organization of either. In the ages before the Industrial Revolution a culture based on status existed. Every man had his place, and the organization was such that few aspired to rise above the place they had, and the cohesion of society held them from falling below it. The Industrial Revolution broke up the old organization, and for a time put nothing in its place. Then for some eighty years the inhabitants of England suffered the maximum of inequality and the minimum of security. We are still, to some extent, suffering from the psychological damage inflicted on the state at that time.

The organization of both security and equality is closely bound up with private property, and with the status system connected with it. So long as a man has property, may do what he likes with it, and receives differential treatment from the community in accordance with the amount he accumulates, there will be inequality. So long as there is not enough for all, and so long as the possession of a certain type of property confers powers over other people, there will be insecurity. The great difference, in this respect, between our capitalist civilization and that of more primitive peoples is that capitalism confers vast powers on the owners of the means of production. They can, by the wages and conditions of work they offer, profoundly affect the lives of those who must use their machines in order to live. They can thus produce differences between one man and another far in excess of those possible in a simple culture. The pattern of capitalistic society allows for greater inequalities than any other.

Among primitive peoples security is achieved by a community of goods so that resources are equally divided, and the enlarged family, so that the individual will not lack the psychological assurance of the group. We have abandoned both of these, though we are returning rather furtively to the former. Society has, as we have said earlier, adopted the small family, and concentrated the holding of property, the management of children and the display of love in the small group. We have done nothing to intrude into the emotional isolation of the family; and the frequent loneliness, the sense of helplessness when anything goes wrong, the expectation of a solitary or unhelped old age, still remain with us. On the other hand, we have begun to feel that the state should provide a minimum of material security. The old poor law which offered so little has been developed by stages we shall talk about later. From increased differential taxation we have created a fund that is administered in various ways to provide better conditions for the poorer citizens: subsidies to keep down the price of certain foods, unemployment pay, sickness benefit, and so on. This has been done while leaving the structure of private ownership untouched. The provision of hospitals, of meals in schools, of old-age pensions, has brought into a system that was felt to be intolerable, many of the benefits of communism. With ever-increasing social legislation material security will increase till acute, serious fear—the fear of starvation, or being left to die unattended—should cease to be part of the psychological experience of the nation. Whether we shall ever revert to the emotional security of the extended family it is difficult to say. At present the idea is too firmly held that an extended family lacks privacy, and that it must inevitably be dominated by the crochets and conservatism of the old, and that the only way for youth to have its chance is to break away into separateness.

The problem of equality is much more difficult and complicated. There are, roughly, four kinds of equality: political, economic, cultural and social. All must be achieved in the perfect state; for the denial of any, unless that denial is so absolute that no idea of change enters the mind of the deprived, produces tensions and frustrations in the community that must make for harm. Unfortunately men themselves are not equal. They are born unequal, and in our society this inequality is increased, or even produced, by the unequal treatment that is accorded to them. The only equality that is possible is one that allows each man to develop as far as he is able, and then finds him a place in the state suitable to his abilities. During this development all must experience the same cultural training, and whatever place they finally fill they must be treated with equal social consideration. Under our present system, something very different takes place. In our status system, although based on considerations of wealth and birth, some considerations of personal merit have a place. The system is not completely rigid, and personal effort, intelligence and force can fairly easily alter a man's position. On the other hand, those who start in a less-considered position are offered a very different cultural training from that given to those whose parents are more highly considered. Thus even if a man does rise by his own merits he is apt to lack some of the qualities that his more fortunate competitor possesses; and he can only rise if his abilities are well above the average. It follows that those who fail to rise are usually inferior in many respects to those who are born to or achieve higher position. The phrases 'upper' and 'lower' classes imply a definite inferiority in those of lower status, and this inferiority frequently really exists, either through natural endowment or through the cultural education they have received. The deficiency in cultural training offered in England to the children of the poor accentuates in many cases a poverty of natural endowment. It does more. It prevents their full merits from developing, and thus, while stunting their growth, produces a sense of inferiority that makes them less good than their abilities would have permitted. On the other hand, a member of the governing classes learns while very young the attitudes of mind, the manners and behaviour that become his position, while the sense that all is possible leads him to develop his powers to the utmost.

All forms of inequality, except those due to innate endowment, are closely bound up with our ideas of status. The idea of status, the conventional superiority of one man to another, is not universal through the world, but it is very widely spread. In some cultures it is based simply on accumulated wealth, or the need to have some leader of the group for certain purposes. In others birth or race is important, usually in association with wealth. Where there is a system of primogeniture the inheritance of property is as integral a part of the system as is the inheritance of titles and honours. Aristocracies and monarchies are associated with great wealth, and it has frequently been considered a 'wrong'—something contrary to the proper organization of the world—when wealth passes from the hands of the aristocracy to the industrial magnates.

The extreme form of the aristocratic and monarchical tradition of status is the divine ruler, and Perry in his *Children of the Sun* has shown how the idea has passed round the world, possibly from a cradle in Egypt. In later ages kings have claimed to rule by 'divine right', and more than one monarch had to lose his head on the scaffold before they shrank almost to the status of ordinary men.

In much of Europe in the past the basis of the status system seems to have been a difference of race. The ruling caste were invading warrior tribes who made themselves masters of the land and wealth among a subject population of serfs of different race. Even to-day, in those places where the status system is at its strongest, there is a difference of race behind it. The treatment of negroes by white men in various parts of the world, the behaviour of Englishmen in India, the attitude of the old-established American families to the newer immigrants of different race, all show clearly the effects of this sense of superiority. When status depends solely on economics, there is still a feeling of a difference in kind. The rich woman, well preserved, well groomed, assured, seems to be very far removed from the shapeless, soiled charwoman. Up to the end of the last century a gulf that was edged by "condescension" and "respect" existed between the landlord and the agricultural labourer. The same gulf, only deeper and wider, opened with the Industrial Revolution between the large employer and his workmen. Here again the starveling, overworked, stunted, ignorant brat in the mills hardly seemed a member of the same species as the owner's well-fed, well-taught children. This idea of status and difference underlies all forms of social inequality and will find further expression as we discuss them separately.

So long as the society based on status was fully accepted the damage it produced was of a certain kind. It did not include much that we group under frustration. The serf suffered, but he had never expected anything else: he did not think himself fit for a higher destiny. Each man knew his place, and saw God's work in his position.

> The rich man in his castle,
> The poor man at his gate,
> He made them, high or lowly,
> And ordered their estate.

When the fabric of this order was torn—when poor men rose suddenly to high wealth, when slaves were free north of the Mason Dixon line—then the damage became more complicated. For, to actual deprivation, was added the sense that a wrong was being done, and the oppressed struggled to attain, and hated those who restrained them.

The easiest form of equality to secure, once the idea has been accepted, is the political. Give everyone a vote, make the ballot secret, and there you have the first step. Pay members of Parliament, remove the tests of religion and race, command the judges to show no favour, and the thing is done. When, in the past, the idea of political equality first presented itself, it was at once seen to be bound up with status, and this and the corresponding deficiencies of the underprivileged were used as an argument against it. How can we give the vote to the ignorant? Why should these new industrialists intrude into the political preserves of the landed gentry, their betters? How inconceivable that a negro should sit in the parliament of the Union of South Africa to represent the interests of his own race! So long as political equality is refused, especially when there is an aristocratic or absolutist form of government, traditional vices develop in the rulers, traditional oppressions are inflicted on the ruled. When we are discussing later in this chapter the corruption of a governing class this will be clear.

Under a static social system of the aristocrat and serf type, political inequality is hardly resented. It passes the imagination of the oppressed to think of any other order. When opposition does arise it comes, not from the chief sufferers, but from those who are only just excluded from power; young men and women, educated, culturally the equals or superiors of the governors, who feel no doubt of their ability to take their full share in the political life of the community. Because they are themselves excluded they adopt the cause of the more completely oppressed, and revolutionary movements develop. Political inequality may be of different degrees of severity. There may be absolute political oppression of a whole nation as was the case in Tzarist Russia, or a much milder and more selective kind as in Prussia-dominated Germany. In Russia there was steady agitation of the discontented intelligentsia trying to bring about reform and their own inclusion in the governing classes. In Germany there was much political acquiescence, due partly, as will be discussed later, to the invention of the great German myth, and partly to the tendency of the abler men of independence to emigrate or of the more conformable to take to industry and find in its organization the scope for their powers that was denied to them politically.

Since it is the educated and the superior who lead the demand for political equality, this demand will only arise when there is a sufficient body of people able to make it. Critics of British policy have said that the British always have trouble in every country they govern because they educate the natives and teach them that they have a right to self-government. This is largely true since the leaders of the agitation against the ruling race are those who have received the education of their rulers. The force of the revolt comes from two sources: the educated youth is not absorbed fast enough into the machine of government, and it is while unemployed that they find revolution so attractive. When, in addition, the supreme frustration of the social colour bar is added, the hostility is sharpened. This is all the more poignant as the educated native has generally been at school or college in England where he is accepted as an equal. He only meets the worst form of the status system when he returns to his own land, trained and, as he thinks, ready to take his full share in political control.

This type of frustration may not be strictly political. It may simply be the deprivation of a place in the governing ranks of society, whether that society be the state or some smaller body such as a school. One of the more recent psychological discoveries, that has entered education some considerable time after it has been recognized in politics, is the realization of the power that office and integration with the group exercises on the human mind. Schools are too fond of giving office only to those children who already show a high degree of conformity. They are not prepared to trust the rebel with the sobering delights of power. Other organizations are wiser, and it is being generally realized that many a child opposes the government merely because he does not hold office in it. This of course is one of the discoveries that England made long ago in the political sphere. While Russia was sending courageous and high-minded students to the gallows or Siberia for whispered words against the regime, England was reporting their Union speeches in *The Times*, and speculating what constituency they would best represent in a year or two. This policy broke down with certain pacifists in the first German war. Many young men, who were then imprisoned for their views, and were treated with great harshness, have never forgiven the state or put their talents fully at the public service. They have always remained in hostile opposition. Had their treatment been more intelligent they might have made a great contribution to the national welfare, for they were very clever, and possessed courage and strong convictions. It is to be hoped that the more lenient treatment of conscientious objectors in the second war will prevent this waste of talent.

The person who feels himself excluded from a reasonable place in society is always ready for revolution. It was the unemployed, the officers from the 1914-18 war, and the young graduates who saw no hope of work equal to their abilities, who formed in large measure the revolutionary core of National Socialism in Germany. The same revolutionary feelings are common to young people in all lands before they have made a place for themselves in society. For the ambitious, the years from about twenty to twenty-five are most exacting. He feels he has reached his full powers—that he is generally mistaken makes no difference to his feelings—and society refuses to listen to his voice or to let him establish himself. It is clear, therefore, that society is rotten and should be swept away. In time, when success is achieved, or failure accepted, the order of society, as it exists, seems preferable from its familiarity to any unknown dangers.

Economic inequality is a far more difficult problem. Until there is a complete reshaping of the social pattern there will always be economic inequality so long as men are born with different abilities and interests. Without complete communism, extending to the smallest personal possessions, this inequality must remain. But under capitalism it assumes monstrous forms. At first the agricultural serf-labourers, then the urban proletariat were depressed and degraded. Inequality passed beyond anything based on natural abilities and became a social poison. This state of affairs, at first accepted as inevitable and natural, has now been seen to be intolerable, and a complicated system of taxation has been devised which, like Robin Hood, takes from the rich to give to the poor. At the same time a multitude of labour laws and practices have so changed the conditions of employment that the poor, instead of being overwhelmed by their misery, now have the strength and vision to resent it. When economic inequality is at its height both the richest and the poorest suffer. The very rich either develop the characteristic vices of excess or they become anchorites of wealth. Anatole France, in his description, at the end of *Penguin Island*, of the capitalist hell describes the martyr to his own riches.

"Like all true aristocrats these powerful men affected a great severity in their habits and customs. They were the ascetics of wealth, several denying themselves all happiness, all pleasure, and all rest, spent their miserable lives in rooms without light or air, furnished only with electrical apparatus, living on eggs and milk and sleeping on camp beds. By doing nothing except pressing nickel buttons with their fingers, these mystics heaped up riches of which they never saw the signs, and acquired the vain possibility of gratifying desires they had never experienced.

"Among the rich, all were devoted to the social order. Some felt the severity of their position cruelly, but they endured it either from pride or duty."

It is not necessary to take Anatole France in anything but the spirit of satire, but there are millionaires, world famous, who are not so unlike his description. Without anticipating a later discussion, it can be said that as regards women, the extent to which great wealth precludes any useful or satisfying activity can be matched only by the disabilities of high rank in the eighteenth century.

The effects at the other end of the scale are much more important because so many more persons are affected. For every hundred who suffer from excess of wealth, a million or two suffer from its deficiency. Much is of course obvious. Lack of food, lack of comforts and conveniences of life, lack of beauty, cleanliness, quiet: all these deficiencies are apprehended directly and have a direct moral effect. In children they produce hooliganism, destructiveness and violence. In adults, drunkenness, theft, cruelty, suspicion, hostility. The pattern of behaviour when a slum population is moved to a new housing estate is always the same, and the improvement in behaviour follows regularly, in about three years, when conditions have had time to influence the mind.

The effect of property frustration is very obvious in some cases. A child brought up in a property-owning society without personal possessions seems to develop very unsatisfactorily, and the figures for petty theft are higher for domestic servants than any other class of the community; although the number of cases that appear in court represent only a small fraction of those that occur because, in most cases, employers do not prosecute. The domestic servant's position is one of the maximum temptation and the maximum frustration. She is so employed that she can take almost anything in the house, and she sees her mistress, no more deserving than herself, with silk stockings, linen handkerchiefs and scent—all the prestige-bearing accessories that she herself needs for her social life. When it is realized that, in addition, the ordinary domestic servant is not very intelligent, and that she is not likely to be able to control her actions by foresight, it is the comparative honesty that is remarkable.

In England, the frustration of poverty has been as much through its limitation of opportunities for development as through the deprivation of possessions socially desired. The English system of education is only just beginning to throw open opportunities to the really poor. In this we have been far behind such countries as America. With the passing of the Education Act of 1902 and the establishment of the secondary schools, the poor child for the *first* time had some chance of an extended education. But it was a very faint chance. In 1904 the secondary schools accommodated less than 3 per cent. of the children of the relevant age. The percentage has now risen to 16 or 17 per cent. of the age group; but of these only 1.6 per cent. of the age group stayed till the age of eighteen. Admission was by examination at eleven and a child had to be unusually quick and industrious to reach admission standard. Moreover, the really poor, if the retardation due to poverty was not enough to exclude them, were often prevented by their parents from taking the places offered.

The poor child, under-educated, has nothing to look forward to when he goes to work. He needs money and has to take the job with the highest pay that offers, irrespective of its prospects. He does not receive a training, and a future of unskilled or semi-skilled work opens before him, with the realization, if he is really intelligent, that his children will be condemned to much the same life. It is, of course, the intelligent who are most conscious of this frustration, and who feel most bitterly against society and who are most inclined to rebel.

When unemployment is added to these constant factors the social bitterness is increased. Unemployment has been a regular factor in our type of industrial civilization, and has come to be regarded by many as a necessary and almost salutary element. Its effect on the psychological character of our culture has been very marked. The damage inflicted by unemployment is complex. A man loses so many of his sources of satisfaction. His work means more than money. It means society, self-respect, interest, a settled way of life; even a place to go to where he is warm and dry. The terror of starvation that hung over the unemployed before unemployment pay was thought of has gone, but a period of unemployment still has serious financial consequences and puts an end to hopes of prosperity and prestige. But it is in its other aspects that unemployment brings most suffering. The industrial worker has been trained to one type of life, one type of activity, and he has never learned, or has forgotten, other ways of spending his time. The unemployed man who takes a knapsack and sets out to use his enforced leisure in travel is very unusual; most of them can think of nothing to do. Without his work—its comradeship, its interest, its sense of being a valuable part of the community—a man has nothing, and stands unhappily at street corners waiting for the day to pass. The fear that this will happen, the damage when it does happen, has had a very great effect on the mentality of the poorer sections of the nation.

Opportunities for cultural development were for many ages dependent on rank or wealth. In Europe only the Catholic Church realized the importance of throwing open education to those who could profit by it, irrespective of their position in society. Thus the Church assured for itself a supply of talent, and introduced into a society that was dominated by ideas of status a democratic element. To-day the state is at last learning how dependent it is on ability, and is making an effort to secure the able and to train them in the most socially useful way. Though education will provide a man with technical skill or with certain kinds of knowledge it will not of itself bring full cultural equality. In England we now offer a passable schooling to all, irrespective of wealth, but we take great care to maintain a quite separate system of fee-paying schools to which the richer parents may send their children to learn ideas and manners which will distinguish them through life from those who have not had so much money spent on them. On a university staff, where all the members have attained high academic success, there are noticeable differences of manner and speech which are due to early training, and have persisted in spite of all subsequent experience.

The able man who receives an inferior cultural training often suffers considerably. Questionnaires seem to show that what he particularly resents is his inability to converse on an equality with others. He is conscious of his difference in accent, in style and his lack of general knowledge. Since he feels undeservedly inferior he is suspicious and hostile. He is apt to exaggerate his shortcomings to make it appear that he is not ashamed of them. The characteristic type of self-made man shows clearly how the lack of cultural equality poisons a man's outlook on society.

On the other hand, the less-educated defend themselves from invasion from above. A university degree is in England an *absolute* disqualification for one seeking employment as an unskilled labourer, and, unless one is a refugee, as a housemaid. No such disqualification exists in America, where the level of cultural equality is higher. The garage hand, the window cleaner, the tourist-coach driver or the girl in the hairdresser's may all be products of the local university. In those parts of America where the status system is strong, some other criterion than general culture has to be adopted: often the membership of some secret society.

The worst type of cultural inequality arises when it is associated with a colour bar. In nearly all countries where white and dark races live together the white attempt to maintain their superiority by varying devices. In South Africa the antagonism between the minority white population and the majority negroes is particularly keen, and the difficulties and frustrations are perhaps worse than in any other country. The legal position of the natives is very bad, the facilities for education extremely small, and at work they are confined to the least-skilled tasks. A city like Johannesburg with its millionaires' palaces and its dog-kennels in which large sections of the native population are segregated; its streets of luxury shops and its utterly insanitary squalor; its schools of art and its refusal of any kind of culture to the negro in the mine compounds, produces a sort of social nightmare. It is no wonder that the murder rate in the city is probably the highest in the world.

Social inequality, which is in a way the summary of all these types, is one of the most outstanding characteristics of our civilization. In England the two chief qualifications for high social status are birth and wealth. Culture, except the knowledge of special attitudes, is far less important. The rich and well born intermarry and mingle their social qualifications. The children of the aspiring are given the "best" education, and make their way to wealth and rank through their school friends. The upper stratum during this century has become much more fluid. The history of British snobbery has not yet been written. It is a topic that should provide the social historian with an absorbing field, but, before we say a few words about it, we will give a very brief description of the American version which is simpler, has on the whole a less complex history, and has been the subject of recent investigation.

In the modern community studied under the name of *Yankee City* the social stratification was most precise and definite. There were three main social classes, each subdivided into an upper, a middle and a lower. In the main, wealth was the criterion of division, but this was complicated by considerations of birth. The upper-upper class was almost entirely defined by birth, though sufficient money was necessary to support the position. This most superior group was highly intermarried, and used the language of kinship (i.e. cousin) of members who were not really related but had grown up as one of the set; and it maintained its distance from the other groups. This maintenance of distance was partly physical. Most of the houses were in the same part of the town, they were large with large gardens, the children were not allowed out unattended, and as all the families had cars they did not use the trains or buses. Also for social purposes there were exclusive clubs.

The members of the lower-upper group were not accepted by their social superiors. They came from outside the city, and had risen from still lower groups through ability and wealth. Although the children went to the same schools, and therefore were gradually assimilated, the parents were cold-shouldered. This produced a family situation sufficiently common in America, where the children feel themselves superior to their parents, and are therefore not subordinate to them in matters in which, in more stable groups, the elders claim control. Eccentricities of behaviour tend to characterize the climbing member of the lower-upper group, such as comparatively large sums spent on motor-cars, or an excessive devotion to an aeroplane that seemed to the owner to symbolize the higher life.

In addition to this, further social discrimination was introduced by cliques. "So-and-so belongs to the X set", and So-and-so was at once established in his proper social niche. This tendency of American society to break up into countless cliques or secret societies seems strange to an Englishman. In particular, the semi-secret greek-letter societies, with their initiation ceremonies and their high prestige, introduce a field of social rivalry unknown in England. In this field the stratification of society can find its fullest expression and a sense of dignity and prestige be fully satisfied.

When the aspirants to a higher sphere threaten to intrude on the ground of their betters they are repulsed. "Those poor Starr children (lower-upper class) were here", says the hostess after a political vote-catching party. "Katherine was looking a little beaten and worn, and Johnnie was apparently suffering from a hangover. I don't care what people say, I can't dislike them. I always remember, when they were growing up, how badly they were treated. I've actually seen people turn their backs on them when the two of them came in. It's not as bad as that any more, but it's bad enough. This town does something to children like that."

If in society the two upper classes maintain such relations, the same sort of behaviour is repeated down the social scale. In *Babbitt* the corresponding social life of a middle-west town is dramatized with its seeking for superiority and its disappointments and frustrations.

The same social prestige is sought among the young in America. They learn early to play the social game and to strive to achieve high rank among their contemporaries. The system of "dates" is one way of assessing social prestige. The high-ranking girl or boy is always in demand, and learns to show off her charms or his power. At dances the girl who cannot command admiration suffers more even than the English "wallflower" of a generation ago. As part of this social competition the American "teen-age" girl learns to dress far better than her English contemporary—who is still, probably, in a gym tunic—to make use of cosmetics, and to take part in smart social back-chat.

In English snobbery, with its infinitely longer history and more picturesque quality, the same thread runs. The importance of some arbitrary standard of merit, such as birth; the maintenance of social distance; the right of the superior to resist by rudeness encroachment from beneath; the limiting of the activities of the superior classes. In a book on manners published just over a century ago a story is related with approval, showing how this distance and rudeness was employed.

"A certain Mr. Timms of the Treasury who fancied himself a great man chanced to dine next to a noble lord at the club and got into conversation, which the lord allowed out of his magnanimous desire to set the poor fellow at his ease. A few days later they passed in the street and Mr. Timms endeavoured to presume on the casual acquaintanceship, explaining who he was.

" 'We dined together at the club the other evening. I'm Mr. Timms of the Treasury.'

" 'Then', said the noble duke, turning on his heel, 'Mr. Timms of the Treasury, I wish you a very good morning'."

Mrs. Sherwood shows how the same principle was early introduced into the minds of children. Mrs. Fairchild in her youth climbed a tree in company with the gardener's daughter, and was sternly rebuked.

" 'Besides', said my Aunt Emily, 'the shame and disgrace of climbing trees in such low company, after all the care and pains we have taken with you, and the delicate manner in which we have reared you'."

The language of social inequality is interesting. Perhaps the most important word in one connection is *condescension* and in another is *presumption*. One is *conscious of one's position* or one *demeans* oneself by some foolish act. The *lower orders* are *respectful* or *insolent*. *Consciousness of worth* is but a step from *pride*. *Dignity* and *humility* are the graces of the master and servant.

Caught in this net of ideas both parties suffered: those who had achieved and those who desired to enter the golden gates. The upper had their activity restrained in countless ways and were compelled to many types of ostentation, one of which was idleness, to emphasize their wealth. On the other hand, the excluded suffered much in their efforts to establish themselves in a higher circle, and were in many cases driven to behave even worse to their inferiors in their attempts to avenge the snubs they had themselves suffered and to re-establish their dignity.[1]

The most important of the disabilities that social prestige inflicted was the denial of adequate occupation. We shall discuss this more at length later, but we can say here how completely idle and secluded the women of the upper classes were. In Poland not so long ago no nobleman was *allowed* to engage in trade or handicraft. England had no law forbidding such a thing, but social custom was so strong that it was unthinkable. The women even more than the men were held in the bonds of decorum. From babyhood upwards behaviour must

[1] Cf. Fagg: Sheridan's *Rivals*, Act II.

be *ladylike* and with a *sense of one's Position.* In the pages of Jane Austen we meet Lady Bertram or Lady de Burgh and both are fully conscious of their position and effectively restrained by it from any activity. Judging by literature, this repression of natural activity, and the necessity of keeping up one's own position by depressing others had a number of effects. In some it bred fear of the world and a dread of the "lower orders". The woman who could not go out alone—though to judge by Dickens the streets of London were far less safe than they are to-day—had developed these fears partly to justify the social exclusiveness she had been taught. It was not snobbery, she felt, but a proper appreciation of the dangers of her position that kept her in her carriage, or only allowed her to go abroad with maid or footman in attendance. She did not resent her lack of freedom, she had never imagined a world in which she could enjoy it.

Pride, a characteristic strangely common in the past, to judge by literature, was also part of the defensive mechanism that supported the restrictions of gentility. Pride established distance between the upper and lower classes, and it also consoled the higher for his failure to reach the highest places. The emotion was based on considerations of wealth or birth, and, if wealth was lacking, even an impoverished nobleman could afford to scorn the most prosperous linendraper. The proud duke who rebuked Mr. Timms of the Treasury was only doing on the grand level what every whipper-snapper with a silk coat on his back was doing to his inferiors. The proud woman who spent her time in snubbing or insulting her inferiors was one of the standard characters in Victorian literature. She is frequent in Charlotte Young, as for example Lady Price, wife of the rector of Bexley. This pride was taught to children along with the other restrictions of their position. The unpleasant Augusta Noble in *The Fairchild Family*, who was, as might be expected, burned to death, shows how this teaching was given.

The Fairchild family had been invited to spend the day with the Nobles, and the children had been sent off to the school-room for dinner and play.

"It was now getting dark and a maid-servant came in with a candle, and setting it upon the table said:

" 'Miss Augusta, it is time for you to be dressed to go down to tea with the ladies.'

" 'Well', said Miss Augusta, 'bring me my clothes, and I will be dressed by the fireside.'

"The servant then went into the closet I before spoke of, and returned with a beautiful muslin frock, wrought with flowers, with rose-coloured sash and shoes, and a pearl necklace. Emily and Lucy had never seen such fine clothes before, and when they saw Miss Augusta dressed in them they couldn't help looking at their own plain frocks and black shoes and feeling ashamed of them.

"When Miss Augusta was dressed she said to the maid-servant, 'Take the candle and light me down to the hall.' Then turning to Emily and Lucy she added, 'Will you come with me? I suppose you have not brought any clean frocks to put on? Well, never mind. When we get into the drawing-room you must keep behind your Mamma's chair, and no one will take any notice of you.'

"So Miss Augusta walked first, with the maid-servant, and Henry and Emily and Lucy followed. They went along the great gallery, and down the stairs, and through several fine rooms all lighted up with many lamps and candles, till they came to the door where Sir Charles and Lady Noble and Mr. and Mrs. Fairchild and a great many other ladies and gentlemen were sitting in a circle round the fire. Lucy and Emily and Henry went and stood behind their Mamma's chair and nobody took any notice of them, but Miss Augusta went in among the company, curtsying to one, giving her hand to another and nodding and smiling at another. 'What a charming girl Miss Augusta has grown', said one of the ladies. 'Your daughter, Lady Noble, will be quite a beauty', said another. 'What an elegant frock Miss Augusta has on', said a third lady. 'That rose-coloured sash makes her sweet complexion lovelier than ever', said one of the gentlemen. And so they went on, flattering her till she grew more conceited and full of herself than ever, and during the rest of the evening she took no more notice of Mrs. Fairchild's children than if they had not been in the room."

There is a third moral type that results from the restriction of activity characteristic of social rank: the over good, who were actively and zealously engaged in passing on to the new generation what they had themselves suffered. The force and defiantness of character which these people possessed made them eager to rule, and if they were women, the restrictions of their life often left them no empire but the children of the family, who could not escape from their domination. Mrs. Fairchild ruled in the name of religion; the matriarch of a Chinese household holds sway in the interests of age-old custom.

When the basis of social status is wealth it is necessary that this quality should be emphasized. The principle of "conspicuous waste" is found among Europeans and more primitive peoples alike. The Kwakiutt hold feasts and burn valuables as a demonstration of abundance. One of the authors, as a child, used to watch fascinated while her mother's tea-time guests ate incredibly oily bread and butter without removing their white kid gloves. The world is full of examples, but from the moral point of view, by far the most important is idleness. Idleness can be carried to exquisite lengths. In sections of society where it is fully practised, a man or woman cannot possibly put on his own clothes without help, or pour himself out a glass of water. The degeneration affects the mind before the muscles. The fox-hunting squire might normally be a very fine horseman, but only a very remarkable man can be both a Rothschild and a distinguished scientist.

Wealth can be emphasized by other things besides waste. For many people, social status is bound up with the ownership of prestige-bearing possessions. Almost every society has objects to which it attaches importance. Sometimes these possessions are for use, immediate or potential, as when a man's status depends on the multitude of his flocks and herds. In other cases the more intrinsically useless the objects are, the more they are felt to raise a man's status. Gold and jewels have long been the marks of kingship and royal rank. It has been suggested that gold has had attributed to it, from the most remote ages, a magic quality that makes it supremely desirable. Its symbolic importance has remained strangely constant. The King of England has his crown and jewels, and if he lost them the whole nation would feel bereaved. For the service of religion vessels of gold and silver have, from remote times, been felt to be necessary or suitable; and the Christian Church, though there is no warrant in the New Testament for such precious metals, has always, when once it was rich enough, felt them to be desirable aids to devotion. Among people at large, jewels, furs, silks, mansions, and cars are possessions that confer social distinction. Among the lower income groups, washing machines, refrigerators, permanent waves or grand pianos are desired for the sense of dignity that they bring.

In different countries the sense of social stratification probably varies. There is little doubt that it is less keen in some parts of America than others. To a traveller in the "wild west" of Arizona or Colorado there appears to be little. In New England the stratification is largely determined by birth, and perhaps in other areas wealth is more important. In Germany the remains of the old status system seem stronger than in England, though the industrialization of Germany has gone very far. This is suggested by the comparatively small numbers of students reaching the universities (about 5 per cent.) who come form the lower income groups, although the university fees are very low. In England a far higher percentage reach even Oxford and Cambridge, so long the exclusive province of the rich. It is suggested also by the respect that titles or army rank can command, or the continuation of the custom, condemned in England as vulgar in a book of etiquette of 1836, of putting Mrs. *Col.* Figgins on one's visiting card.

So long as the status system was absolute it had bad effects, but not perhaps its worst. As political inequality is not resented so long as there is no rift in its oppressiveness, so a society in which each man knew his place did not produce a certain kind of bitterness. When Eppie in *Silas Marner* refuses to be made a lady and prefers to keep her working-class status and marry her Aaron, all is well. It was quite different for those who felt themselves fit for higher things and wrongfully excluded from them. The conviction that one has a right to rise in the world may come solely from the possession of wealth, when it produces the kind of social climber so often parodied in the more snobbish novels; or it may come from a consciousness of ability. Humbert Wolfe in his autobiography, *The Upward Anguish*, describes with humour and candour the struggles of a young man who, soon after the Boer War, determined to transfer himself from the lower middle classes to the professional. From his Manchester Grammar school he had long worshipped in imagination the Public School boy as something infinitely superior. On his way to Oxford to sit for his scholarship he firsts meets these demi-gods in the train.[1]

"The persons in his new compartment on the train, might they not be Public School boys? These to him were still legendary figures. Chiefly they were known to him in the pages of *Stalkey and Co.* and more remotely in those of *Tom Brown's School Days.* But in either case they were invested with the peculiar sanctity that all those outside the pale tend to attribute to those within. They had, it seemed, 'studies' where they were up to the most spirited things; they swore splendidly and drank deeply; they had compulsory games; they were divided into distinct houses and had long-standing inter-house feuds; they 'crammed' for the army; they constantly won scholarships at Balliol; and above all they did not have to go home for mid-day dinner and high tea.

"The Public School Boy remained an untarnished ideal for the scholarship candidate. It was with an emotion that he was to experience later on seeing his first real mountain, that he learned from the conversation in progress between the two heroes that not only were they on their way to Oxford for the Queen's College group, but that they were Rossall boys. He gazed at them with ardent satisfaction. . . They had a particular vendetta, it seemed, against two schools, St.

[1] P. 5.

Paul's and Manchester Grammar School. 'Grimy swots' they held them to be, equipped, as a result of their sneaking addiction to books, with an abominable capacity for snaffling scholarships. To all intents and purposes they put lead in the gloves—and blacklead on their finger-nails—added the hero with the moustache. The stroke was deadly to the fellow traveller. He examined his own fingernails for the first time with attention. Black as stygian night they glowered defiantly. . ."

Wolfe was clever, and unremittingly directed his efforts, while at Wadham, to the task of entering the aesthetic set and forming intimacies with the superior beings who inhabited Magdalen and Christ Church. The chronicle of his achievements is touching in its naive seriousness. His pleasure when he is asked out to lunch, his eager discipleship of the rising poets, his pursuit of those with undergraduate reputations. This labour of social snobbery was the real work of his years at college, and he received his reward.

Too often barriers could not be overpassed. At Oxford, while Wolfe was making his way into social esteem, the wife of a tradesman was cut off from any intercourse with the wife of a don—and this quite irrespective of the merits and culture of the two women, or the wealth of their husbands. In those smaller country towns in which a public school is located the least efficient of masters used, till almost to-day, to feel himself quite entitled to insult the cleverest and most successful of the shopkeepers. It was a social wrong that the latter should own property and horses, while his superior lived uncertainly on a low salary, tyrannized over by a headmaster.

Taking all these types of inequality and all these forms of insecurity together, we can see that our society is full of strains, full of situations that produce frustration and hostility, exclusiveness and cruelty, envy and destructiveness. Where status is of the greatest importance, and where it can be obtained by wealth, it is obvious that the pursuit of wealth will be made more vigorous and relentless. Where there is little security, either emotional or economic, it is natural to cling too closely both to persons and things. Where work is unpleasant and the rich idle, there is a tendency to seek to join the unoccupied. Where there is, in addition, little definite attention directed to the group as an object of interest and care; when, in fact, the social system, as well as its current morality, lays stress on the individual rather than his place in an association, then most of the vices common among us fit naturally into their place as the products of our way of life.

The frustration and hostility that arise in our social system spread widely. One of the most interesting observations on the effect of frustration is the way the resultant hostility may be displaced from its original object. This happens particularly in two types of cases: where the real cause of the trouble is too powerful to be safely attacked and where the frustration depends on circumstances that are too complicated to be properly understood by average people. Typical of the first kind are cases where children are kept under too strict discipline. In one example a child kept in an institution at first attacked the grown-ups who thwarted him. Punished for that he attacked the other children. Punished again, he took to banging his own head against the wall and pulling his own hair till he had done himself so much injury that he had to be transferred to a milder regime. In the second type of cases the frustration usually arises from the economic or social ordering of the community. The lynching mob in the southern states of America is generally protesting, in reality, against the low price of cotton, or a social order which considers the white man much superior to the negro and then leaves him with lower pay and poor prospects. Elaborate studies have been made of lynching mobs and there is a close connection between the price of cotton and the frequency of lynchings. There is also a clear pattern in the events and the leaders. It is the lowest grade of white man who acts as leader, and those most directly in competition with the negro who form the bulk of the participants.

Of all the great transferences of anger the persecution of the Jews in Nazi Germany is the most remarkable and appalling. The extraordinary spiritual, political and economic conditon of Germany between the wars must have produced a state of acute frustration and utter bewilderment. When a nation is bewildered and frightened it is very suggestible, and for a long time there had been anti-Jewish feeling in Germany. Germany, as we have said, had retained its rigid status system more strongly than some other countries. The Jews had the status only of wealth, and that is a very invidious position. In consequence it was easy to augment the existent feeling and direct it to the Jews. Thus the Jews became responsible for all the inexplicable ill that affected the state, and all the anger and dismay could be vented on their heads.

Books referred to in this chapter:

Abram Kardiner, *The Individual and his Society.*
Ruth Benedict, *Patterns of Culture.*
W. J. Perry, *The Children of the Sun.*
Anatole France, *Penguin Island.*

Warner and Lunt, *Social Life in a Modern Community* (Yale University Press); *The Status System of a Modern Community.*
Sinclair Lewis, *Babbitt.*
Mrs. Sherwood, *The Fairchild Family.*
Jane Austen, *Pride and Prejudice; Mansfield Park.*
Dickens, *The Uncommercial Traveller.*
Charlotte Young, *Pillars of the House.*
Veblen, *Leisure Class.*
George Eliot, *Silas Marner.*
Humbert Wolfe, *The Upward Anguish.*
Hadley Cantril, *Psychology of Social Movements.*

Chapter 2

SOCIAL INEQUALITY

We have discussed in the last chapter the general pattern of society and its general effect on the moral type of the State. The difficulties are not of modern growth, nor are they the fancies of the modern psychologist. Long ago Plato, considering the Greek State which already showed most of the characteristics of ours to-day, saw the dangers to morality.

"Then it seems we have found some other things against whose secret entrance into the city the guardians must take every precaution."

"What are they?"

"Riches and poverty", he answered, "for the one produces luxury and idleness and revolution, the other revolution and meanness and villainy besides."

Although the dangers and frustrations of wealth and poverty are spread all through the state, those most affected are, naturally, the highest and lowest sections of the community. In this chapter we wish to give a sketch of the aristocrat as he appears when circumstances are favourable for his full development, and, on the other hand, of the industrial worker who is the most characteristic product of our age. The aristocratic society is now already in decay. In its full form it was always associated with a monarchy, usually a despotic one, and monarchs are much diminished in numbers and even more in power. In most of the countries of Europe the pattern of kings and emperors, of aristocratic power and privilege, has been shattered. France, Russia, Austria, Germany, Poland and the Balkans have all felt the liberating hand of revolution or war. If in many countries a new despotism has arisen its characteristics are so different and so unexplored that they must wait for a new psychology to unravel them. The industrial worker, on the other hand, is a comparatively new social factor. He is different from his predecessors in toil, whether they were serfs or craftsmen. He has been with us hardly 150 years, and has been a conscious problem for only 100. As the world becomes more and more industrialized his numbers will increase. Thus he has a far greater importance than the rich aristocrat whom we shall soon know only from literature, or as inhabiting those backward countries where supertax is not 19s. in the £. Yet it is worth studying the moral characteristics of the aristocrat because they illustrate tendencies in human behaviour that it is important to guard against. Moreover, the last genuine examples were with us as late as the 1920s. They had no political power and could not order an inattentive waiter to be flogged, or send a social rival to the Bastille; but in folly and personal degradation they were not unworthy successors. It is also a corrective to those sentimentalists who look back to a glorified past and fail to see that many of the changes that have taken place have been very definitely for the public good.

If we consider the tyrant and the aristocrat there are two basic causes for their moral condition—excessive power, and lack of an adequate occupation. The power is either directly political, or economic through their riches; the lack of occupation may simply be due to the fact that their riches come from safe investments, or it may be due to the organization of a society that refuses them the right to work. If we divide aristocracies of birth or wealth into two groups, the division is the possession of direct power over the lives and deaths of others. In the group of the powerful, cruelty is the most strikingly developed of the vices. It is a curse that spreads through all society, from unrestricted power at the top down to ignorance and misery at the bottom. In the second class idleness is the chief quality, and sex instability, an excessive and perverted sensibility, drunkenness and gluttony are the chief results. As the powerful group is also largely unoccupied they also share the vices of their less dangerous peers.

There is little difficulty in understanding why cruelty is so common with tyrannies. Cruelty has two attractions for men in a certain position. It is vicarious revenge for what they have themselves suffered—and even the heads of absolutist states have suffered frustration in being cut off from ordinary life—and it is also a form of excitement that makes a strong appeal to various types of people. Where there is low intelligence, little imagination, and scant sense that the sufferer is of like nature with oneself, then cruelty and the sight of suffering provide a thrill that rouses the emotions when all else fails. There is no doubt that the Spanish bull-fight has an attraction for many people. D. H. Lawrence in *The Plumed Serpent* describes vividly the effect that the bloodshed had on the crowd of Mexicans. It is also certain that if hangings were once again held in public in this country there would be a large audience. Even as it is, people gather outside the prison gates when there is an execution, and try to derive, from this slight proximity and the tolling of the bell, some of the emotions they are anxious to enjoy.

Idleness, whether with or without power, brings the vices of drunkenness, gluttony and sexuality. Drunkenness is a method of escaping from an intolerable situation, gluttony is the conversion of one of our most fundamental impulses into a substitute for other interests in life; sex amusement is the regular resource of the idle, as we see any day on a pleasure voyage.

Almost any aristocracy could provide examples of this constellation of vices. We could take examples from Imperial Rome, pre-Revolutionary France, Habsburg Austria and from the cosmopolitan rich at the end of the nineteenth century—though they were not quite up to the earlier standard; but perhaps we possess the most recent and fully documented accounts of Tzarist Russia. There, till the middle of the last century, a society existed based on slavery, maintained by a ruthless police tyranny and governed, surprisingly often, by a ruler who, if not mad, was far from sane.

That we can see Russian society so comparatively clearly is due to the burst of literary glory associated with Chekhov, Dostoevski, Gorki, Tolstoy, Turgenev, Gogol—with such avowedly revolutionary writers as Leonid Andriev or Prince Kropotkin. The historical figures of Ivan the Terrible, Peter the Great, Catharine and Paul throw into relief the characteristics that were only partially modified by the passing of time, while the freeing of the serfs mitigated but did not remove the degradation of the peasants.

Perhaps the clearest brief pictures come from Prince Kropotkin's memoirs. We see from these the tyranny under which the aristocracy lived as well as the tyranny they passed on to the serfs.

The date is about 1840 and Nicholas is Tzar.

The first extract is in a description of a general who came to offer Kropotkin's father a second wife.

"General Timofeef was a terrible man. He would order a soldier to be flogged almost to death for a mistake made during a parade or he would degrade an officer and send him as a private to Siberia because he had met him in the street with the hooks on his high, stiff collar unfastened."

Military service was for twenty-five years. Even an officer cadet was liable to be flogged. The soldier suffered incredible things. When a serf was sent to be a soldier, chained and in a cart, his friends saw him off with the chants for the dead. Any serf who showed too much initiative was liable for this fate.

But it was not only the poor men who suffered. As in the reign of Nero, a message from the Emperor was only too likely to be a sentence of death.

"One night while all the household was asleep, a three-horse carriage, ringing with the bells attached to the harness, stopped at our gate—a man jumped out of it, loudly shouting, 'Open, an ordinance from His Majesty the Emperor'.

"One can easily imagine the terror this nocturnal visit spread in the house. My father trembling went down to his study. 'Court martial, degradation as a soldier', were the words that rang in the ears of every military man; it was a terrible epoch. But Nicholas simply wanted the names of the sons of all officers in order to send the boys to military schools, if that had not been done."

Or lastly, the disturbance when a servant had to go to the master of the house and announce that the supply of tea was nearly exhausted. Shouts, abuse, a summoning of one servant after another. Finally the full wrath falls on Makar, the piano-tuner and sub-butler, who had broken a few plates the day before.

"Of a sudden there was a lull in the storm. My father takes a seat at the table and writes a note. 'Take Makar with this note to the police station, and let a hundred lashes with the birch rod be given him.' My father was not among the worst of landlords. What we saw in our house was going on everywhere, often in much more cruel forms. The flogging of serfs was a regular part of the duties of the police and the fire brigade."

Here we have in companion descriptions the nervous terror of a member of the aristocracy, and the way in which he releases his pent-up emotions of shame and fear by the injustice and savagery of his treatment of those in his power. The mechanism of displaced aggression could not be more clearly shown. At the same time in these memoirs Kropotkin describes how his father, a general without command, used to issue written marching orders and instructions to his family when they were making any expedition, and finish up the document by giving permission to his wife to use her discretion in any unforeseen circumstances. As nearly all the actual circumstances eventually proved to have been unforeseen, the orders only served to amuse an unoccupied man.

The picture of Russia given by one writer after another, from historical times to Alexei Tolstoy's *Darkness and Dawn*, stresses the element of brutal cruelty that exists in it. The whip is the symbol of authority—whoever that authority may be—and any regard for human life is absent from the minds of the rulers. This cruelty spreads all through the nation and finds strange perverted forms. In *The Brothers Karamazov* is the story of how a child eight years old was stripped naked and hunted to death by hounds before the assembled serfs, including his mother, because of some accidental injury he had inflicted on one of them.

Nor is Russia at all peculiar. In Roman history it is remarkable how the bloodiness of the gladiatorial shows increased when the Republic became an Empire, and, as the Empire became more tyrannical, so more blood flowed; till, with Nero's massacre of the Christians and slaughter of beasts, a standard of destruction had been reached. Spain is one of the few countries in Europe that has retained its aristocratic organization, and there the bull-fight with its disgusting slaughter of animals continues. Across the border, in republican France, it has been changed to an almost harmless Sunday afternoon amusement, in which the worst damage likely to happen is that a youth might sprain his ankle. It is perhaps worth noticing that in English history it needed one beheading to establish the Commonwealth, and a great many hangings to effect the Restoration; but whether we should blame the King or the Presbyterians for the latter is not very clear.

Gluttony and drunkenness are not a rewarding subject. The larks' tongues of the nobleman recorded by Kropotkin, who stripped himself of a large patrimony to pay for food alone; the drunken orgies of every petty princeling throughout Europe are recorded in every book. It has to be a very constitutional monarch who can give a state banquet on gold plate (the honeysuckle service) and have for menu:

Consommé jardinière
Mayonnaise d'homard
Perdreaux rôti sur croûtes, légumes de saison
Soufflés, glacés aux fruits, petits fours
Café.

Sexual depravity such as made the Court of Catharine the Great so fertile a field for the biographer is characteristic of absolutism. A mere commoner could hardly command the rights of princes, but the rich societies of Rome and London, even at the end of the nineteenth century, were completely without sexual morals. Among many books two can be mentioned: d'Annunzio's *Child of Pleasure*, written with a frankness that Wilde in his *Picture of Dorian Gray* did not dare to equal. In *Dorian Gray* the sexual adventures are hinted at. The voluptuous and excessive concentration on beauty, which is only the secondary theme in d'Annunzio, is made more important.

It is hard to tell if d'Annunzio meant his hero to excite pity—one almost thinks so—or if he were holding him up to horrified contempt. The young man makes his uncertain way through a maze of intrigues, swooning in love.

But love, for this society, is not something natural and straightforward ending in satisfaction. Both the women loved are married, and neither has any intention of leaving her husband. Moreover, though the hero does not share it, sexual perversion plays a large part. This too is a characteristic of such a society. The great names of sexual corruption, such as the Marquis de Sade and Sacher-Masoch, belong to degenerate aristocratic societies, and the homosexuality of Wilde's circle is well known.

Here is the passage from d'Annunzio[1] describing the connoisseur gloating over his collection of the obscene:

"Lord Heathfield opened the great book-case containing his private collection and turning to Sperelli:

" 'You should design the clasps for this volume', he said. 'It is in quarto and dated from Lampsacus 1734—the engravings seem to me extremely fine—What do you think?'

"He handed Andrea the rare volume, which was illustrated with erotic vignettes.

" 'Here is a very notable figure', he continued, pointing to one of the vignettes. 'Something that is quite new to me. None of my erotic authors mention it.'

"He talked incessantly, discussing each detail and following the lines of the drawing with a flabby white finger.

" 'This Dutch edition of Petronius is magnificent, and here is the *Erotopoegnion* printed in Paris, 1798. Do you know the poem attributed to John Wilkes, an Essay on Women? This is the edition of 1763.' The collection was very complete. It comprised all the most infamous, the most refined of sensual works that the human mind has produced in the course of centuries to serve as a commentary to the ancient hymn in honour of the God of Lampsacus, 'Salve, sancte Pater'.

"The collector took the books down from their shelves and showed them in turn to his 'young friend', never pausing in his discourse. His hands grew caressing as he touched each volume bound in priceless leather or material. A subtle smile played continually round his lips and a gleam as of madness flashed from time to time in his eyes."

And then, too, in the conclusion of the book, perhaps ironic, perhaps sincere, when Andrea, having finally lost both his mistresses, attends the sale of a loved-one's possessions, the exaggerated sensibility, the senseless desolation, are the typical ending of this type of sexual excitement:

[1] *The Child of Pleasure*, Bk. IV, Ch. 6 and 10.

"Andrea could not make up his mind to cross the threshold of that house, but wandered about the street a long time, weighed down by a horrible sense of lassitude, a lassitude so overwhelming and desperate as to be almost a physical longing for death.

"At last, seeing a porter come out of the house with a piece of furniture on his shoulder, he decided to go in. He ran rapidly up the stairs; from the landing already he could hear the voice of the auctioneer.

"Andrea felt stifled—the sale proceeded rapidly. He looked about him at the low faces of the dealers, felt their elbows pushing him, their feet touching his, their horrible breath upon him. Nausea gripped his throat.

" 'Going, going, gone.'

"The stroke of the hammer rang like a knell through his heart and set his temple throbbing painfully. He bought the Buddha, a great carved cabinet, some china, some pieces of drapery. . .

"He forced himself a passage through the closely-packed bodies, repressing his disgust as well as he could, and making the most strenuous efforts to ward off the faintness that threatened to overcome him. There was a bitter and sickening taste in his mouth. He felt that from the contact of all these unclean people he was carrying away with him the germs of obscure and irremediable diseases—physical torture mingled with his moral anguish.

"Half demented, Andrea fled, and at last reached his home. There the porters were just taking his purchases off a cart and shouting loudly. Several of them were carrying the cabinet up the stairs with a good deal of difficulty. He went in. As the cabinet occupied the whole width of the stairs he could not pass. So he had to follow it slowly, slowly, step by step, up to his door."

In *Dorian Gray* the exact nature of the book which so entrances him and which opens his eyes to such a range of exquisite vices is not disclosed. Perhaps it had a place on Lord Heathfield's shelves. In any case vices are more attractive when undefined. Instead, for many pages Wilde tells us about Dorian's exaggerated sensibility and his morbid contemplation of the beautiful.[1]

"He felt keenly conscious of how barren all intellectual speculation is when separated from action and experiment. He knew that the senses, no less than the soul, have their spiritual mysteries to reveal.

"And so he would now study perfumes and the secrets of their manufacture, distilling heavy-scented oils, and burning odorous fumes from the East. He saw there was no mood of the mind that

[1] *The Picture of Dorian Gray*, Ch. XI.

had not its counterpart in the sensuous life, and set himself to discover their true relationship, wondering what it was in frankincense that made one mystical and in ambergris that stirred one's passions, and in violets that awoke the memory of dead romances, and in musk that troubled the brain.

"At another time he devoted himself entirely to music and in a long latticed room, with a vermilion and gold ceiling and walls of olive-green lacquer, he used to give curious concerts in which mad gipsies tore wild music from little zithers, or grave yellow-shawled Tunisians plucked at the strained strings of monstrous lutes. The harsh intervals and shrill discords of barbaric music stirred him at times . . . yet after some time he wearied of them, and would sit at his box at the Opera listening in rapt pleasure to 'Tannhäuser' and seeing in the prelude to that great work of art a presentation of the tragedy of his own soul.

"On one occasion he took up the study of jewels. . . ."

We need not follow him through the intricacies of precious stones, but all the time he was lolling on embroidered cushions and inhaling the fragrance of scented lamps and turning the pages of that book, so intoxicatingly wicked, which obsessed his mind.

The aristocratic pursuit of art is something of this type. To *practise* art requires hard work, thought and effort. This sensibility is a product of idleness, of the blocking of the ordinary outlets of emotion in action, and of the absence of intellectual thought. Further, the art generally preferred by aristocracies is that which imposes the maximum of labour on the craftsman. The elaborate goldsmiths' work done for the Popes by Benvenuto Cellini can be paralleled by the incredibly intricate ornaments made for the Russian Imperial Court by Fabergé or the cloisonné enamels of the Palace of Pekin. In dress, too, the elaboration of the Ancient Régime dissolved into the comparative simplicity of the Republic, and the great age of architectural baroque was also the age of despotism.

A despotism inflicts, even on the aristocracy, additional ills, and the chief is the denial to them of responsibility and the right to work. In some cases this produces nothing worse than acute discouragement. The characters who people the world of the Russian novelists are, at their best, hopeless and without purpose. They have no conscious part in the scheme of things, and they can find no proper outlet for their talents and energies, so, unused, these fester and turn to unceasing introspection. On the other hand, lack of responsible occupation may take away the last reason for restraint. The French nobility, excluded from the time of Louis XIV in any participation in government, had no counter-balance to their vices, and were hated with a violence that never fell to the lot of the English lord. For the Englishman kept his place in the state. The aristocratic and wealthy in eighteenth-century England were on the whole drunken gamblers and much lacking in the finer shades of morality. When all their ordinary dissipation failed to amuse, they held parties at Medmenham Abbey and celebrated the Black Mass; yet these revellers held political positions and a considerable part of their lives was occupied with affairs of state. They could not sink below a certain level, as they had to use their intelligence and maintain a certain standard of self-control. They were not the highest and wisest, but they stood in quite a different category from others who were allowed no political life.

On the whole, the English upper classes have at all times maintained the right to work—so long as the work was unpaid. The work has been done from motives in which a love of power, a desire for prominence, an eagerness to exercise their talents and a sense of social obligation have been mingled. Nothing could be more complete than the difference between the English country gentleman of the eighteenth and nineteenth centuries with his work as J.P., his duties in Parliament, his care of his estates, his management of the fox-hounds, and the Russian nobleman as he appears in Dostoevski and writers of that age.

Women were later in gaining this right. After the managing mistress of a castle, who, in the olden days, was quite capable of organizing its military defence, or the Elizabethan lady who was for ever in the still-room or dairy, the Victorian wife was allowed to do nothing. Her husband paid the servants and the housekeeping bills. The ordering was done by the cook, and the organization by the butler. Nurses and governesses attended to the children. Exhausted by idleness she developed a complaint and retired to her sofa. It was only through much ridicule that she fought her way to committee-room and parish council. Even to-day the very rich find it difficult to be allowed to work, and their lives are characterized by a restless wandering from one house or district to another, emotional crises, and a sense of the vanity of the world. The matrimonial adventures of one American heiress, who appears fairly regularly in the papers, suggests very well the hectic unhappiness, the mercenary interests and the lack of judgment that prevail.

The first condition, therefore, for a healthy society as well as for individual morality, is that there should not be too much power at the top, and that there should be the right and duty on all to work. What form this work should take depends on circumstances, but an unoccupied class is always a source of social danger and moral corruption, especially because its members, being the owners of prestige, are envied; and their vices, as being the most imitable, copied.

The poor, in the form of poverty that constitutes a social threat, exist as part of the pattern that produces the very rich; but their numbers and characteristics have changed in the last 150 years. The beginning of the nineteenth century saw, probably, the greatest mass of misery that the peace-time history of Europe records. The old forms of oppression and political tyranny were suddenly reinforced by the new horrors of industrialism. Even where the old fabric of society remained apparently unchanged the real nature of the relationship was altered for the worse. The political system of serfs and nobles might still be there, but the passage of time had increased the distance between serf and ruler. What had been difference of status, very real, very important, but mitigated by a certain common background of belief and custom, had become, through changes in the nobles, almost a difference in nature.

In England the industrial worker had not even this historical union with his employer. He was a new social phenomenon. The agricultural worker, or the village craftsman from whom he was sprung, had a recognized place, and an established association with the parson or the squire; but when he came to the town this was all gone. His state was not ameliorated by any established custom or pre-formed charitable dispositions. He was what he was, because purely financial considerations were the only ones that counted.

The state of the English urban worker in the early nineteenth century can be learnt from the impeccable pages of government reports, or the more vivid writing of Engels, and the picture of horror and degradation is almost unbelievable to the modern reader.

Engels, writing of conditions in Ancoats, Manchester, has hardly need to describe the inhabitants, the conditions of life are so degrading:

"Here a horde of ragged women and children swarm about, as filthy as the swine that thrive upon the garbage heaps and in the puddles. The race that lives in these ruinous cottages behind broken windows mended with oilskin, sprung doors, and rotten door-posts or in darkest cellars in measureless filth and stench, must really have reached the lowest stage of humanity. In each of these pens containing at most two rooms, a garret and perhaps a cellar, in the average twenty people live. For each one hundred and twenty persons, one, usually inaccessible, privy is provided; and in spite of the excitement into which the cholera epidemic plunged the sanitary police in this year of grace 1844, it is in almost the same state as in 1831."

From such starvation, exhaustion and misery nothing good could come; it was hard for any being to maintain the mere pattern of human life. So, too, the Russian serf, the agrarian counterpart, suffering from the same theory of private property, represented a depth of misery that is hard to imagine. Both of these classes had a great effect on the state. In England disease and misery, degradation and hopelessness, were manifest in Chartist riots, and plain to any man sensitive enough to feel it. But the proletariat was cut off from the governing body by the curtain of the middle classes. There was a section still left of the skilled artisan, and the lowest grade of worker was so low that he could often be ignored. In Russia where there was only a small middle class the effect was more serious.

The Russian serf, and later the peasant, formed a vast sea of ignorance and degradation that had an effect on the whole nation. The monstrous growth of superstition, the mass of conservatism, all drew their power from the peasant. In so far as a child of the upper classes was nursed and tended by women drawn from the land, he imbibed the beliefs of his nurse. Thus from the Tzar downwards right up to 1917 there was a firm belief in the prophetic power of idiots sheltered by the Church. If any man of ability appeared among the serfs, he was either left quite uneducated or, if he were taught enough to make him useful to his master, he would be sent off to be a soldier, or killed supposing he began to show any sign of questioning authority. Thus the nation's supply of ability was wasted and the government remained in the hands of men frequently naturally stupid, and always corrupted by the ignorance below them and the tyranny above. Thus the incompetence of Russian government reached an extraordinary pitch before it was shattered by the Revolution.

The whole movement since 1850 has been away from the unrestricted use of private property. A man may no longer do what he likes with his own. Certainly not if his action affects others. Thus the great reforms to be discussed later all decreased the power of the propertied classes, while giving opportunities of development and advancement to the poorer. But this has not by any means completely removed the problem of the industrial worker. It is true that he is now educated until the age of fourteen, is paid a reasonable wage, is protected from the worst effects of unemployment and is provided in many cases with a decent house. Even the slums of our industrial cities are better than would have been thought possible fifty years ago. But though the worst elements of the system have gone, what remains is still bad enough and there is still a social problem of considerable gravity. As the industrial worker forms more than half of our urban population, he is of the greatest social importance, and on his behaviour the moral state of the community largely depends.

In an industrial society workers fall roughly into two categories, the skilled and the semi- or unskilled, and the psychological conditions of the two differ completely. To the skilled man his work is generally a pleasure. He has probably chosen it because there is something about it which makes an appeal to his nature—one man likes wood, another iron—and there is an almost universal pleasure in the successful employment of our powers. He is probably intelligent enough to understand the purpose of his work and he has generally accepted this purpose as satisfactory. The perfect pattern of a craftsman is given in a medieval picture of St. Eloi working as a silversmith. The bishop sits in his robes in a high chair, his work before him; behind is an open window with flowers on the sill. His assistants are working in front and his pet monkey is walking about among the tools. Silversmithing is a very difficult craft and as such makes an appeal to a high type of man. It is also a craft used directly in the service of God and the Church, and it has numbered St, Dunstan as well as St. Eloi among its practitioners. Here everything is gathered together: skill, an aim that is satisfactory artistically and morally, and, for the Saint, conditions of work that are pleasant.

Not all craftsmen have as good a room as the Saint, but they share, to some degree, his pleasures, and the skilled craftsman is generally a very good type of man as an individual and a citizen. He is intelligent, self-respecting, trained and generally public-spirited. But under modern conditions he is only a small fraction of the population. In an American survey, of a New England town, 11 per cent. are classed as skilled workers. This would include many men of far less skill than silversmiths. Fifty-one per cent. are given as semi- or unskilled. An English estimate given in a survey of the needs of youth suggests that only one in three can expect to enter work that will require skill.

"It appears reasonable", says the pamphlet,[1] "to suppose that one-third enter an occupation which offers an opportunity for learning a skill or for advancement in this particular occupation."

Thus they anticipate the figure for semi-skilled work to be 66 per cent. Out of the remaining one-third of the population come the distributive workers, clerks, professional and managerial groups.

The mentality of the unskilled worker is directly shaped by his work. Suppose a child of fourteen enters industry, he is given some

[1] *Youth's Opportunity*, H.M. Stationery Office.

simple repetitive task to do which makes a greater or less call upon his physical strength. The commendation so often heard of a job—"Oh, it's easy"—implies that the physical strain is light and that there is no thought required. On the other hand, some jobs as a result of noise, dirt, inconvenient posture at the machine and so on, are very fatiguing and leave the worker at the end of his shift tired out. The mental attitude of the repetition worker can be either a concentration on the job, in which the mind turns round and round at the point of the cutting tool; or it can be day-dreaming, in which the hands perform their function and the personality is elsewhere. Sometimes the attitudes change from hour to hour; but, in either case, the effect is bad, especially in young people. It is possible that those over forty do not take much harm, having completed their development; but for the young there must be an arrest of development and an acute exaggeration of the sentimentality and unreality of adolescence. When, in addition, the unskilled worker expects no promotion, has no hope of further training, is bullied and even pressed at his work, frustration can only be drowned in apathy.[1]

There is a further bar placed to natural development. The unskilled worker is not allowed to think. Supposing he is operating a machine that needs small adjustments between one job and the next, he is not given a spanner and allowed to make them himself; he must wait for the foreman or the tool-setter to do it for him. Thus he cannot feel himself in any way master of his environment; he cannot experience even the mild pleasure of a small piece of independent work. It is, of course, contended that to give a very stupid man a spanner is to invite accidents and mistakes. The answer is that frequently people develop as far as they are *allowed*, and a refusal of any responsibility prevents development.

The adult industrial worker, subdued from youth to the task, has developed a life exactly patterned: the hooter in the morning, the break for lunch, the blessed hour when the shift ends and the machines sigh into immobility, tea, and a half-pint. If he breaks the routine and goes to a Wednesday football match, it is absenteeism. If he thinks, he becomes discontented; if he demands better conditions, it is a strike. He has not developed his powers either of thought or action. He has no resource or alternative way of life. He has little power of individual enjoyment.

In her book *The Weeping Wood*, Vicki Baum gives a picture of the motor-tyre industry. This is her description of building the tyre and the labour of the man bound to the task:

[1] *Democracy and Industry*, Reavley and Winnington.

"Then comes the turn for the tyre-builders. Each man stands behind his revolving drum and gauges the speed of it with a foot pedal. You cement the drum and run on the first band, and heel it down tightly with the heel of your hands . . . you stitch it down and put on the next ply, swab it with solvent, stitch it down, one ply after another, and the last goes over the bead and has to be hooked under. Then you put on the breaker and the chafing strip and stitch these down; so now you buff off the carcass with a wire brush and swab it with solvent. Then you've got to finish your tyre, that is put your tread on. It doesn't come in bands, but as a belt and has to be pulled over the drum. That's the toughest moment and takes all of a man's strength. Next you slip on the side walls to match the tread and roll them down. You trim off the bead with a knife and stop your machine. You collapse the drum and pull off your tyre, and that's tough work too, believe me. . . . To-day you can't take more than four minutes all together for building a tyre, and that's damn fast time, believe me. . . ."[1]

This intense repetitive work exhausts a man mentally and physically. He has nothing over for himself and his own life.

This is not, of course, strictly true in all occupations. If, when waiting for a train, one gets into conversation with the porters on Oxford station, one finds that the ticket collector's hobby is building dolls' houses, that the man with the luggage trolley plays in a band, and that most of the others are gardeners; but in the great industrial areas there is far less activity of this sort, the slum houses have no room for plants or sensible human interests.

Though there has not been any general study of the psychological effects of industry we can see a type that is more or less imprinted on the individual by his circumstances. It is a type lacking in knowledge, in skill, in responsibility, in dignity, rendered stupid and careless by the manner in which work is arranged, and encouraged to day-dreaming and in need of escape from a life that is ugly, fatiguing and lacking in variety.

This is not all. In the past, unemployment hung as a continual dread over all industrial workers. It became an integral part of the system, and was deified into a fundamental principle. The doctrine, still occasionally seen to-day, of the "labour reserve" meant that at least 10 per cent. of workers were expected to be always unemployed. This served two purposes. In boom periods there were men who could be quickly engaged; and, except for a very brief period at the peak of the boom, all those with jobs were kept docile and industrious

[1] *The Weeping Wood*, page 231.

by the fear of losing them. This actual insecurity by engendering fear exhausted the nervous energy of the workers and their wives. They could never develop self-confidence or self-respect. If they were skilled workers the shock of losing a job was even greater. They were deprived not only of a wage but of the right to use their skill and for many a man this was the greatest damage. They were deprived of their hold on life. When they were discarded so that cheaper, half-skilled labour, often female, might be taken in to work machines, they felt that the meaning had gone from existence.

This again is from Engels and tells of the skilled artisan reduced to the depths of shame. He is found by a friend also in search of work:

"There sat poor Jack near the fire, and what did he, think you? Why he sat and mended his wife's stockings with a bodkin: and as soon as he saw his old friend at the door-post he tried to hide them. But Joe had seen it, and said, 'What the devil art thou doing? Where is thy missus? Why is that thy work?' And poor Jack was ashamed, and said: 'No, I know this is not my work, but my poor missus is in the factory: she has to leave at half-past five and works till eight at night, and then she is so knocked up, that she cannot do aught when she gets home, so I have to do everything for her, what I can, for I have no work, nor had any for more nor three years, and I shall never have any more work while I live', and then he wept a big tear. Jack again said: 'There is work enough for women folks and children hereabouts, but none for men; thou mayest sooner find a hundred pounds on the road than work for men, but I should never have believed that either thou or anyone else would have seen me mending my wife's stockings for it is bad work. But she can hardly stand on her feet: I am afraid she will be laid up and then I don't know what is to become of us, for it's a good bit that she has been the man of the house and I the woman: it is bad work, Joe', and he cried bitterly."

A further result of all this complex of conditions and characteristics is that there is a sense of hostility against the social system or some part of it. It has, in fact, treated the man badly, and, as we have said already, a sense of wrong issues in all kinds of unsatisfactory behaviour. Not only does a man feel that he has been badly treated, he also lacks a sense of his place and importance in the social scheme. Hence many of the characteristics of the lower-grade worker. His unreliability, his carelessness, his tendency to unofficial strikes or go-slow methods. Hence, too, the forms of recreation most common among the industrial workers; in particular the cinema and certain forms of gambling. The cinema provides emotion, excitement and the sensation of luxury. It requires no effort on the part of its patrons and it takes them away to another world where the things they long for are commonplaces: houses, yachts, motor-cars, marvellous lingerie, romantic love. The gambling on horse and dog races or in football pools dangles before their minds the dream of sudden wealth, a solution of all their problems, a realization of their dreams achieved without more effort than filling in a betting-slip. The man at the machine, his mind vacant but for day dreams, has something to base his dream on if he has posted his sixpenny coupon.

Karl Mannheim in *Man and Society* complains that the greatest danger of the modern world, the characteristic, in fact, that has produced the disasters of the last twenty years, is the liability of the industrial masses to attacks of unreason. In spite of the apparent patterning of their lives, within there are unorganized emotions and minds that are not sufficiently developed or trained to take charge of their conduct. This is largely true; the industrial worker is not master of his fate and he does not understand the forces that act upon him. He has no alternative way of life that he can adopt when things go wrong. He is thus a prey to fear as soon as the pattern of his life is upset. Thus, when unemployed, he is a prey to anyone who propounds an apparent solution of his problems; and with this stored frustration, that the organization of society forces upon him, he is liable to choose the solution that offers violence. The relation between the type of work a man does and his social behaviour was well illustrated in Liverpool just before the war when prolonged unemployment among the dockers had led to mild rioting. A ticket-collector, asked what he thought of it, replied, "Damn fools, woke me up at four o'clock this morning". In the growth of that attitude and that class lies social stability.

It is a testimony to the natural goodness of man that so large a part of the industrial population, the moment they are provided with an even passable way of life, are so law-abiding, so full of domestic virtue, for they are very far, even under the best modern arrangement, from achieving the two conditions that seem among primitive man to make most for virtue: security and equality. There is no doubt that the status system, the class system, has decreased a little in England in recent years—it seems hardly to have decreased at all in the parts of America where it is strongest—but it still exists to a degree that makes any claim to real equality nonsense. A system of social security as imagined by recent legislation will decrease to some extent the element of fear in life, but it will not remove it, or set individuals or individual families free from the necessity to compete and struggle. The industrial worker will still be in private employ, and, though most men develop a surprising loyalty to "our place", the fact that both their conditions of work and the profits of their labour are the affair of others prevents a full development of corporate feeling.

In the past the incentive to work was either fear or private gain; mainly the latter for the employer, and the former for the employed. Now, when fear is to be largely removed from the industrial worker's life, what is to be the motive that makes him put out his full powers? This will be one of the practical as well as one of the moral problems of the new order of society. It is a problem that cannot be solved at once. Before a man will work his best without the stimulus of much fear or much reward he must get rid of the feeling that work is something to be avoided, he must learn to feel responsible for his part in the social order, his work must be so arranged that it makes no demands on him that cannot be met with dignity, he must have learnt the pleasures of activity and the use of his skill. Most of all he must be put to work that suits him both mentally and physically. This training of the individual to work his best with a sense of his part in the whole community is one of the tasks that will fall on the educators of the years that are immediately approaching.

In Russia this problem of incentives has been particularly important. A population, unaccustomed to industrial work, has had to be trained, and that at a time when output was of the greatest importance. Some of the deterrents to effort that exist in England do not exist in Russia. The sense that the harder one works the more profits for the employer, the fear that if one hurries, the piece-rate will be cut or the job finished, has made the worker unwilling to do his best. The stored historic anger and frustration from the worst days of the last century do not affect the minds of the inhabitants of more newly industrialized countries. Moreover, the Russians have relied to a much greater extent than some other countries on suggestions from the workers themselves. The two best-known devices for increasing production, the organization of socialist emulation and of Stakhanov groups, were both allowed to grow up from the bottom, and received only encouragement from above. The competing groups in the same factories or in similar works were stimulated by praise, public demonstrations of their skill, and intelligently chosen rewards. The Stakhanovite methods again, which in America would come under such a heading as scientific management, were devised by the most intelligent workers, tried out by them and taught to others, with the sympathy, but not the coercion of the management. They were thus received willingly and without opposition. In addition, every effort is made to increase the skill of the workers, to upgrade the able, and to give incentives in the form of differential wages to the most productive and efficient.

The trade unions, debarred from political activity, have been given a new role.[1]

"Trade unions must now take a more active and direct participation in economic construction and an increased interest in production. Socialist emulation, and its offspring, the exemplary workers' movement, are decisive and fundamental factors vitalizing and improving trade-union work and bringing wide working masses to participate in industrial management. . . . That is why socialist emulation and shock brigades must become the corner-stone of trade-union production activities in enterprises and workshops. Exemplary workers are called upon to become the backbone of production conferences. Trade unions must combat personal management of a faulty bureaucratic type, and any disregard on the part of the management of the productive initiative and self-expression of the masses."

When this positive ideal of associating the workers in the planning and management of their own work is joined with severe penalties for labour indiscipline, it is clear that there is a very complete system, both material and psychological. The capitalist system, especially when it is working under the full influence of competition, and relying mainly on the threat of dismissal as an incentive, will have great difficulty in carrying out the psychological part of this method.[2]

[1] *Soviet Economic System*, Baykov.
[2] *Democracy and Industry*, Reavley and Winnington.

Books referred to in this chapter:

Plato, *Republic*.
D. H. Lawrence, *The Plumed Serpent*.
Prince Kropotkin, *Memoirs*.
Dostoevski, *The Brothers Karamazov*.
Alexei Tolstoy, *Darkness and Dawn*.
d'Annunzio, *The Child of Pleasure*.
Oscar Wilde, *The Picture of Dorian Gray*.
F. Engels, *Condition of the Working Class in England*.
Edwin Chadwick, *Report on Sanitary Condition of the Working Classes*.
Kay Shuttleworth, *Moral and Physical Condition of the Working Classes*.
H.M. Stationery Office, *Youth's Opportunity*.
A. Pearl Jephcott, *Girls Growing Up*.
Reavley and Winnington, *Democracy and Industry*.
Vicki Baum, *The Weeping Wood*.
Karl Mannheim, *Man and Society*.
Baykov, *Soviet Economic System*.

Chapter 3

MYTHS

We have discussed very briefly in the preceding chapter the actual organization of society from the point of view of political and economic equality, and indicated some of the ethical effects of inequality. A material fact is never quite divorced from a psychological event, and there is almost always a belief which accords with a material arrangement. Private property, as an institution, is intimately connected with many beliefs, and these beliefs would seem very strange to a member of a culture based on communism. To trace all the beliefs that are necessarily involved in private property would be difficult. They concern the nature of things as well as the nature of man. Ownership, as we understand it, gives the right to use or destroy. It enables the possessor to refuse to another the benefit of a thing he does not himself want. It allows one man to resist the needs of the community or to claim and take preferential treatment in his group. It grants to a man control over the lives and happiness of others, not on grounds of his own worthiness, but simply because of economic possessions. We are accustomed to such a system of ideas and they seem natural, almost inevitable. To others of a different culture they would seem incomprehensible.

The idea of ownership may be carried farther. In the past serfs and slaves have been owned as much as chattels and animals, and so long as slavery persisted the ideas that went with it were so natural that it was hard to discern them. In time ideas changed, and then a struggle took place to adjust institutions to new ideas, and after much effort and suffering slavery was abandoned by all the nations that took their ideas from Europe. To-day we can see the gradual decay of another idea. As we have said earlier, parents still legally own their children, up to a certain age. It is true that the state has considerably invaded this right. They may no longer destroy them, exposing them on the hills to die, as did the Greeks. They may not neglect or maltreat them beyond a certain limit; they must allow them to attend school. This limitation of full ownership, important as it is, still leaves the ideas on which the law is based strongly operative in many cases.

This mental structure in a community, corresponding to its

material structure, is a very important thing, and the two must be in accord. If they are not, there must either be a change in one or the other or there will be a permanent disharmony that it is very difficult to resolve. An example of this disharmony may perhaps be taken from the religion of Western Europe. A very proud, pugnacious and militaristic race, with strong institutions of private property, adopted a religion that was in essence meek, pacific and communistic. At different periods one or the other of these irreconcilables has been dominant. We have seen Renaissance popes and warrior bishops, as well as St. Francis and the Quakers. We have never yet succeeded in fully reconciling our beliefs and our institutions. Archdeacon Grantly has always had a place in our Church, to the scandal of those who desire a more apostolic type of Christianity. In other cases ideas have been more successful in moulding institutions. The abolition of slavery is one example, the gradual building up of social security is another. Of course the institutions that actually exist are continually forming our ideas. Merely to grow up in a certain order of society is to learn the ideas on which the institution depends. That is the reason why it requires such an effort to bring about a change. It is not even easy for an individual to imagine a change. But when he has done so he must transfer his vision to other minds so strongly that they will make the effort to build anew the ruins of what they themselves have cast down. That is why sudden revolutions are such things of toil and passion, and gradual ones so slow.

In addition to this mental structure that is so closely bound up with institutions, there is another of the greatest importance. It is freer from direct material ties, though in most cases it has an embodiment in material form or in the structure of the community. It is not easy to find a word that will serve for the system of ideas we wish to describe. We have called them myths, using the word in a sense rather closer to the modern German use than to that of Max Müller. We shall distinguish several types of these myths, and perhaps examples of what we mean will be clearer than a definition.

The simplest type of myth, the form that for us has little importance, is a story that offers in concrete or dramatic form an explanation or description of some phenomena of nature—Demeter, the corn goddess, giving her gifts to man, or rosy Dawn rising blushing from her lover's bed. The shining nymphs of grove or stream have an aesthetic rather than an ethical importance. For us the important type of myth is a belief that has importance for action.

In logical terms a myth is a universal, but it can be one of various types. The simplest myths are in their nature similar to a common noun. They gather together, for purposes of convenience of thought, certain common characteristics that exist in different individual examples. A myth of this kind that we shall mention several times is the myth of the Victorian Woman. She was a construction, a counter for thought. Her nature, her capacities, her proper functions, were all embodied in the myth. When the myth was given an imaginative reality, as in Dickens's heroines, we are surprised at its poverty. It was none the less powerful. It moulded women to its form, and it provided the complete answer when individual and eccentric women demanded votes, education or a career. The myth of the Father was equally potent and had a more complex origin. Both these myths have been largely destroyed by social and especially economic changes.

An even more important type is that which, while claiming historic truth in form, is really an embodiment of abstract truths applicable to all men in all times and places. It is in fact a rationalization of psychological experience. Many of the most potent of these myths are religious. The great circle of ideas that includes the Revolt of the Angels, the Fall of Man, and the Devil, has had a very wide influence on the beliefs and practices of Western Europe. It is a myth that was centuries in growing. It seized the imagination of men because it appeared so well to explain human experience. When once the myth was established, then, in its turn, it motivated action. The followers of Satan, though they may have been followers of a pre-Christian cult, had accepted so much of the dominant beliefs that they saw their own practices largely through the eyes of their persecutors, and felt their own courses to be evil. The belief in the Fall of Man and Original Sin dominated education and the treatment of children both in the Port Royal schools and in the nursery of the Fairchild family. Theology and the social system alike bore the imprint of this body of ideas.

Probably the most important type of myth is that covered by the German definition, "a symbolic idea of life-renewing force". This type has been called a concrete universal, meaning that it is a pattern through which all parts receive a firmer integration and a fuller meaning. In most cases the myth is pseudo-historic, and the examples that the German author gives are the myths of Frederick the Great, the Blood and the Reich. Jung, discussing myths, says the same thing in different words: "This rule applies to mythical tradition. It does not set forth any account of the old events, but rather acts in such a way that it always reveals a thought common to humanity, and once more rejuvenated." We would include other myths of a somewhat different type, in particular the myth of the Englishman. In most of these myths a state of society or a state of mind is made into a system, and each part of the system is enriched by its incorporation with the whole.

This incorporation and enrichment was undoubtedly one of the functions of the Nazi myth. The nation was broken and humiliated and individuals were left without adequate sense of security and dignity. The myth, by giving to each the membership of a great idea, made them partners in the glory of the whole. This is a common function of organizations. The humblest member of a great organization, if the unity of the whole and his part in it is kept before him, feels his dignity increased. When the clerk in a great firm, when a recruit to the army, or the new boy at school, feels himself absorbed and accepted he experiences an increase of dignity. So a myth which is able to make all partakers in one glory, whether they are general or private, party boss or crossing-sweeper, has a great effect.

We can perhaps classify these myths in another way that will show their importance to society yet more clearly. Myths may idealize, they may explain, or they may serve as compensation. These categories are not exclusive, and a figure is held up to admiration even when, as was the Victorian Woman, she is mainly a rationalization of a social system. So too the person of the Führer acquired Messianic dignity as well as the power to save his people.

We have discussed in the later part of this chapter two of the idealizing myths: Chivalry, which idealizes a social order, and the Englishman, which is concerned with personal qualities. Myths of this type play a very important part in the culture of a nation. They are the ideals towards which the young strive, and towards which their elders guide them. Every developed culture has these ideal figures, and they are one of the most influential parts of the tradition. They accompany a man through his seven ages. The mother has her image of what her baby should be and treats it accordingly, while the child responds to the treatment it receives. The prodigy of the Victorian age, learning Latin at three, has given way completely to the sturdy ruffian who does not know a single letter. The schoolboy of fiction who takes a "licking" rather than betray a comrade in mischief, the young warrior of Rembrandt or Rupert Brooke's poems, the lover, the lawyer, all have their predestined ideal forms. Even the grandfather sitting in the sun, watching the children sporting on the green, fills his proper role. This being so, the ideal figures that are impressed on the nation give it much of its character. It is interesting to trace the changes of such figures. One such, the Industrious

Apprentice, has his rise and fall, and now is in somewhat of a social eclipse.

The rationalizing myths take many forms. Some, as we have said about the Devil and Original Sin, give form and reason for what man feels about his own nature. Others are economic or political. One that had a great influence on our history in the last century was the myth of the Economic Man. This being, obsessed solely with his material needs, deaf alike to the lure of art or patriotism, was the form in which a commercial age could best understand its own activities. It gave form to the guiding lines of action and rendered argument simple and thought direct.

By far the most spectacular myths are the compensatory, and they possess in a special degree the characteristics of unreason. They arise, naturally, when things go wrong, either with an individual or a nation, and their power depends on the urgency of the need that is being frustrated. In this chapter we discuss the German myth that has overshadowed Europe for the whole of this century. Elsewhere we shall give examples of the way irrational ideas, translated into concrete myth form, are accepted and, for a time at least, determine action.

It is not enough to point to the existence of myths, it is also important to ask why they exist, and why in this form. It is fairly clear that the human mind prefers to think in concrete symbols. Personification is one of the commonest devices of literature or politics, and corresponds to an actual psychological fact. Most people use visual images in their thoughts, and these images assume definite, often personal, shape. For most men *Justice* is not an abstract noun, it is a figure, blindfolded and holding the scales. When we dream, the unconscious turns abstract thoughts to concrete and dramatic form; and the visions of seers have always been in terms of sense perception rather than revelations of pure ideas. The myth, the symbolic figure or the pseudo-historic tale, thus conform to the nature of man's mind; and if an idea is to make a wide appeal it is necessary that it should take this form. Plato long ago realized this, and instructed his Guardians to make their regulations palatable to the uninstructed populace by clothing their reasons in mythical form.

At the same time there are certain symbols and figures that seem to possess the power of arousing emotion generally, and thus are most suitable as the basis or as a part of a myth.

Jung has pointed out that there seem to be symbols connected with a racial unconscious, so widely spread are they, and so general is their power to evoke emotion. The myths that have grown up through the ages often seem to have drawn these elements to themselves.

For Jung the use of these symbols is a regression to earlier, childish, fantastic thought, a retreat from the abstract controlled thought of science. But it is a stage that many people never outgrow, and to which almost all return when confronted with an emotional difficulty in their personal or social thinking. Thus the myth is characteristic not only of children but of nations. We would go further than Jung and see in the symbolism of myths a normal method of conveying ideas.

A symbol that appears in many religions and systems of thought is the snake. To the snake is attributed wisdom and sexual potency, and the wide distribution of the symbol, its appearance in literature and dreams, seems to indicate that for great portions of mankind it possesses a special power of evoking emotion.

So, too, the position of the Father or Leader in myth and story carries with it an emotion that rarely fails to have its effect.

When a myth is invented more or less deliberately, as was the German myth, it is possible to see various elements being added to it. Thus Wagner brought the old Teutonic stories before men's minds with the utmost vividness, and added the powerful figures of Odin and Siegfried to Arminius and the slightly anaemic female figure who gathered flowers in the forests of Hölderlin. Later still Nietzsche and Rosenberg contributed elements that had a strong emotional significance. So, too, the myth of chivalry incorporated elements and symbols of great emotional potency.

It would be interesting to compare the great myths, using the word in this sense, with the folk-lore and stories of the nations and periods that produced them. One could see then what elements were common and what symbols belonged to both constructions. Our knowledge of Grimm's fairy tales is generally limited to some half-dozen that have been rubbed and polished by constant re-telling to a childish audience; but the corpus contains a number of a singularly unpleasant character and with an emphasis on cannibalism that shows a very different state of mind from the urbane pages of even our Red Riding Hood. The *Nibelungenlied* in its more primitive form has an emphasis on race and a brutality of detail that is startling. The choice of *Blood* as the basis of a myth, the fierce spirit that pervaded Nazi training, clearly connects closely with older stories.[1]

A myth when it is achieving its full perfection usually appears in literary form. The proud poet who was prepared to let any one make the nation's laws provided he was allowed to provide the songs was telling a literary truth. The intelligentsia could have capped his boast

[1] Contrast modern fairy stories, e.g. those by de la Mare.

by saying that any one else could make anything they liked so long as they made the myths. The part that the individual plays in the formation of the legend is clear from a few examples. In the myth of the Fall of Man the main parts of the myth have different origins and were brought to their complete form at different times. The story of the Fall of Man and the curse put on him occurs in Genesis and from this has been deduced the doctrine of original sin, but this doctrine grew up only by degrees. Later Jewish thinkers made a fundamental change in the story when they made the sin *moral*; in Genesis it concerned man's material existence. The apple that Eve took was more truly the fruit of the "tree of the knowledge of good and evil", a scientific knowledge of what conduces to man's material welfare. Once the apple has been plucked the development of cultural and technical skill involves mankind in ever-increasing disaster, as can be seen in the very next chapter of Genesis where we read of Tubal-cain "the forger of every cutting instrument of brass and of iron", and listen to the grim shout of his father Lamech exulting in the bloody and extravagant vengeance that his son's discovery has enabled him to exact for a trifling offence.

For I will slay a man for wounding me,
And a young man for bruising me.
If Cain shall be avenged sevenfold,
Truly Lamech seventy and sevenfold.

We can hardly help being reminded of the modern discovery of the atomic bomb.

The story of Adam and Eve has thus originally no connection with a doctrine of hereditary Original Sin. The allusions to that doctrine in the New Testament are scanty, and the sole authority for connecting it with Adam is to be found in half a dozen passages from the Epistles of St. Paul to the Romans and Corinthians. It was not until St. Augustine that it reached its classical form, and began its baneful career in the world.

The story of the Revolt of the Angels, as we know it, does not occur in the Bible. It existed vaguely in many forms in many traditions. The Devil, as man's adversary, had been the subject of centuries of thought. For Western Europe the whole story was put together and given form in the *Works and Days* of du Bartas, and from him it passed to Milton and *Paradise Lost*. The astonishing popularity of du Bartas, whose work was translated into all European languages, showed that in giving form to the myth he was supplying a need in men's minds. The domination of *Paradise Lost* in English thought shows the same need. But *Paradise Lost* adds to the myth, for it is clear that Milton accepted the sexual element in the symbolism, and that when Adam accepted the apple Eve offered it was a surrender to female attractiveness as well as a desire for a knowledge of good and evil. In the same way literature fixed the form of the chivalric legend. The German myth in particular grew up from authors and musicians, and the legend of the Englishman has received expression from a host of writers from Chaucer and Shakespeare to P. C. Wren, Eric Linklater and Anthony Hope.

These myths are taught to children; in fact, a very large part of moral education consists in this teaching—as we shall see later. It is very important, therefore, what quality these myths have, for they are one of the chief factors in the moral and social development of a nation. If one can say that a whole nation is virtuous or vicious, if one can characterize the moral quality of the age, it can be done only in relation to the prevailing systems of myths which exist at the time. It is true that in all complex societies there are sectional beliefs, but nearly always there are overriding beliefs that set their impresson on the whole nation and colour the thoughts and behaviour even of those who reject or only partially admit the merits of the dominant myth.

Thus the intelligentsia, whether they actually create a myth or only transmit it, giving it in passing a new application or a topical turn, bear one of the chief responsibilities for the moral state of their society. In certain countries where this has been realized the state or the church has assumed control of literature, and through a censorship or the deliberate commissioning of work has attempted to keep the presentation of ideas under control. In other communities the populace has been left free to choose between different ideas presented to it. In these cases the responsibility of the intelligentsia is greater, or perhaps the responsibility of each individual reader. Both author and publisher must consider the moral effect of a work on the community as well as the degree of financial profit to be made from it. In England many people take this responsibility very seriously.

The intelligentsia cannot, of course, force a myth on a nation that is in an unsuitable state to receive it. When for some special reason some section of the community wishes to add to or change a current myth, it is often difficult to get the new to blend with the old and to get the new accepted. For example, the Grail legend with the stories of Galahad and Parzival was probably invented by Cistercian monks who wished, by the glorification of chastity and faithful conjugal love, to counteract the tradition of courtly *amour* that played so large a part in so many tales. The two sets of ideas have never fully blended, and a reader of Malory may well find the transition from one to the other bewildering. He will be all the more grateful to Professor Vinaver who, in his recently published edition of the Westminster manuscript, has shown that the tales included in the Morte d'Arthur were originally conceived and executed as separate "works"; and that these inconsistencies of characterization and ideal exist between one work and another, but not within a single work. Each work reflects the tone and temper of Malory's French source. Moreover, it is probable that the new stories had little effect at the time when they were composed. It needed the Victorian age with Tennyson, Watts and Burne-Jones to bring Galahad to the forefront of the story—and even then Lancelot never quite lost his pre-eminence.

In another way the myth reflects the society in which it grows; it gives in concrete form the interpretation of the universe as it appears to the age. Nature and its powers appear in pleasant or terrible guise. The beings who haunt the woods are nymphs or ogres, fairies or trolls. The gods recline cheerfully on Olympus or crave their dark mede of human blood. It is to be noted that tribes with harsh customs feel themselves surrounded by demons, and those of milder manners have a less terrifying spiritual environment. A man's inner nature is projected outwards and determines the characteristics of the universe as it appears to him. The Christian claim that God is Love, that the universe is ordered and benevolent, is something very different from the doomed and war-torn realm of Odin. In Christian thought the emphasis on Hell and punishment varies from age to age, and the thought of damnation has been most vivid and influential when wars without and social injustice within kept men's minds hostile and fearful.

In the pages that follow we have discussed three typical myths which illustrate the various forms myths take and which indicate their importance for the individual and society.

The myth of Chivalry shows very clearly how a state of society becomes transmuted by imagination and ideals into something noble. The struggles of Christendom against the threatening might of invading Mohammedans, the endless strife of feudal barons each coveting his neighbour's lands, the insecurity of life and the low position in many ways of women, all became the substance of a high story and a way of life that men felt to be in accordance with the finest ideals.

Chivalry grew up as part of the wider concept of Christendom of which St. Augustine says: "That most glorious society and celestial city of God's faithful, which is partly seated in the course of these declining times, wherein he that liveth by faith is a pilgrim among the wicked, and partly in that solid estate of eternity, which as yet the other part doth patiently expect. . . ."

This Christendom was the union of all Christian men on the earth, with the Church on earth and in heaven, and it stood over against the Pagan world that threatened to engulf it. So its members were warriors, bound by the Christian bond, and ever alert to do their duty. It grew up to form the feudal relationship of lord and vassal, warrior and squire, and "chivalry was the fine flower of honour growing from this soil, embosomed in an abundant leafage of imagination."[1]

The combination of the mutual faith of lord and follower with the duties of a Christian meant, since Christendom was constantly at war, that the first virtue of a knight was valour. Next, since life and property hung on mutual aid and troth, the virtues of truth-speaking and troth-keeping took their places in the chivalric ideal. Another useful quality and means of winning men was generosity.

Valour, troth, largess, had no necessary connection with Christianity. It was otherwise with certain of the remaining qualities of a knight. According to Christian teaching, pride was the deadliest of the sins, so haughtiness, boasting and vainglory were held vices by the Christian knight. He should show a humble demeanour, save towards the mortal enemies of God. Humility entered knighthood's ideal from Christianity; and so perhaps did courtesy, its kin, a virtue which was not among the earliest to be admired, and yet reached universal recognition.

Since knighthood's ideals took form in crusading times, the slaughter of paynim became the supreme act of knightly warfare.

If such elements of the knightly ideal were of Christian origin others were even more closely part of medieval Christianity. First of these was faith, orthodox faith, heresy-uprooting, infidel-destroying *fides* in the full Christian sense.

From faith knighthood advances to obedience to the Church, a vow expressly made by every knight on taking the cross.

This pattern of behaviour is summarized in Chaucer's description of the Knight in the *Canterbury Tales:*

> A knight there was and that a worthy man,
> That from the timë that he first began
> To riden out, he lovëd chivalry,
> Truth and honour, freedom and courtesy.

[1] H. O. Taylor, *Mediaeval Mind*, I. 522.

Full worthy was he in his Lordës war
And thereto had he ridden, no man ferre,
As well in Christendom as in heathenesse,
And ever honoured for his worthiness. . . .

And though that he were worthy, he was wise,
And of his port as meek as is a maid.
He never yet no villainy ne sayd
In all his life, unto no manner wight.
He was a very parfit gentle knight.

Chivalry was not only an ideal that has influenced succeeding ages. It was a pattern upon which actual lives were moulded. Godfrey of Bouillon, the hero of the First Crusade, fought his way to the walls of Jerusalem and into the temple in a glorious welter of infidel blood, but when he was made king he refused to wear a crown of gold where His Lord had worn a crown of thorns.

St. Louis of France, in the pages of de Joinville, appears as a warrior, a most devout servant of the Church and a careful king, who listens faithfully to his subjects' problems and treats them and all others with the most scrupulous courtesy and good faith. The pointless and futile crusade in which he engaged contains battles that are exactly like pages of the *Morte d'Arthur*.

"While I was coming back thence the Turks bore down upon me with their lances: my horse was brought to its knees by the weight and I went over the horse's ears, and I got up as quickly as I might, shield at neck and sword in hand, and my Lord Everard of Siverey who was with me came to me and said that we should betake ourselves right to a ruined house, and there await the coming of the king. And as we went, a great rout of Turks on foot and on horseback came to strike at us and bore me to the ground and went over me and tore the shield from my neck.

"My Lord Everard of Siverey said to me, 'Sir, if ye think that neither I nor my heirs will be held to blame, I will go and seek aid for you from the count of Anjou, whom I see yonder in the fields.' And I said to him, 'My Lord Everard, meseemeth that ye will do yourself great honour if ye go to seek aid to save our lives, for your own is likewise in jeopardy'."[1]

These delicate considerations of honour really had a part in life. The knights-errant who in the *Morte d'Arthur* ride somewhat pointlessly through forests and deserts, had their rather more efficient counterparts who went buccaneering in the pages of Froissart. The

[1] Joan Evans, *History of St. Louis*, Bk. II, Ch. XLVI.

legend and facts were near enough to each other for there to be action and reaction the one on the other.

A knight was a true servant of the Church, but chivalry contained an ideal of courtly love that was far from approved by the ecclesiastics.

Lancelot and Gwenevere, Tristram and Isolde, are the romantic portrayal of a life that was prevalent in all the courts of the period. The knight gave devoted and absolute service to his mistress and all deceits were lawful when love must be defended. Neither pair loved in the bonds of lawful wedlock and there was a saying of Marie de Champagne—one of the acknowledged arbiters of love matters—that there could be no true love between husband and wife, and that the possession of an honoured husband or beautiful wife did not bar the proffer or acceptance of love from another. But this unwedded love was not disgraceful, it was ennobling, the giver of might and purpose, the overriding power of life. It was not a thing to be ashamed of. Rather the knight wore the lady's favour in his helmet and maintained her superiority against all comers.

In Malory's *Morte d'Arthur*, as in many of the later romances, there is a marked opposition between Lancelot and Sir Galahad: the one representing the courtly love, now beginning to be thought wicked, the other chastity. It was religious influence that added chastity to the chivalric virtues: and Parzival, whose rather dreary career occupies so long a German epic, stands forth as the example of the constant husband. The story of the Holy Grail is, in all the collections of tales, the most difficult to follow and possesses the least human interest. Parzival's dealings with the Grail King, his unasked question and his wanderings, leave merely a sense of wonder in the mind. It has neither the robust battles of some of the stories nor the romantic love of others. It is in fact an attempt to graft on to chivalry an ideal which does not properly belong there, and which never really established itself—not even in Tennyson, whose omissions are always clearly felt.

The importance for us of this chivalric legend is that it sprang out of the circumstances of the time, and gave to facts an ideal nature. It shows how an ideal can be created without abandoning the realities of a situation, and in assessing moral theories it is necessary to remember that mankind will respond to an ideal representation. Wordsworth has suggested this in his *Happy Warrior:*

> Who, doomed to go in company with Pain
> And Fear, and Bloodshed, miserable train!
> Turns his necessity to glorious gain:
> In face of these doth exercise a power
> Which is our human nature's highest dower,

Controls them and subdues, transmutes, bereaves
Of their bad influence and their good receives.
By objects, which might force the soul to abate
Her feeling, rendered more compassionate. . . .

'Tis he whose law is reason, who depends
Upon that law as on the best of friends;
When, in a state where men are tempted still
To evil for a guard against worse ill,
And what in quality or act is best
Doth seldom on a right foundation rest,
He fixes good on good alone and owes
To virtue every triumph that he knows.

The chivalric ideal did not die with the passing of an age. It appeals to something in most men and the stories of King Arthur, in their popular form, are astonishingly popular with children. It has, in fact, come right down to our day, though somewhat changed, and at the end of this chapter when we are discussing the myth of the English, we shall see how much of it determines our moral thought to-day.

The second myth we shall discuss is very different. The men who created the myth of chivalry thought well of themselves and used the myth to bring them nearer heaven. The German myth is a typical product of frustration and fear, and grew up in the national life by the same mechanism as crime springs in the mind of the individual.

The myth began as a defence mechanism in the shame and confusion of the Napoleonic conquest. For the first time Germans began to consider themselves as a single unity and to claim, though somewhat hopelessly, a future of importance and dignity. They had hardly yet begun to talk of power. The first stage in this was the invention of an ideal Germany, beautiful, gentle, pure. Hölderlin, writing in the midst of war, thus described the pure child Germany:

The priestess pure, that gentlest child of God,
She even fairer in deep simplicity, silence to guard. . . .
She who with widely open eyes did gaze as if, quite unaware,
When late a storm, death-omened, roared, and raged above her head,
Because the child presaged some better thing,
Until at last throughout wide heaven there grew
Amazement, that one was full of faith and greatly so,
Like to the blessed power of might divine.

Since days of yore, when hidden midst the woods
And nodding poppies, full of sweetest sleep . . .
So now my breast is full of grief and love,
And full of happiness and full of peace.
Germania, priestess thou of future rites
At which, unarmed thou wilt deal out advice
To all the kings and people round about.[1]

Schiller, at much the same time, was imagining a similar consolation and proclaiming as inevitable the final spiritual glory of Germany:

"It is the destiny of the German to perfect in himself general humanity. To wreathe the lovely blossoms of other people into a crown, the centre of Europe's peoples, he is the kernel of mankind, as they are the flower and the leaf.

"He has kept and preserved the treasures of the centuries, all the precious things that arose and declined in other times and peoples.

"It is not the task of the German to strive for the moment and play out his part, but to achieve the great fulfilment of Time. Every people has its day in the history of the world, but the day of the German is the harvest of all time.

"Dominion must in the end belong to him who is formed and ruled by the spirit, for finally, at the goal of time, if the world has any plan, if the life of man has any meaning, finally morality and reason must conquer, crude violence must submit to form, and the slowest people will overtake the rapid and fleeting people. The other people would then appear as the flower that fadeth. When the flower is faded the golden fruit is left, it grows and swells toward the harvest.

"This precious possession of ours, the German language, expresses everything, all that is deepest and most fleeting, the spirit, the meaningful soul.

"At this moment when the German ingloriously abandons a sorrowful war, when two arrogant peoples place their feet on his neck and the victor decides his fate, can he feel his own worth and glory and rejoice in his name? Can he raise his head and step with dignity in the ranks of the peoples? Yes, he can! He was unsuccessful in battle but he has not lost what constitutes his true worth. The German Reich and the German nation are two different things. The majesty of the German never rested on the heads of his princes. The German has found his specific value elsewhere than in the political sphere, and even if the *imperium* collapsed, the dignity of the German would remain unassailed.

[1] Prof. Butler's translation.

"Theirs is a moral greatness, it dwells in the culture and character of the nation, which are independent of its political fortunes. This Reich flourishes in Germany, it is in full growth and in the Gothic ruins of an ancient barbaric constitution the shattered pieces are taking shape."[1]

A third voice was also raised. Fichte, lecturing when Napoleon's spies sat at the back of the room, declared that the German language gave to the German nation its unity and its highest justification. The Germans, united by their language, were the supreme race, gathering the flowers of other inferior races and tongues, and weaving from them, as Schiller said, a garland for her own head. Germany must rise by spiritual power. The spirit of Arminius, exhorting his countrymen, called them to stand firm:

"For if the German race were destined to go under and become a part of Latin civilization, then it would have been better to succumb to the old than to the new. We withstood the old and conquered it. You have been scattered to be chaff before the new, and now, as things stand, your duty is not to conquer them with bodily arms, it is your spirit that must rise up erect against them. The greater destiny has been allotted you to found the absolute empire of the spirit of reason, and to destroy brutal material might altogether in a sovereign power in the world. If you do this you are worthy to be descended from us."[2]

And in his own peroration he presents the Germans as the hope of the world:

"You among all modern races are the one race in which the germ of human perfectibility is most decisively present, and to whom the progress of its development has been entrusted. If your race becomes extinct, then too all hopes of the salvation of humanity from the depths of its evils will be extinguished with you. Therefore there is no other way out, if you go under the whole of humanity will perish without hope of a future rebirth."

The paths to this new Germany were through a new concept of education:

"The new education would have to consist solely in this, that, in whatever soil it undertook to till, it would totally exterminate freedom of will. . . . If you really wish to have any influence over your pupil then you must do more than merely admonish him. You must remake him, and make him in such a fashion that it will be quite impossible for him to will anything but what you wish him to will."

[1] Prof. Pascal's translation.
[2] Prof. Butler's translation.

Here, then, at the time of Napoleon was a dream and a method. The method waited a little but the dream took a hold at once, for Germans are good at dreams and singularly ready to forsake reality especially if it is inconvenient. As Heine said,

> The Land belongs to the Russians and French,
> The English own the sea,
> But we in the airy realm of dreams
> Hold sovereign mastery.
>
> Our unity is perfect here
> Our power beyond dispute:
> The other folk in solid earth
> Have meanwhile taken root.[1]

The difficulty about the dream was only that it rather lacked content. The peaceful virtues of the Germans, their beautiful original language, the promise that they alone carried the moral future of the world, and if they perished all perished, might be consolation in the hour of their defeat, but it was not strong enough to matter when things began to improve and Napoleon was defeated and Germany once more free. The years of suffering had, however, already set their mark on the nation in the myth. A successful nation rarely claims that it is the hope of the world or threatens that if it is destroyed all will perish with it. This is the self-exalting cry of the defeated, and this cry, once heard on German lips, has been echoing ever since. We hear repeated again and again Schiller's claim that if history has any meaning Germany must be victorious. The change is that material destruction, military victory, have been substituted for the spiritual triumphs that earlier writers claimed. For spiritual triumphs, apart from the material ones, do not make much appeal to the mass of the people; and a fervent patriotism acclaimed in one sphere can very easily be interpreted in another. This happened in 1848 when the German liberals, asked to decide whether the dominion of Germans over Poles and Italians should continue, decided on imperialism because of the belief that they were the master race, the designated rulers of the world.

So, too, the unity of much-divided Germany through language and race rapidly becomes fantastic. "The compass of our language", writes Arndt, "is not merely confined to Germany and its dialects. We are joint heirs to the Netherlands, Denmark, Sweden and Norway, just as we concede to them joint inheritance as far as

[1] Prof. Pascal's translation.

we are concerned. Indeed we are also joint heirs to England and Scotland...."

This again is the fantastic voice of the defeated, the day-dreaming of a schoolboy who will be revenged on his tyrannical master.

The last vital ideal of the formative period of the German myth is the idea of the "folk"—that primitive entity out of which grow the language and the spirit of the people. The "folk" was the ultimate group which gave Germany its superiority over all other nations.

While the philosophers and poets were preparing a theory, Prussia was forming an army, and the next fifty to sixty years were dominated by that army. Thus, though Germany became great and unified, she did not become free internally. The defeat of the Liberals in 1848 drove the dominant spirits once more to dream. The campaign that ended in the defeat of Austria at Sadowa was also a defeat of the other German States, and still in the minds of many Germans the sense of defeat and frustration persisted in spite of victories without.

There was also the humiliating fact, clear to most intellectuals, that Germany was behind other nations in ideas, industry and general culture. She still borrowed the "flowers" of other nations. Wagner, in exile, performed two services for the Germans. He gave them in opera, the most overwhelming of the arts, a mythology and legend of their own. No longer had they to rely on Arminius and the vague praises of the "folk"; they had Siegfried, and Wotan, Thor and the Valkyrie. This sudden glory of the imagined German past was overwhelming to the Germans, but it was not quite enough alone. If the past had been so wonderful why was the present defective. It was the Jews. "The Jew . . . took over the German headwork, and we see to-day a disgusting caricature of the German spirit held before the German people as its supposed reflection. It is to be feared that in time the people will believe that it sees itself in this reflection. In that case one of the fairest manifestations of the human race may perhaps be killed for ever."

If a German wanted to get rid of his carping sense of inferiority there was one supremely successful institution to which he could turn, the Prussian army; and this by its triumphs against France offered Germany the only morsel of success that it had been able to enjoy in common for a long time.

In the period 1870–1914 Germany had developed a supremely successful industry, a high standard of social life, a powerful army and fleet, and still the sense of inferiority existed. There was no real political freedom, and that was perhaps the cause of the growing sense of failure which was expressed in the oft-reiterated plaint that she was excluded from colonies, lacked living-space and was encircled. When she tried to get rid of this sense of inferiority she imagined the conquest of Central Europe, and rattled the sabre against the arch-enemy, England.

The defeat of 1918 was a great shock to the Germans, so great a shock that it was never really believed. But disorder, inflation, occupation, a democratic government that could never establish itself, and the slump, all together brought the frustration and confusion to the highest peak. "Things can't go on like this", was the cry in the 1930s, and when the Germans sought to find an outlet there were the old myths of German greatness, the myths of blood and race, strengthened by the ravings of Nietzsche, and the invocation by Spengler and Van der Bruck of the divine leader. A sick, neuropathic nation turned to the one way of greatness it knew and understood and revived a chorus from the Napoleonic war,

> Strike him dead, at Judgment Day
> None will ask your right to slay.
>
> *Schlagt ihn tot! das Weltgericht*
> *Fragt euch nach den Gründen nicht,*

oblivious that once such a view is generally believed all sorts of conduct becomes possible. The Nazis seized their opportunity. The German belief in themselves as a master nation, designed for the role of overlords, set them free from a great number of the restraints that control the behaviour of ordinary men. If it was destined that Germany should be the centre of Europe, possibly the world, then the concentration of power, the elimination of competitors, the construction of a grandiose State was natural and almost inevitable. If, too, the domination of the world by Germany was decreed, and this domination was to come through the Party, then anything which weakened the party was a crime, and anything which aided it, a virtue. Moreover, the destined lords of the world could cultivate an arrogance of personal behaviour and a contempt of others that fitted their august role.

Into this myth of German domination fitted the theories of blood and soil and the glorification of the leader as the embodiment of the victorious forward march of Germany. Fitting as naturally to this pattern is every kind of greed and cruelty, of repression and aggression, of deceit and lying. The story of Japan is almost exactly the same, but the scientific apparatus of destruction and torture was less well developed.

The German myth has been about a century and a half in growing and it seems to have drawn its strength from the weakness and enslavement of Germany itself. A country that never developed a liberal government, where power always resided in the army and the army caste, inevitably bred in its citizens a sense of frustration and anger. This, through the myth, was turned outwards and the hostility directed against other states rather than its own government. The defeat of 1918, partial though it was, rendered thousands of young people desperate and helpless. Again the myth came to their aid—they would rise again and prove their overwhelming might. The myth was in its essence the day-dream of an unsuccessful schoolboy, in its practical outcome the most dangerous thing in the world.

The psychology of the German state led inevitably to the concentration camps, the destructiveness, the grandiose imaginings. These manifestations of frustration are as characteristic as the hooliganism of the slum child and the cruelty of the unloved. The practical problem for moralists and administrators alike is how to prevent a new myth forming which will be stronger and with more bitterness than ever after this new and more terrible frustration.

The myth of the English is different again. It is not in essence a political myth, nor does it idealize a certain structure of society. It is a code of personal behaviour, and it is taught to children by exhortation or punishment, or by the presentation for admiration of heroes in the past. It is, of course, far more difficult to analyse those ideas among which you live than those that can be viewed from outside. The English myth is rather subtle. It is compounded of self-satisfaction along certain lines, and the complete acceptance of failure in others. It is vainglorious in its humility, and it is very seldom talked about. The Englishman all too often appears to the foreigner as insufferably arrogant, and very stupid; yet the tale of the English achievement in art and science is splendid, and "swank" is the quality most completely banned by the tradition. An Englishman hardly ever mentions his moral standards, and yet they are generally of an almost terrifying rigidity and firmness. Private Angelo, that most attractive of poltroons, comments sadly on the English sense of duty when he comes up against it:

" 'I think it is the proper thing for you to do', said Simon stiffly.

" 'Oh my God', said Angelo, 'now I am doomed indeed!'

"For he knew that when the English say 'It is the proper thing to do' the inexorable laws of nature are supplemented by another that they discovered, and only they can understand. Oh their wild notions of propriety! They are like sunspots, he thought, for they cannot be explained or foretold, and their effects are incalculable. Whether they have indeed certain absolute standards of behaviour, or merely a tribal instinct, or perhaps an hereditary taint, a sort of itch—it is impossible to say. But you can no more argue with them than you can dispute the law of gravity. Yes, he admitted, I am doomed, and there is no escape.

" 'Lucrezia', he said, "will not be pleased when she hears that I am to become a soldier in the infantry again. It will be very difficult to say Goodbye to her.'

" 'Well', said Simon, 'that sort of thing is always difficult. But it has to be done, of course'."[1]

Not long ago all England went to see the film of *Henry V*, and that, in the most marked way, illustrated the moral standards of the average Englishman. The film followed Shakespeare's play closely, and what was so interesting to the moralist is that the standards of Shakespeare, three hundred years old, are still absolutely valid to-day. We have not changed from then till now, and the standards were not new when Shakespeare wrote. They were the chivalric standards made a little more robust, purged, on the one hand, of excessive ecclesiastical influence and, on the other, of the courtly love that tied a knight to his lady. There is less of the saint about the king and more of the hero; there is less romance and more practical military science; marriage, a good common-sense marriage, is the end of courtship, and not fantastic devotion. The path of arms leads to the conquest of a realm, not to the slaying of dragons and the relief of damsels.

If we consider the morality of the nursery, still more the morality of the public school, we can see how a large part of it coincides with *Henry V*.

We find in the play the hero-king in the typical form he has retained ever since. Slow to wrath, anxious that his cause should be just; unwilling that a man should be punished for drunken words; gallant in battle, merciful in victory, bold and courteous in love.

The French are boastful and cruel and deserved their overthrow. The British army was as gloomy as usual on the eve of battle, and as staunch on the day. It was not ashamed of its fears, nor did it give way to them. Kings might be kings, but they should treat even Falstaff with politeness; and if they went about disguised must receive tolerantly the rebukes of the common soldier.

Even the King's failings are traditional. He speaks French abominably, and he has no skill to turn a love-sonnet. And the squabbles between the various national groups in the army go as far as such squabbles traditionally do, but have no effect on the general will to

[1] Eric Linklater, *Private Angelo*, p. 215.

win. There is fair play for all, and it is wrong to deride Fluellin for his accent, and it is wrong for Pistol to boast and brag. It is far worse, worst of all, for the French needlessly to slaughter the camp boys, and it is this deed of cruelty more than all else that angers Henry.

This "ideal Englishman" as represented in *Henry V* is part of the national myth. It appears in many forms in children's books, in films, and above all in the version of history taught to the young. If most people cast their minds back over their knowledge of the past, certain episodes stand out—the defeat of the Armada, Alfred burning the cakes, Magna Carta, and it is these events that are in accordance with the myth as shown in *Henry V*.

We even re-enact it to-day. Montgomery visiting every camp before "D"-day, and making a speech in each, is a very obvious successor to Henry V. The clearest difference is that Shakespeare did not write Montgomery's lines, and so the speech, as we shall show later, reads less well in retrospect; though it appears to have been highly successful at the time. Montgomery is not the only commander who has kept alive the myth. Alexander, taking off his greatcoat, when he was inspecting troops in the rain, so that he might get as wet as they, is exactly in the same tradition.

This is one part of the English myth. There is another that has impressed itself strongly on our history and our thought—our connection with the sea.

This English ideal is treated by R. L. Stevenson in a well-known essay as embodied in the English admirals: "The admirals are typical in the full sense of the word. They are splendid examples of virtue indeed, but of a virtue in which most Englishmen can claim a moderate share, and what we admire in their lives is a sort of apotheosis of ourselves. . . " "We do not all feel warmly towards Wesley or Laud, we cannot all take pleasure in *Paradise Lost*, but there are certain common sentiments and touches of nature by which the whole nation is made to feel kinship. A little while ago, everybody, from Hazlitt and John Wilson down to the imbecile creature who scrubbed his register on the fly-leaves of *Boxiana*, felt a more or less shamefaced satisfaction in the exploits of prizefighters, and the exploits of the admirals are popular to the same degree, and tell in all ranks of Society."

He describes how this national myth affects the Englishman's attitude towards the sea: "The prostrating experiences of foreigners between Calais and Dover have always an agreeable side to English prepossessions. A man from Bedfordshire, who does not know one end of the ship from another until she begins to move, swaggers among such persons with a sense of hereditary nautical experience.

To suppose yourself endowed with natural parts for the sea because you are the countryman of Blake and mighty Nelson is perhaps just as unwarrantable as to imagine Scotch extraction a sufficient guarantee that you will look well in a kilt. But the feeling is there and seated beyond the reach of argument." That this is not altogether without foundation in fact, that a taste and an aptitude for "messing about in boats" really is a widespread characteristic, and one of considerable importance, is suggested by the success of sea-scout troops in many places where there is hardly water to float a gun-punt. The rescue of our army at Dunkirk with the aid of pleasure yachts, as well as, of course, more regular craft, gave that disaster a strange air of triumph. The national characteristic had proved its worth again.

The British are, on the whole, over the course of their history, a very successful nation. They have avoided major disasters, they have achieved a political development that has been exceptionally free from the more shattering upheavals. Their industrial development was early and brought with it great suffering and destruction, but the sense of social obligation developed early also, and the ultimate damage, though severe, is more aesthetic than anything else. They have hardly ever lost a war, and their most serious defeat was by their own kinsmen. They have spread all over the world, and their policemen are marvellous. This record gives them pleasure and a great self-confidence. They believe in themselves, but on the whole they visit the scenes of their defeats rather than their victories. They take their successes so much for granted that they pride themselves on other qualities than mere success. Above all they dread the stigma of professionalism.

" 'Life is war,' says Angelo, meaning something else, 'and we who are virtuous may well lose every battle but the last one.'

" 'That,' said Simon's friend with noticeable stiffness, 'is the prerogative of the English.'

" 'Because you are good?' asked Angelo.

" 'That is an attractive hypothesis,' said Simon.

" 'There was a time when we aspired to goodness,' said his friend, 'and the world regarded us as hypocrites. Then we decided to pose as realists, and the world said we were effete.'

" 'But why do you win your last battle?' said Angelo.

" 'We are amateurs,' said Simon's friend with a heavy yawn, 'and the amateur lasts longer than the professional'."[1]

This claim that we can do it without taking too much trouble is the excuse of the schoolboy, and the pride of the dilettante.

[1] Eric Linklater, *Private Angelo*.

For the Englishman in India there is one shrine, holy above all others—the Residency at Lucknow. There the Union Jack was never hauled down, even at night. There a group of English men and women held a completely indefensible position against inconceivable odds, till their leader was killed and the survivors rescued. The English are prouder of that defeat than of Plassey or Arcot or any of their victories. The Indian Mutiny should never have happened. It was provoked by bad statesmanship and red tape. It could have been checked at the outset by reasonably vigorous action by the officer commanding at Meerut. It is a sorry story in many ways, yet the British are not particularly ashamed of it. They never thought they were very good statesmen or that the senior officers in 1850 were particularly efficient. But they did think they could show constancy and bravery, and their pride grows when they stand where these qualities appeared in the highest degree.

We have, on the whole, forgotten the First Afghan War. It was the most complete military defeat possible, following a story of political folly and military ineptitude; yet even that, when we remember it, causes us no shame. We think of the incredible march from Ferozopore to Kabul across the Indus and through the Bolan Pass; and of the fortitude of the prisoners after the defeat. In the same way we shall remember Singapore for the bravery of the men in the Japanese prison camps. It is a strange treatment of history and adds the final touch to our national arrogance.

The fact that our moral code has been roughly constant in its elements for three hundred years and more, gives a firmness to British morality that cannot be possible in countries where changes take place. There have been variations of standard, as when the Victorian commercial morality held sway, and there are differences between sectional groups; but over the mass of the population the standard has been constant. The young man entering the Colonial Service has roughly the same ideas about " fair play " as the lad entering industry. Both have, rather roughly, the same ideas of honesty, honour and letting the other fellow have his say. There is not, in essence, a class morality. Nor, when either reads of the past, does he find standards that differ. They do not have to say, " Things were different then. We have changed all that." The Englishman is, in some ways, nearer to the stability of the primitive community. It is thus that he does not have to talk about his morality. That makes him difficult to understand.

From the moral point of view the stability of the myth is of the utmost value. It makes the land almost a closed society. When, in addition, the average Englishman really despises all foreigners, or rather, if he likes them personally, pities them for their unfortunate institutions or deplorable history, he is not likely to be susceptible to antagonistic suggestions from abroad. Practically in England the only nation that has this suggestive power of prestige is the U.S.A. We may admire French art, enjoy German music, but we are not tempted to imitate either. We learn our slang from the Yanks.

Books referred to in this chapter:

Margaret Murray, *Witch Cult in Western Europe; The God of the Witches.*
Edward Langton, *Satan.*
Bernard, *Little Schools of Port Royal.*
D. M. Emmet, *Nature of Metaphysical Thinking.*
Franz Kafka, *The Castle.*
Jung, *Psychology of the Unconscious.*
Grimm, *Household Tales.*
Walter de la Mare, *Collected Stories for Children.*
Rosenberg, *Myth of the Twentieth Century.*
A. R. Chandler, *Rosenberg's Nazi Myth.*
St. Augustine, *Anti-Pelagian Treatises.*
Du Bartas, *Works and Days.*
Milton, *Paradise Lost.*
N. P. Williams, *The Idea of The Fall and Original Sin.*
P. C. Wren, *Beau Geste.*
Anthony Hope, *Rupert of Hentzau.*
Eric Linklater, *Private Angelo.*
G. P. Gooch and others, *The German Mind and Outlook.*
Malory, *Morte d'Arthur.*
Vinaver, Edition of *Morte d'Arthur.*
Tennyson, *Idylls of the King.*
St. Augustine, *Civitas Dei.*
H. O. Taylor, *Mediaeval Mind.*
Chaucer, *Canterbury Tales:* Prologue.
Joan Evans, *History of St. Louis.*
M. Richey, *Parzifal.*
Froissart, *Chronicles.*
Wordsworth, *The Happy Warrior.*
Shakespeare, *Henry V.*

Chapter 4

THE CLOSED SOCIETY

We have described how each society has a structure which is in essence its material form of political and social life and the ideas which belong to this material form. In addition there is the mental structure, that we have called its myths. Between them these structures determine the form of society and provide the framework in which the moral life of its inhabitants develops. If we take morality on the lower, practical level of conformity with the customs of society, we can say that the successfully moral man is one who has been moulded to the form and has absorbed the beliefs of his group, and shows forth in his life the ideas of his day. If morality were merely this, then, once a pattern was firmly established, all would be well, each generation growing up would learn to conform and each would be virtuous. In practice, for many of the societies of the world, this is not so. In each society there are conflicting elements, and, further, the dominant ideas of the whole society change from time to time. In England, although much of the moral tradition has been quite stable, as we have said, there have been many sectional variations. The commercial morality that found its most complete embodiment in Dr. Samuel Smiles is one; the harsh treatment of children another. In these fields, when the moralist of to-day looks back sixty or seventy years he shudders at the picture of virtue that was then presented. From the distance of, say, two generations in England it is not the vices of our forefathers but their virtues that seem horrible—our ideas have so completely changed.

From the ethical point of view these changes of standard are of the utmost importance. It is their existence that lends substance to the discussion of the ultimate good and that makes an understanding of the whole process of ethical teaching so important. In later chapters we shall consider these changes as they take place in a society that is naturally liable to change. They can occur in two ways. There is the change that is in accordance with reason and is felt to bring society more into conformity with noble ideas, and there is the change that is irrational and often for the worse. In the present chapter we want to consider those elements which make for a stationary society and those which allow of change.

We can roughly divide societies into two groups: those that possess a homogeneous set of ideas well established and little liable to change, and those that are altering rapidly. In this last class there is also a sub-division into those societies that change as a result of pressure from without, and those that carry their diversity within themselves.

The static type of society exists at different levels of culture. It can be seen among very simple people where there is hardly any social discrimination and very little contact with the outside world. It can also be seen in a complex, isolated community such as China during the long period when it was cut off from the western world. It could also be seen in some European states that for periods had a certain form of social organization. The other type of society, that which is liable to change, exists both in the complex society where there are groups with differing interests and moral types, and in the simple society invaded by ideas from outside. The changes that come to these types are very different. When changes arise within the society itself they are likely to be beneficial. When they are imposed by the disruptive force of an outside power they very seldom are.

It is possible to convert a changing society to one at least temporarily static by closing it to new ideas. This must, of course, be accompanied by the suppression of all internal ideas antagonistic to the system preferred. From time to time, states have done this with more or less success. The establishing of a closed society has always been the desire of a certain type of educationalist—as we shall see later—and it is a necessary device of some reformers or innovators, as well as of some conservatives. As a political and moral device it is of the highest importance.

Such closed societies appear very valuable to certain minds. In them there is no conflict; duty is clear, and habituation ensures that in most cases duty is done. Moreover, to those who are convinced that they know the truth, nothing could be more beautiful than to see truth lived and thought in a paradise where error never rears its ugly head. Such maintainers of an established system are particularly common among those engaged in religion. Priests are usually a conservative force in a state. The reasons are various. Where religion is practised for such purposes as assuring rainfall or the health of crops, any change in the ritual may be dangerous. What has proved successful in the past must be perpetuated, lest any innovations should make the ritual less effective. Again, when religion is based on a revelation in the past, part of the function of the priest is to guard the truth that has once been committed to his charge. When this truth is very important to each individual, when a proper faith is necessary to the soul's health, then it is the duty of the priest to keep his flock safe from the contagion of alien ideas, and to educate them in the truth through the whole pattern of society. This idea of a closed society in which both myth and institutions lead to faith has been well expressed by a modern Catholic writer.[1]

"The idea of educating by environment is not peculiar to Christian theory, but it is specially suitable to it because of the value which Christian thought attaches to objective truth. If once a way of life can be evolved which can be looked upon as giving concrete expression to definite principles, it will have an incalculable power of leading the mind, imperceptibly but surely, to the full acceptance of those principles—*Christendom* is a corporate society in which, ideally every man, and especially the less well instructed, will draw from the social conditions in which he lives, from the institutions, customs and buildings which surround him, knowledge and love of dogmatic truth."

On the other hand, the critical mind, trying the social order against some standard, if only our standard of happiness, finds much to improve: and the moralist who feels that true excellence can only be achieved by and through struggle, wishes men to be exposed to the temptations that may strengthen their understanding and will.

There exist, in reality, naturally closed societies which, for some reason or other, are cut off from ideas that conflict with the prevailing system, and which can apparently continue indefinitely without radical change. These naturally closed societies are characteristic of a primitive way of life all over the world, and seem to depend mainly on geographical isolation. In those societies to which immigrants have come, immigrants that is of a higher culture, sudden developments have taken place, as for instance among the Maya of Central America. But those that have remained isolated have apparently been able to retain an almost constant pattern for a long time.

Whether, from our point of view, such a static, closed society is ethically good or bad depends entirely on the characteristics it has developed. Many primitive societies have a way of life that seems to us horrible in whole or in detail, others can present a very pleasant picture. We have already briefly mentioned the Chuckchee who inhabit Siberia and have as unpleasant a pattern of life, both material and moral, as could be imagined. On the other hand, many writers, especially about communities in the Pacific, describe, if not a paradise, at least a community that appears to be happy and virtuous. There

[1] M O'Leary, *Church and Education*, p. 27.

is a virtual equality that pervades the group, there is an observance of customs and taboos, there is little friction or cruelty. Physical conditions of life are pleasant, and man has not invented torments for himself.

Under the circumstances the characteristic ethical qualities of the closed static society appear. There is no questioning, hardly any resistence. Each child sees before him a pattern of life, he sees no alternatives and he follows the pattern without complaint or hostility. This acquiescence is, in fact, the most characteristic mark of the static society. We shall point out later how ready man is to settle down to any routine that is tolerable. Where there is apparently no alternative this tendency is strengthened.

This type of acquiescence appears even in a complex society where there is great differentiation of way of life and much suffering. The condition is that the community should be guarded from outside influences that might offer another pattern of behaviour. In China this condition was fulfilled for many centuries. There, in spite of groups with conflicting interests, the essential social structure which inflicted poverty and wretchedness on large sections of the population, and domestic customs which were often cruel in the extreme, maintained themselves in isolation. The *Autobiography of a Chinese Girl*, by Ping-Ying, shows how suffering was born by the elder sister and how only outside influence helped the younger child effectively to escape.

"My elder sister had started to embroider when she was only eight and she seldom left her room but worked all the time, and when she was eighteen she was married. In her tiny room from six o'clock in the morning until six o'clock in the evening she did nothing but embroider. In the evenings she had to spin. My poor elder sister, even when she was very weary and fainting, would not dare to say anything about it to her mother—all she did was to sigh in secret. . ."[1]

Ping-Ying, herself a natural rebel, tried to avoid having her feet bound, but was overpowered. She could then only think of forcing her family to send her to school by the threat of suicide. Her first school was the one the village boys attended and she was still in the grip of tradition. It was only when, aged twelve, her family sent her away to a boarding-school among Americans that she found girls of her own age with unbound feet, and threw her bandages away and never allowed herself to be recaptured by the spirit of custom.

The coolie and the peasant have in the past been as uncomplaining as poor elder sister and for the same reason. They have not been able

[1] *Autobiography of a Chinese Girl*, by Hsieh Ping-Ying. (Allen & Unwin.)

to imagine any other way of life. Now that new ideas have been forced on China by the terrors of war that even they cannot ignore, the grievances of the past are leading to the long-drawn-out bitterness of civil war between those who wish to keep as much of the past as possible and those who wish to remodel society on a much more extensive scale.

The long refusal of the Chinese to accept ideas from outside shows clearly that for the spread of ideas in this way prestige is necessary. The innovating foreigner, if he comes with engines of destruction sufficiently powerful, will usually be listened to. He can also get a hearing if he can cure the sick or show new methods of prosperity. The isolation of China has finally been broken down by the bombing aeroplane. The breach was made in her isolation by the medical missionary and the successful trader.

These communities are of natural growth, but it is possible to establish them artificially. If in such a society the rulers are able people, the way of life pleasant, and external ideas excluded, they flourish very well. The charm for many people of running a boarding-school is the establishment of such a private state; and the tendency to place these schools at rather isolated places in the country, or to refuse to allow the children to go out if the school is in a town, represents the attempt to keep the children isolated from more than the danger of physical infection. The devisers of Utopias from Plato downwards have enjoyed the same dream. It is not often that such imaginings can take shape and reality. Because it is perhaps a unique case, and because it is nice to contemplate anything so happy, it is worth quoting a passage from Willard Price's *Japan's Islands of Mystery*, which concerns the island Kusaie, an island near Ponrape. It was visited by American whalers in the last century who shanghaied the men and brought disease till the population fell to 200. Then the missionaries took over.

"For eighty-four years the missionaries of the American Board of Boston have been at work on Kusaie. Let us go ashore and see the results. We are met by Arthur Herrman—lone American planter. Evidently Kusaie agrees with him—he is portly and jovial. On the copra-scented pier we meet Mrs. Herrman, a native of Kusaie, more jovial and more portly. One look at her beaming and enlightened countenance and we conclude the missionaries have done a good job.

"Her face is not unusual—we walk down the village street through a sea of seraphic smiles. There are low bows and soft good-mornings. All the inhabitants are in long white robes as in the realms of the blest. Houses are so neat they ache. Music drifts about—whistled and hummed and twanged—hymn-tunes familiar in New England churches."

Attendance at church was exemplary.

"We find the white church on the shore already occupied by a thousand people. The King leads the singing. The most blasé visitor must feel a tingle run along his ribs as these thousand trained voices take the air. The volume and beauty of it are so great that one would not be surprised to see the sheet-iron roof go sailing off into space. Then the native minister in high-collared white drill suit and bare feet preaches. Through the open windows we can see the ship, waiting for stevedores (who are all in church). The service is long. When the last prayer is finished and we make to rise, the King, who sits beside me, whispers 'Now, Sunday School'.

"No one leaves. It is not until nearly one o'clock, after three hours of services, that we pass out and some of the men answer the insistent whistle of the steamer. But they must work fast for there is another service at three and another at five. Double pay cannot induce them to miss a service."

The general state of the island is equally satisfactory.

" 'How many murders a year now?' I asked the King. He smiled, 'There has not been a native murder in my lifetime', he said. The King was sixty years old. 'How about minor offences? How many cases of detention in your jail in a year?'

" 'Jail', exclaimed the King. 'But there is no jail.'

" 'Well', I said, 'whatever you call it you must at least have some place to put the tipsy ones until they sober up.'

" 'But there is no drinking in Kusaie.'

"There is no house of ill fame, there is practically no disease. There are no native medicine-men, no charms or other superstitious devices to ward off illness, and the Japanese doctor goes fishing. Native physique is splendid."

And it had all been accomplished by Miss Hoppin, "a cunning little old lady" who mingled moral exhortation with large meals, and the two Miss Baldwins aged seventy-six and seventy-eight who taught school and insisted on dresses that took six yards of cotton cloth.

The imagination is stimulated by such a picture of virtue. If one could only find such an island and there in isolation with the prestige of knowledge and force of will build up the ideal community!

These static communities are not only isolated, they lack the power of speculative thought. The simpler communities have no advanced education, and would be unable to appreciate abstract ideas even if they were presented to them. Of China it is more difficult to speak, but for centuries education has largely consisted of learning the classics. Science with its habit of disturbing established customs has played little part. Moreover, a national pride and self-sufficiency have disinclined the leaders of thought to seek new ideas, while the concentration of family power in the hands of the aged has prevented youth attempting new paths.

When we turn to the type of society that is stationary not because of geographical remoteness but because of social conditions, we find that it is the same element of thought that is lacking. These are communities that consist essentially of two classes, nobles and peasants, and there is no middle class between the two. We shall speak in the next chapter of the intelligentsia, it is enough here to show how their absence affects a state. Communities which are in touch with their neighbours may be as effectively isolated by ignorance or prejudice as by impenetrable wilds. In Europe, long after the western states had developed a middle class—both skilled artisans and traders as well as members of the professions—Russia and Poland retained the old organization, and with it an essentially unprogressive society. In this grouping neither party can initiate change. The nobles have all they wish, and their way of life seldom allows them to cultivate their minds or possess ideas. The serf is kept from education, as we shall see in the next chapter, often by severe laws, and thus is as unable as the Chinese coolie to imagine a better way of life. Intercourse with other countries is very limited and in any case the stranger comes into contact only with the upper classes.

When Peter the Great attempted to modernize Russia, his difficulties were largely due to the structure of society. He had, of course, to draw his ideas from outside.[1]

He left Russia ignorant of all but the rudiments of learning; he returned a highly skilled technician, a philosopher, a reformer in religion and a man who believed in efficiency. He was met by apathy, a determination to cling to the ways of the past, the whole weight of an obstructive, ignorant priesthood, and the incompetence and corruption of the nobility. He built his new city to turn men's minds to the sea, and to break the associations centred around Moscow. Every difficulty he met with, every destructive flood, every trick of the marshes, was hailed by the opposition as an act of God confounding Antichrist. There was no group in the state on which he could rely. Even his son, brought up by the conservative elements, was only waiting his father's death to revert to the old customs. If there had been an educated middle class to be employed in Government work

[1] Merejkowski, *Peter and Alexis.*

or even a group of skilled artisans who would have put their services at the disposal of the ruler and been ready to learn new arts, the changes he desired might have taken place.

The middle class of artisans and small traders, useful as they are, is a very different thing from an educated middle class. From this class springs the intelligentsia, and their part in the development of ideas will be discussed later. It was not until a century or two after Peter, when an educated middle class had developed in Russia, that it was possible for the ideas to be produced and disseminated which led to the break-up of the conservative system and the accomplishment of a revolution.

The history of Poland has been made much more difficult because it was the Jews who were sucked into the space between the nobles and the peasants: and as aliens they were allowed no political part in the state. Thus the changes which a normally powerful middle class introduces became impossible.

In Freya Stark's book, *East is West*, one of the points she is engaged in making is the great importance, political and social, of the growth of an educated middle class in the Arab States. She calls them the "young effendis". In the past the unprogressive grouping of nobles and peasants existed, and this was maintained by the ignorance and insecurity of Turkish rule. Now in the security that British guardianship has given, the educational labours, mainly of Americans, have produced a new class; educated and progressive rulers, intelligent civil servants and a whole body of men whose minds are open to ideas. Little, almost forgotten, states, like Transjordania have towns with the amenities of civilized life, have hospitals and schools, have increased their population from 2,000 to 43,000 in twenty years and have succeeded in afforesting a large part of the waste land of the country. There is also peace, and the desert tribes have ceased to raid the cultivated land. This and such-like development has been made possible by the production of men able to apprehend and transmit ideas, who have been introduced to completely new standards, moral and material.

Without such an educated middle class that will receive ideas and can understand them, a state may be preserved for long periods in a static condition. Thus the provision of education is neglected wherever it is to the interest of the rulers to keep things as they are. It is useless for the reformer, if he can escape the police, to preach if those that listen fail to understand his words. It is equally possible to surround men with a wall of prejudice that prevents any new ideas penetrating to their minds. Aldous Huxley has satirized in *Brave*

New World the teaching that from an early age murmurs into the drowsy ears of children contentment with their lot. "I'm glad I'm a beta", says the hypnotic voice, and the waking mind retains unshakably the conviction of good fortune. So too in many societies prejudices are taught from an early age; social taboos, moral and political judgments. They are taught without reason given, at an age when a child is incapable of questioning—and once firmly established make a man as safe from new ideas, though he walk the streets of London or Paris, as if he were on a Pacific atoll. It is these prejudices that keep the rulers of the unprogressive countries as secure from change as ignorance keeps the peasants.

Change may come to a country from without, or it may arise from conflict within. When it comes from without it is nearly always brought by people with prestige, and it is often fatal to the indigenous culture. The contacts of Europeans with primitive peoples have frequently been unhappy.

The changes that occur when a more developed civilization comes into contact with a primitive static one have often been described and deplored. The European comes armed with immense prestige. He comes too often as a conqueror or exploiter. He spreads disease, economic ruin and serfdom. The primitive tribe has no means of resistance, either material or spiritual. The important point for our argument is the destruction, not only of the material basis of life, and of the institutions by which the society has lived, but also of the spiritual, the myths. Pitt-Rivers has described the spiritual desolation of a Maori village that was in process of being absorbed into European culture. They were not being exploited economically, they were being offered education, but they were cut off from their past. They felt lost in a world that had been deprived of its meaning, and they were pathetically anxious for help to rebuild the system of beliefs by which a man lives.

"Anxious to know how much the younger generation knew about their own history, I asked one of T——'s nephews what he learned at school. He had learned about the Wars of the Roses, and was now learning about the Hundred Years' War, but had never been told about the fighting along the Wanganui River during the Maori War that had taken place sixty years ago.

"The younger generation are guarded from all knowledge of their pagan past. Three European religious bodies contend for their souls. They know little or nothing of the pride, ambitions, hopes and outlook of men who lived in another world incompatible with the present one. There came a time after the disastrous and depressing catastrophe of the Maori War, when the rising generation refused to look back into the past, refused to learn the names of the long line of ancestors, which their fathers, now the old men, took pride in remembering. . . .

"When the Europeans criticize the present-day Maori for having acquired all the vices of civilization they should remember that it is our proselytizing culture that has arbitrarily sought to stamp out the meaning and therefore all the virtues of the past."

The following is a translation of the speech of welcome by an old Maori:

"Once numerous, now diminishing and dispirited, our women no longer bear many children and our villages are emptying. The white man brought us Christianity and we accepted it; but he broke our *tapu* and our *mana* left us.

" When our *mana* was destroyed the whole world became dark, and if this work of yours will bring light it will do good and we will gladly assist you and welcome it.' "

The strange inability of the white man to realize what is relevant to the development of a more primitive group was curiously illustrated on a *mesa* in the Hopi Indian reservation of Arizona, where one of the authors found an American teacher showing her class how to draw sunflowers and thatched cottages.

During most of the history of Western Europe change has been accepted voluntarily. There have been conquests of one State by another of slightly higher culture, as the conquest of England by the Normans, but in most cases the mechanism of change has been very different.

One of the peculiarities of the civilization of Western Europe is the diverse origins of its culture. Hardly any of its culture was strictly indigenous. It all came from farther east. Rome took it from Greece, or later, Palestine—France from Rome and Germany. The religion was Jewish, the law Roman, the philosophy Greek, the tradition of chivalry Germanic, the art at one time based on Rome, at another on Greece.

The sectional elements in the social system were well marked. Church, army, nobles, craftsmen, peasants, each had its own way of life, each its own interests. For almost everybody there was a choice of life and an alternative standard of excellence. The anchorite starved and shivered for his soul's salvation, the nobleman gathered all the spoils of Greece and Rome and enjoyed the verse of Lucretius. The workman imprinted on the cathedrals his own rather lewd humour, while the clerk berated the sins of the flesh. The cloister and the court were equally open to a princess, and the noble who did not fancy a military life might become a bishop.

The consequence of this diversity has been that European civilization has never been stagnant for very long. Nor has it ever refused ideas from without, rather it has expected them. The Crusaders brought back more than plague from the East. When the Moors introduced a higher civilization to Spain ideas were readily accepted; when growing knowledge and the fall of Constantinople made the Greek past available it was received with passionate enthusiasm. At a later date Europe accepted fashions from all parts, and at one period Chinese furniture and at another Japanese prints were the rage.

This mixture with its possibilities of change has never been pleasing to all parties. The unfolding Renaissance was hated and feared by some, and bitter and bloody attempts were made to maintain the "purity" of religion against change. Writers such as O'Leary, quoted above, look back to a more or less imaginary "age of faith"—and contrast it sadly with the lack of control to-day. But such an age, if it ever existed, did not exist long, certainly not in any but districts too remote to hear the voice of the orator. Wherever men gathered, ideas were never long without change, and this was made more obvious by the rapid growth of an educated middle class. In the earlier periods book-learning was largely confined to the Church, and from among the various ranks of "clerks" not only priests, but statesmen, civil servants, teachers and others must be found; but in England from the reign of Elizabeth a complete secularization took place. Nicholas Bacon, the first lay Lord Chancellor, inaugurated a new age in politics. While France was still governed by cardinals, England enjoyed the rule of country gentlemen.

Moreover, Western Europe has had many stimulating experiences. The discovery of America set men's minds to new thoughts: the development of science never allowed one set of ideas to persist too long; invaders such as the Moors enriched the lands they conquered; realms, more fertile in men than crops, have thrown out adventurers to seek their fortune, and return with strange knowledge. When Olaf Tryggvason returned to Norway after his service with the Varangian Guard he brought back with him Christianity as well as a scarlet cloak and gilt helmet.

It sometimes happens that a state deliberately attempts to establish a closed society. It may do so for one of two different reasons. It may wish to preserve the present against the ravages of change, or it may wish to establish a new system and to let it grow up unharassed from outside. In both cases it must close the frontier to ideas and keep control of the expression of opinion at home. The censorship of imported books and news is one of the commonest methods, the persecution of the intelligentsia and the prohibition of education is another. When an attempt to preserve is made, it is seldom successful for long. Individuals slip out, learn and return. Dangerous literature lurks in the false bottoms of trunks. The Czar's government fought, for a century, a losing battle against enlightenment.

When it is a new idea that has to be established, there is, perhaps, more chance of success. From the chaos of the 1920s in Germany arose the Nazis, offering a new idea of life that fitted perfectly many of the desires of the nation. But it had to encounter opposition from a minority, mainly older people. The problem was a double one, to keep out ideas from other lands, and to prevent parents influencing their children. Russia in 1917 had just the same problem, complicated by the immediate aftermath of a disastrous war. Perhaps the Russians had fewer parents to contend with.

During 1917 and the following years a large part of a generation perished. There were thousands and thousands of children to be brought up by any one who could catch them, and had the means of educating them. Moreover, the government had at its disposal both fear and hope. It used the fear ruthlessly in the persecution of those with anti-revolutionary tendencies, or to break up groups—such as the Kulaks—that delayed plans of reorganization. On the other hand, it offered hope of favour and advancement to those who joined in its work. The Nazi system was not very dissimilar. It tried to keep its children occupied; silenced its parents by giving the domestic spy every encouragement; threatened with the concentration camp those who dissented and rewarded those who adhered with office and power. In both cases it practically closed the frontier to ideas or information.

From the present point of view it is important to notice that with the remodelling of institutions went the creation of myths, and that with the provision of suitable myths the nation's spirit revived. All writers about Germany in the period just before the rise of Nazism have emphasized the loss of belief and of confidence. No one knew what was going to happen next, and all felt that the present state of things could not continue. The moral decadence that afflicts primitive peoples deprived of their beliefs affected Germany. Sexual perversion was exceptionally common, and open violence was general; a financial scramble raged in which all classes struggled for their own profit without the slightest regard for the public good. Books such as Christopher Isherwood's *Goodbye to Berlin* or Arnold Zweig's *Calf of Paper* show how the general disorganization appeared to different observers.

The difficulty with Germany was that the whole policy of the state was aggressive, and sooner or later there was bound to be a conflict. If Germany was unsuccessful then the whole structure was bound to collapse. In Russia the aim was more pacific, and but for the Nazi attack should have developed, and may yet develop, in peace till it feels sufficiently strong to discard its closed position and join openly with other groups, permitting the normal interchange of ideas, confident that its beliefs are as stable and integrated as those of any one else.

It is interesting to consider briefly the parts that the different elements in the state have taken in the process of change. Organized religion, from the days of Akhnaton, has generally been opposed to change. A ruling class has generally striven to maintain its position, though men as wise as the Duke of Wellington have often decided to lead the reform when it has become inevitable. Traders and merchants pushing out to new lands have received new ideas, and at home have resisted those elements in their own society that have hindered their activities. The professionalization of the army has had curious effects on the development of different states. When the army became a separate caste, and not merely the nobility engaged in war, it has been a conservative influence in some directions and an innovator in others. No country shows this more clearly than Germany. The Imperial General Staff has maintained ideas and myths that have prevented much intellectual development. They have, on the other hand, encouraged all types of industrial development and all kinds of research that might have military uses. England has been, by contrast, curiously hostile to the growth of military power and influence. The oldest regiment in the British Army—the Grenadier Guards—dates only from 1660; and, as late as 1800, the army was kept to the smallest possible size. In consequence, the British Army has never had, or claimed, any political power. It has believed itself the servant of the state, and done what it was told. If Cromwell ruled, or the Duke of Wellington became for a time the most important man in England, it was their personal qualities rather than their leadership of an army that distinguished them.

Inevitably, the real instrument of progress in a state is the small group of people who have received an advanced education and who are interested in ideas. This group, the intelligentsia, we shall discuss in the next chapter.

Books referred to in this chapter:

Samuel Smiles, *Self-Help.*
O'Leary, *Church and Education.*

Milton, *Areopagitica.*
Ping-Ying, *Autobiography of a Chinese Girl.*
Willard Price, *Japan's Islands of Mystery.*
Merejkowski, *Peter and Alexis.*
Freya Stark, *East is West.*
Aldous Huxley, *Brave New World.*
Pitt-Rivers, *Clash of Culture.*
Christopher Isherwood, *Goodbye to Berlin; Mr. Norris Changes Trains.*
Arnold Zweig, *Calf of Paper.*

Chapter 5

THE INTELLIGENTSIA

THERE is one group in the State that is particularly antagonistic to the principle of the closed society. This is the intelligentsia, who may roughly be divided into two classes: the scientists, the men of learning and the teachers who increase the supply of knowledge; and the writers, journalists and publicists generally who are interested in ideas rather than in facts, and whose special part it is to transmit those ideas in forms that can be rapidly understood. This distinction is not absolute. There have been many men who belong to both classes. There are scientific men like Huxley who have a great gift of popular writing, and there are writers like Gilbert Murray who combine scholarship with high literary ability, and take a prominent part in the world of social and political ideas. This does not prevent the distinction serving roughly to divide the two groups.

A visitor from Mars, or even from another country, might not be very impressed by the intelligentsia. The more strictly academic portion—the scientists, the philosophers, the teachers—do not receive very high salaries, they do not dress particularly well, and their manners are generally mild and unemphatic. The publicists have been unkindly described by Dr. Schiller[1] as a group whose "members are not held in personal esteem and are rarely to be found among the ruling circles of modern society". This is perhaps hardly fair to such eminent persons as Dr. Schiller himself, still less to the controller of a great newspaper. The editor of the *Thunderer* can make cabinet ministers quake, but, because there are exceptions, it does not prevent Dr. Schiller's description applying fairly well to many members of the class, perhaps more to the Continental type than to the English.

The word intelligentsia comes to us from the Russian and to many minds it suggests shabby men in cafés engaged in interminable discussions about God, Truth, the Soul and the Theory of the State. The women are suspected of wearing sandals and having views on "free love". Such talkers might be thought insignificant and the university professors merely useful. That would be a mistake. The social function of the intelligentsia is most important. The one group

[1] *Social Decay and Eugenical Reform.*

supplies facts, the other ideas, and, between them, they drive the state on its way. The facts, as we shall say later, maintain the material development of the community, the ideas develop the facts and offer solutions of the problems that beset the state. The discovery of facts is a difficult specialized job. It can only be undertaken by those who have had an advanced education, and possess in a high degree qualities of mind and character that are rare in the general body of a nation. The dissemination of ideas is not much easier. The publicists, whose duty it is, must also be men of special ability and training. Ideas need to be taught, and they must be expressed in certain forms before they can be fully apprehended. The progress of an idea, from the study to realization, is by way of books, reviews and the popular press. Before it can be acted on it must generally be so well known that all have heard of it, and a body of support been formed strong enough to force action on overburdened officials, or to persuade politicians that the gain is greater than the risk. It is very informative to follow a newspaper campaign through its various stages and see how an idea is presented to the nation at large.

These ideas may be of different kinds. Many are philosophic ideas such as the great principles of the eighteenth-century socialists who saw the liberty and equality of mankind as the means of transforming the world. The first duty of these writers was to imagine a society in which conditions were so changed that liberty and equality became possible, and to make this imagination part of the general ideas of the age. In doing this they had to break down beliefs that had formed the skeleton of the social structure for hundreds of years. They had also, as time passed, to realize that political equality was not enough, and to devise means for giving economic equality also. An idea of a more immediately practical type, though one that ultimately depends on a philosophic concept, is the nationalization of various industries. An idea such as this must be explained and expounded, repeated and justified till it is known to all and believed by some. In a yet more restricted field such an innovation as one-way traffic in the centre of a big town has to be argued and made the subject of articles and a newspaper campaign.

In another sphere publicists and scientists combine. The invention of new machines and processes is the part of the scientists, but the publicists explain them to the world. The intense interest taken in, say, the development of a new drug owes much of its force to the articles that appear describing its powers. But even more important than this, the publicists help to unfold the ideas that are implicit in new discoveries. Such an invention as the atom bomb might shock the world at the moment of its explosion, but the full significance is only gradually being worked out.

This exposition of the ideas that are implicit in a fact is one of the most important functions of the intelligentsia. Without their aid it would take far longer for the modifications of thought and belief that are due to the scientists' discoveries to take place. The psychologists discovered some time ago that high intellectual ability occurs in all sections of the community and not only in the "well born". This "fact" has gradually been seen to involve a major social change. If ability and poverty can coexist and if the nation needs its able men, then education up to—and even beyond—university level must be provided free, so that the ablest can be trained and, when trained, used for the good of society. The freeing of education which is now taking place represents the final working out in practice of the ideas implicit in the psychological fact. An exactly similar history lies behind the present organization of the Royal Military Academy. It was the discovery of the psychological fact that powers of leadership occurred among all levels of society that made it necessary to remove social and financial barriers to position in the army.

This production and dissemination of ideas is only one aspect of the work of the intelligentsia. It is an important function, but a spiritual one, and as such might be ignored. The scientists hold real material power, and can never again be forgotten. The scientists may work in one of two ways. They may be engaged in pure research—in the search for facts quite irrespective of any immediate use to which the facts may be put—or in applied research that is directed to some preconceived useful end. Both these groups of men supply to the modern state its chance of military dominance or its supremacy in the arts of peace. The pure research started by Rutherford ended in the terrible practical form of the atom bomb. I.G. Farben, the great chemical combine, was the organization that made Hitler's war possible by the development of synthetic substances, explosives and other munitions. In England it was the group of scientists who developed radar that gave us the possibility alike of defence and attack. It is this group of men, perhaps a few hundred strong in any state, that really holds supreme power over the destinies of the group—whatever the politicians may think. Without them a state rapidly becomes helpless; with them it can survive, if their talents are used rightly. It was different in the past. Experts were needed, but not in very large numbers. When Hyder Ali, about 1770, wished to modernize his army to challenge the British in India, the change could be accomplished with the aid of two or three renegade sergeants and a captured expert in gun-casting. Ventura and Avitabile, two soldiers of fortune, sufficed to ensure that Rungit Singh's troops, in 1840, were kept at the proper pitch of excellence. It is only recently that the scientist has become indispensable, and it is only more recently still that he has been accepted at anything like his true value.

Any government that in any way realizes the importance of the intelligentsia naturally wishes to command their services. It needs the scientists to work for it and the publicists to disseminate its ideas. The Nazi regime was as much dependent on Goebbels, Rosenberg and Streicher as it was on the chemists. Our own newspapers, which record and extol the rising output of the nationalized coal-mines, perform services to the government of first-class importance. But it is contrary to the very nature of the intelligentsia to commit themselves to any predetermined body of ideas. They must think things out, they must pursue the Truth. Thus the intelligentsia can play very different parts in the state. They can be a body opposed to the government, demanding reform and importing disruptive ideas from abroad, or they can be government men, arming the powers with thunder and making them speak with pentecostal voices. There thus arise two groups among the intelligentsia themselves. There are those who see themselves as employees of the government, working on government problems for the good of the state. They are patriotic men doing their duty in a special field, as truly as the fighting soldier does his in battle. There are others who feel their allegiance to the world, and would make their contribution general. It is not easy for a government to control the rebels. The scientists, who always need money and yet more money, are in a weaker position. But a grant for apparatus is not enough to ensure conformity, and if too much pressure is employed there is always a fear that by limiting the freedom of thought and communication some fundamental discovery will be lost. There can, in these cases, be acute tension. When the results are of the highest importance, and when the politicians believe they should be the property of a single state, a number of scientists will accept the nationalist point of view. Others will protest and attempt to influence public opinion against secrecy. The moral issue is very complicated, especially as all the arguments are hypothetical and unverifiable in advance. When a specific piece of information is involved, a piece whose importance is clearly realized, it is in many cases possible for a government to force secrecy on the scientists. But if the scientists give way on the one point they still maintain that freedom and publicity in general are essential conditions for satisfactory work.

The teachers and publicists are quite as difficult to control. Some, of course, voluntarily join the government; but a large number will insist on their freedom to acquire and disseminate ideas outside the orbit of those officially approved. When a government attempts to establish a closed society these are the people who will continue to break through the guarding fence. As we have said, their activities will vary according to the type of isolation imposed. Where the government has something real to offer, something to stir the imagination, they will on the whole be loyal. When it is desired to perpetuate injustice and inefficiency, they will risk their lives to defy the ban. Thus to a tyrant they are dangerous in the extreme.

When Hitler overran Poland one of his first acts was to murder all the university professors he could catch and close all secondary schools and colleges. He did much the same in Czechoslovakia. In October, 1940, when the organization of Poland was taking shape, he issued a directive of which the following is part:

"All representatives of the Polish intelligentsia to be exterminated. Polish priests will receive food from us and will, for that reason, direct their little sheep along the paths we favour. The work of the priests is to keep the Poles quiet, stupid and dull-witted."

Now this edict of extermination paid the intelligentsia two compliments. They were supremely dangerous to the German regime, and they were incorruptible. They could not be bought for "food".

Hitler probably feared three things from the Polish intelligentsia. They were themselves people of superior knowledge and ideas, with trained minds. They would supply the brains of an underground movement. They also had trained characters and were far harder to terrorize than others. Then, secondly, they would teach; and it is every tyrant's desire to keep his slaves ignorant. Lastly, when, if ever, a break in his tyranny did come, they would at once step forth as leaders, ready with solutions for the problems of their land.

To say that these are the characteristics of the intelligentsia is to make a claim for learning. That mysterious thing, an advanced education, has results that go very deep into the make-up of the individual. It is not so much what a man knows that affects him, but from his training he gains a sharpening of the wit, a steadying of character, a firmness of attention. Not in every case do the desired results follow; but, as a class, the university graduate or teacher has a certain combination of characteristics that is not possessed by many outside the group. If this is so, a nation is dependent on its intelligentsia in many ways. It is largely on their activity that a state relies for moral growth, as well as for material development. It is important therefore to consider how a satisfactory intelligentsia is formed.

In the first place, the intelligentsia takes its rise mainly from the educated middle class, and needs in its turn a good education. No intelligentsia is possible without an advanced education. Ideas are not easy things to deal with, and though the art-school student talks a lot, the ideas he talks about are rather common ones, and his conclusions are not very remarkable. The valuable part of the intelligentsia has usually had a thorough training in ideas, either those belonging to some special science, or the wider ones given by a training in literature, history or law. On the Continent this study of law is the commonest introduction to the world of ideas. In England, in the past, it was the classics that opened the door to human thought. To produce the best results this education must not be narrow, and it must allow the maximum freedom of mind to the student. Thus *Literae Humaniores* at Oxford gives a better training for this purpose than pure classics, and Gladstone was not far wrong when he referred to the *Republic* of Plato as "that most unsettling work".

If the education is too narrow and contains too few general ideas, or consists almost entirely in learning certain established views, then the effect is not the same. Browning's *Grammarian* represents one type that results from an education that is too restricted in scope; the religious fanatic, reared on the exclusive study of ancient books and in unchanging observance of ritual, is another. However great the writers of the past, however noble the thoughts they contain, it is stultifying to the young to be expected to keep their minds confined to one set of ideas. There must be a sense of novelty, excitement, experiment, and a belief that each man makes his contribution to the life of to-day. Advanced states that have become static, have done so generally by imposing a wrong type of education on their youth. It is thus true that a nation gets the intelligentsia, like the government, it deserves; and a defective system will put a state far behind its neighbours in thought.

No worth-while intelligentsia is possible without freedom of thought and expression, and the possibility of playing a responsible part in the affairs of the nation. We shall speak later of the dangers of an inexperienced intelligentsia. If it is deprived of all outlet in action it will turn to frivolities simply as a means of exercising the mind. When Milton went to Italy in 1638 he found a society cultured, intelligent, completely cut off from politics and living under a system that made any comment on public affairs dangerous. In consequence all their time and thought was devoted to the niceties of classical scholarship. For a year Milton found this charming; but no sooner had he returned to England than he plunged into the bitter life-and-death politics of growing Presbyterianism. The Englishman who went to Germany at the end of the last century found a society occupying itself delightfully with music and literature. He admired German culture and contrasted their sweetness and light with the Philistine darkness at home. He forgot, till the white cliffs had received him, that for his period abroad politics had been banned. In the Russian novels of Tzarist days an intelligentsia, cut off from politics, living under a despotism it despised, struggled hopelessly in a continual emotional fever that could have no issue in action. The young men in Turgenev are defeated at the outset of life because there is no field to which they can properly apply their talents. Those who refused to be rendered useless perished.

There must thus be a suitable political environment for the proper development of the intelligentsia. There are also social needs. The intelligentsia require an audience. No one can teach without auditors or persuade when none listen. He also needs the stimulus of his fellows. He must receive ideas, give them in exchange, and see how, when bandied from mind to mind, an idea develops. He must have the society of his peers, and time to enjoy it. For ideas are shy things, and too steady work will drive them away. England has never had a luxuriant growth of the more obvious intelligentsia. Their characteristic occupation is talking in cafés, and the English licensing laws, which discourage the habit, may be to blame. Certainly we look back on the Mermaid Tavern and White's Coffee-house as homes of wit that cannot be paralleled to-day. Deprived of cafés the Englishman has to fall back on the printed word. This reaches a larger public than café talk, but is less stimulating to the individual. On the other hand, one of the other conditions for a successful intelligentsia exists. It is possible for a clever man to earn his living, and still have the leisure to deal with ideas. University posts are generally so arranged that a certain amount of leisure is available. A poet can earn his bread at the B.B.C., and the less exacting forms of journalism keep many.

Not all the clever members of the middle class join the intelligentsia. Those who are less interested in ideas for their own sake than in their application, go into the Civil Service, the Army or business. They become important in a way that the intellectual never expects and seldom desires. The university lecturer who, during the war, rose to unexpected heights in the Civil Service, has, in most cases, resigned his dignity and emoluments, and returned thankfully to his obscurity.

There is, in fact, a profound difference in the two approaches to the world. Yet the two groups are in reality closely linked. The clever man who makes the Civil Service his career is a necessary counterpart to his fellow who stays outside it. The ignorant and inefficient bureaucrat is intensely hostile to the intelligentsia, but the clever official, trained with the man of ideas, able to follow his thoughts, is usually willing to absorb and disseminate the ideas that the other invents. Even when his first response is hostile, he is generally open to the persuasion of public opinion. In any case he has been taught to respect ideas, and he has no impulse to persecute them.

The most outstanding example of this interaction is the career of Beatrice and Sidney Webb. They boldly set themselves to provide ideas and knowledge for the Socialist party and the nation at large. As independent thinkers, by amiable persuasion such as lunch-parties, or by peaceful agitation, they forced a number of new ideas on reluctant bureaucrats and politicians. From the vantage-ground of a Royal Commission, Beatrice defeated the officials who wished to continue the Poor Law practically unaltered. Through the Fabian Society they made known the basic principles of Socialism to generations of young men who were going to hold ruling positions. Their historical researches into Local Government facilitated understanding and reform. They had the greatest influence on their age, and they had it because they were not interfered with and because they were understood. It is curious to reflect that, had they been born in Russia or some other countries, they would either have died on the gallows or spent the major part of their lives in exile.

The claim of the intelligentsia to the free communication of knowledge is historical, and has been part of the cultural tradition of Europe. The wandering scholars carried ideas from city to city, and the man of learning would take pupils from all parts. The medieval universities were not exclusive. Their very essence was that they were international. When true science emerged, Francis Bacon was perhaps the first to give expression to the idea of a commonwealth of science organized and transcending all national bounds. He saw how the thing would be, and, as in a cloud, imagined the organization of experimental research with its necessary corollary of free communication and discussion. Scientists have made it part of their honour that results should be published and should be freely available to other researchers. This has been adhered to in cases where, to a mind filled with commercial motives, the wildest possibilities of private gain existed. When Pasteur found the cure for hydrophobia he made the knowledge and the treatment generally available, and by so doing gave up the chance of holding the whole world to ransom. The Curies laid before the world all the knowledge of radium and its uses. The discoverers of penicillin had and rejected much the same opportunity as Pasteur, and it will not be for lack of scientific protest if any aspects of atom research are made the exclusive possession of any one nation.

This intercommunication would not be so important if what the scientists made public were facts alone. One of the dangers to a closed system of ideas is the threat to established beliefs that facts bring. On the whole, over a period of time, a belief that claims to concern facts, but yet is not in accord with the discoveries of science, will disappear. But its destruction is not easy, and may not be complete. Beliefs become embodied in myths, and it needs the creation of another myth, as well as the discovery of hostile facts, to destroy a structure of the mind once it is firmly established. The existence of the myth gives an emotional support to the belief, and the tendency of a belief to generate its own evidence gives it a kind of intellectual backing. It is hard to believe the scientists when we have immediate personal testimony to support the opposite belief. Yet when once the system of ideas has been invaded both the myth and the evidence vanish, and facts assert themselves. In the past this has been true of many systems of ideas. The belief in witches and witchcraft produced the evidence on which it was based, and the evidence vanished or was reinterpreted with the failure of the belief. More recently the German belief in racial superiority was unshakable so long as the Nazi system endured. In spite of facts known to every scientist it held men's minds and collected evidence in its own support. The destruction of the Nazi regime meant the end of the belief.

When a system of beliefs has existed for a very long time and has become through the ages a very complex thing, the impact of facts will strip off the less valuable parts. When Galileo used his telescope to demonstrate the movements of the heavenly bodies, still more when William Smith made geology a science and Darwin published his theories on the origin of species, the cry went up that Religion was in danger. It was rapidly found that the spiritual essence of religion was untouched by these changes in belief about certain things, and that there was no need for Joshua to have literally stopped the sun or for the creation to have followed the exact pattern given in Genesis for the Bible to contain the most important truths. With such matters which are spiritual in nature, and are not dependent on particular facts, science has nothing to do.

When, however, a belief is, in essence, false and bad, the discoveries of science may destroy it completely. In the period from

Pasteur to Flory and Fleming the whole attitude to disease has been changed. The old ideas that it was a punishment for sin, or that death was a merciful providence snatching the good from a wicked world, have simply vanished before the growth of knowledge, and been replaced by a sense of human responsibility. The death-bed scenes in Victorian novels, where the child dies surrounded by his whole family in an ecstasy of sentimental grief, probably consoled an age in which cause and cure were alike beyond human knowledge and control. Heaven seemed near, and the white-winged angel stooping to receive the tender wraith, a beautiful image. Now orange-juice and calcium, the healing injection or the blood-transfusion have brought earth more firmly under our feet, and made the Better Land more remote. It is a disgrace if a child dies, and we hardly like to blame God for our own incompetence.

So, too, with the Victorian belief in the beauty of toil. The sweep of technical progress has given us instead the cult of leisure, and not to be able to enjoy oneself freely is now felt to be as discreditable as the earlier age found it not to enjoy work.

Where science has not developed, ideas of remote antiquity survive, often divorced from any memory of their origin. The sweeper caste in India, whose duty it is to deal with sewage, are forbidden to draw water from the well or cook food. This is, clearly, a very salutary precaution against typhoid. But because there is no scientific backing to the custom it cannot be interpreted or modified. The scientist, by providing proper methods of sewage disposal, could set the sweeper free to pursue some other calling and take his place in the general body of the state. Facts will achieve this liberation when they are once fully appreciated and the older ideas shown to have no further value. The mystical uncleanness of a whole caste will vanish before the advance of the W.C.

Not only do facts destroy beliefs that are inconsistent with them, they also form the centre round which new beliefs can grow, like sugar crystals round a string in sugar candy. Any material arrangement has its associated set of ideas—just as a set of ideas will produce material arrangements—and when material things alter, ideas must change too. Thus a new invention inevitably brings with it mental changes. Just as the discovery of America changed the current of European thought, so the wireless, the aeroplane, the atomic bomb all make it impossible for men's thoughts to continue in the old paths. Thus the little band of scientists keeps the world in a turmoil. As the pace of invention and discovery quickens we have less and less time to build up a set of beliefs round any material order.

We are hurried from railways to aeroplanes and from telephones to wireless. At the same time, because men need stable beliefs, we come to rely more and more on those spiritual values that science so far shows no power to shake. Love, human and divine, trust, quiet happiness.

At the same time, since the scientists keep up an international exchange of facts, and since facts, in the end, prevail over beliefs, the scientists alone are enough to defeat most of the efforts to establish a closed society.

The second group of the intelligentsia is very different from the scientists. Ask a scientist what he is concerned with and he says *facts*. His criterion is, "Is it true?", i.e. will it explain phenomena, help to predict new ones and not prove inconsistent with other phenomena and predictions. The publicist is interested in ideas, and, if they are not strictly true in the scientific sense, they are at least interesting and might by sufficient effort be made true. When G. K. Chesterton wrote *The Napoleon of Notting Hill*, he was preaching a little sermon on the idea that the world is as romantic as you make it and that you must be true to your ideals. He is not offering you a fact, but only a spiritual glimpse of one, a point of view, a suggestion. He is near enough to the truth for certain people to find the extravaganza amusing, but that is all.

When the future Napoleon appears before the king he offers the supreme sanctification of blood:

" 'I know a magic wand, but it is a wand that only one or two may rightly use, and only seldom. It is a fairy wand of great fear, stronger than those who use it—often frightful, often wicked to use. But whatever is touched with it is never again wholly common. Whatever is touched with it takes a magic from outside this world. If I touch, with this fairy wand, the railways and roads of Notting Hill, men will love them and be afraid of them for ever.'

" 'What the devil are you talking about?' said the King.

" 'It has made mean landscapes magnificent and hovels outlast cathedrals,' went on the madman. 'Why should it not make lamp-posts fairer than Greek lamps, and an omnibus ride like a painted ship? The touch of it is the finger of a strange perfection.'

" 'What is your wand', cried the King impatiently.

" 'There it is', said Wayne, and pointed to the floor, where his sword lay flat and shining.

" 'The sword!' cried the King and sprang up straight on the dais.

" 'Yes, yes', cried Wayne hoarsely. 'The things touched by that are not vulgar. The things touched by that——'

"King Auberon made a gesture of horror.

" 'You will shed blood for that!' he cried. 'For a cursed point of view'."

This light-hearted playing with ideas is part of the intellectual's contribution to the community, and it may have a great effect. Bernard Shaw's plays were at first ignored, then found shocking, and now the ideas they contain have been largely absorbed and have almost become commonplace. He played a great part in breaking up the set of ideas current at the end of the last century. Without him and his like, standards would change very slowly.

Writers such as Shaw and Chesterton are the bubbles on the surface of thought. They have the power to see and analyse, to invent new ideas or ridicule old ones, but they do not deal in any real way with the fundamental economic or political structure of society. Shaw's *Apple Cart* is amusing because it reveals certain aspects of politics which are felt to be true, but are not generally realized. It is most unlikely to lead to any change in the organization of society.

Yet it is also the intelligentsia who produce ideas of fundamental importance, the solution of the problems of state. When a set of ideas has continued for long without change, it degenerates. What was once reasonable becomes absurd, what was once, in essence, just, because it corresponded to certain differences of function, becomes an abuse. When this degeneration has gone far, almost the whole state may be conscious of the evil, but few can devise a remedy. The power to imagine a new organization of society is not given to many. Yet, once a remedy is suggested, if it appears to offer an escape from an intolerable situation, it is accepted eagerly—the more intolerable the situation the more eagerly. In a later chapter we shall discuss different types of these ideas, some good, some bad. Here it is enough to say how they are devised and how accepted.

Ideas are almost always the work of the individual thinker. It is true that at a given time ideas of a certain sort occur to many people. Conditions may have reached such a state that something must be done, and more than one man may devise the same solution. The same thing happens in science where many inventions have been made simultaneously by two people working independently. In politics it is much the same; but, for all that, an idea, however quickly it spreads, however many take part in its elaboration, has a true father, and is not the offspring of a mere "spirit of the age".

Ideas once born have very different histories. Some pursue a placid course bringing mild blessings in their train, others disrupt society and leave a trail of fire and destruction. Some ideas, as those provided by the Webbs, have been absorbed without any great social disorganization. In other cases, the evils were so fundamental in society that nothing but a revolution was possible. De Tocqueville in *L'ancien Régime* gives a half-surprised account of the part the intellectuals played in preparing the Revolution:

"France had been for a long time the most literary country in Europe, yet the men of letters had never before shown the spirit that appeared about the middle of the eighteenth century, nor taken such a place in the nation. Such a thing had never happened before in France nor anywhere else, I believe.

"They were not, as their counterparts in England, engaged in the daily affairs of the nation. On the contrary, they had never been further removed. They possessed no authority, and performed no function in a society that was overflowing with officials.

"On the other hand, they did not, like the Germans, remain strangers, apart from politics, withdrawn in the realm of pure philosophy and literature. They dealt incessantly with matters of government. Whatever their other differences they all agreed that the time had come to sweep away the complicated customs and traditional rules that governed the society of their day and substitute for them simple fundamental rules derived from reason and natural law.

"How did these men of letters attain their position? They had no rank, no honour, no wealth, no responsibility, no power, and yet they became in fact the chief politicians of the day, almost the only ones, for if the others conducted the government the writers had the power."

De Tocqueville has anticipated Schiller's remark about the personal insignificance of the man of ideas, but he has made it clear that the writers, such writers as Rousseau, Mably or Raynal, were sowing the seeds which must in time bring the harvest of revolution.

Why at this time did they have so much effect? The answer is that what they had to say fell on ears that were ready for it. Their ideas, their solution, seemed the only one.

"The sight of so many harmful and ridiculous privileges, whose weight grew ever heavier as their justification grew less, drove every mind towards the idea of natural equality of conditions. So many strange and discrepant institutions, the children of times past, that no one ever tried to bring into harmony with each other, or the needs of the times, naturally gave to thinkers a distaste for antiquity and led them to plan society entirely anew, by the light of their single reason.

"Every man who suffered under the unjust distribution of taxes warmed his heart at the idea of the equality of man, every peasant whose crops were ruined by the nobleman's rabbits from across the fence, was delighted to hear that all privilege was condemned by reason. Every political passion appeared disguised as a philosophy and the writers, taking control of the direction of opinion, found themselves, all in a moment, in the place occupied in other countries by the leader of a party. No one any longer could dispute the place with them."

The danger and damage, says de Tocqueville, came from the fact that these political writers were not practising politicians and had no idea of the forces they would let loose, or the ways to control them. They had the ideas, the ideas that the age needed, and that was all.

It is interesting to compare this account of the first birth of the French Revolution with that of the Russian, the greatest political event of our age.

The writers who drove on the French were Frenchmen writing in their own land, and they unleashed a power of destruction that swept much away. The makers of the Russian Revolution were exiles, some not even Russians, and their inexperience of actual politics and administration was greater. It is no wonder that even more was destroyed, and yet the pattern of the movement was very similar. If ever a great event was the work of the intelligentsia, it was this. It might almost be called their triumph.

The ideas on which the Russian Revolution was based started perhaps in France at the beginning of the nineteenth century. Saint-Simon, aristocrat who had abandoned his class, insatiate student who had learnt too much, driven by poverty, sat in his garret, cold, with no fire, and copied his writings with his own hand. He received no personal respect, except from a few disciples, and he died without regard, yet his dying words are the idea that gives order to the century.

"All my life may be summed up in one idea, to guarantee to all men the free development of their faculties. Forty-eight hours after our second publication the party of the workers will be organized: the future belongs to them."

The idea was a dangerous one. In Russia his books were proscribed. It might be death to have them.

Karl Marx, the dominating political figure for half Europe, was the typical intellectual of Dr. Schiller's unkind description. A Jew, an exile, a shabby man keeping himself warm in the British Museum while he wrote his denunciation of British capitalism. He threw into the nineteenth century the most powerful body of new ideas that the age received. While Dickens was sentimentalizing over the East End or rousing an indignation that saw no outlet in action, Marx showed how and why things were wrong and where the road ahead must lead. He gave the answer that men could accept. The problem of the age was the horrors produced by uncontrolled, individualistic capitalism, the answer was the destruction of capitalism.

Where the idea went, it led to action. Lenin, then Vladimir Ulyanor, read it and saw the light. Lenin's elder brother had just been executed for a plot against the Tzar. "I saw no other way", he said. The phrase of despair by which he justified his action, was: "This weak intelligentsia can defend its right to think only by the methods of terrorism." Lenin, excluded from the University because of his brother, at home in Kazan, had made himself a little study out of the disused back kitchen. There, sitting on the stove, he first read Karl Marx and at once saw another way, another method for the conquest of the intolerable evil of autocratic government. Exile in Siberia, exile in Geneva, were only steps on the path that led to revolution. From abroad, he published his paper, *Iskra* (the Spark), and, little by little, all was ready for the kindling of the blaze.

A tyrant or a tyrant class can never really feel safe when they know a mine such as this may be exploded any moment. The intelligentsia designing ideas which will destroy the evil that the state suffers from, holding up to others hope and an alternative way of life, are a constant menace. They will never suffer the establishment of the peace of death.

As dangerous, almost, is the determination of the intelligentsia to teach. They teach, as some writers write, from a sense of inner compulsion. "That one talent which is death to hide . . ." seems to them, as to Milton, of overwhelming importance; and, though tyrants threaten, they will still do what they feel called to. Because learning must at all cost be kept from slaves, the threat is very real. From innumerable examples of this passion we can take two examples from the early life of Marie Curie.

She was a young girl in Poland. The Poles were governed by the Russians, schools were most closely supervised by Russian inspectors intent on finding out some infringement of the law, and education outside the accepted system was forbidden.

In the schools, the teaching of the Polish language and history was proscribed; and yet the teachers regularly gave lessons on them. At the first warning of an inspector's visit, books were swept away and the class became apparently engaged in an innocent sewing-lesson. Marie, reared in this atmosphere of illicit learning, never gave it up in Poland, whether she was teacher or pupil.

At that time, women were forbidden to have a university education. So, as soon as she had left her secondary school, Marie joined one of the illegal educational groups. This is her daughter's account:[1]

"Marie was admitted to sessions of the 'Floating University', which is to say, to lessons in anatomy, natural history and sociology given by benevolent teachers to young people who wished to extend their culture. The sittings took place in secret. The disciples gathered to the number of eight or ten at a time and took notes; they passed pamphlets and articles from one to the other, at the slightest noise they trembled, for if they had been discovered by the police it would have meant prison for all of them."

Of the classes, Marie wrote forty years later:

"I have a lively memory of that sympathetic atmosphere of social and intellectual comradeship. The means of action were poor and the results obtained could not be very considerable, and yet I persist in believing that the ideas that then guided us are the only ones that can lead to true social progress. We cannot hope to build a better world without improving the individual. Towards this end, each of us must work towards his own highest development, accepting at the same time his share of responsibility for the general life of humanity, our particular duty being to help those to whom we feel we can be most useful."

These were her principles in later life. She was true to them even before they were formulated. When she went to be a governess in the country she spent at least two hours a day, often five, in teaching peasant boys to read and write, knowing that she was thus putting herself in danger of Siberia.

The subjects taught were not in themselves political, and they were desired for mainly non-political reasons; but the Government feared the spread of knowledge because without knowledge men are helpless, and because, as we have said, with knowledge, of whatever kind, comes a certain strengthening and liberating of the human mind. It is set free from the immediate shackles of tyranny and is given a view of a universe in which man is free to develop his faculties, and, having developed them, to use them. This knowledge is a source of spiritual power which may be used in any direction. For the social reformer, knowledge is the beginning. Robert Owen, believing in the rights of a socially-depressed class, started with education. The ideas of his day did not allow him to go very far, not much farther than the very young children, but even that was something: the first step in the upward direction. The history of democracy is essentially linked with the

[1] Eve Curie, *Marie Curie.*

spread of knowledge. The intelligentsia, or at least one section of it, believes in the power of knowledge, and will run grave risks to impart it.

That there really is some power in knowledge is shown by the anxiety of the tyrant to stamp it out, and the eagerness of men in straits to acquire it. Persecution and imprisonment seem marvellously to sharpen the appetite for learning. In the prisoner-of-war camps in Germany where conditions were not too bad, prison universities flourished with extraordinary success. Among prisoners held by the Japanese, conditions were far worse, and any gathering of more than a few men was broken up. Yet, night after night, the prisoners divided into groups of four or five, and, sitting together on the ground, listened to one of the group giving a lecture. The lectures were repeated time and again. The groups changed so that each in turn spoke and listened, and this simple interchange of ideas brought great help and stimulus to men starved, beaten and often dying.

In occupied countries education was one of the regular activities of the underground movement, and students who escaped to other lands brought with them the belief in the liberating power of knowledge. Again, no tyranny is secure when, at any moment, learning may light her torch.

These are clear functions of the intelligentsia. Have they a further one? Are we to consider them as a kind of *moral* élite, or are they merely a group of clever men and women who have the power of understanding ideas and the trick of imparting them? Further, if they have moral obligations, are these to their own states, in a nationalistic sense, or to the international ideal of mankind? There is no doubt that these questions have been answered differently by individuals and nations. Brains and an advanced education do not, unfortunately, guarantee a high moral standard, though they certainly give hope of it; and certain nations have always felt that national considerations were paramount.

The German intelligentsia has seldom opposed the government. The pattern of German history would have been different if the men who could think had been in effective opposition. De Tocqueville notes that, while Rousseau was evolving new theories of society, the Germans confined their thought to abstract problems or found, like Hegel, the supreme Good in the State. All too many thinkers put their minds slavishly at the service of the government; and men like Marx, exiled from their own land, found their ideas taken up in countries other than their own. Many of the independent thinkers left the country after 1848 and bestowed their power in America. Thus, when the crisis of war and Nazism came to the Germans, the intelligentsia were silent or even co-operative, bemused with the national myth and the numbing shadow of fear.

Benda, in his book, *The Great Betrayal*, speaks bitterly of those authors who have joined in the nationalistic fever of their respective countries and supported the politicians and capitalists in their attempts to stir up passion. "They have failed", he says, "to think clearly. The arguments of the nationalists are so bad that any one could see through them. It is only wilful blindness that makes them acceptable."

This is hardly fair to the intelligentsia. The class has never dedicated itself to an unknown God of Righteousness, or even pledged itself to clear thought. Many individuals may honestly feel that the land that has nursed them deserves their utmost service. What Benda is really claiming is that the world, and each state, needs a moral élite; and he is suggesting that it is the duty of the intelligentsia to perform this function. Clearly the politicians are not morally eminent, nor are the capitalists.

For practical purposes, because of their special gifts, the intelligentsia could serve in this capacity if they would accept their dedication. But the responsibility laid on them is very heavy.

The problem of moral leadership in a state is a very serious one. Thinkers on political theory have always recognized its importance, and Plato,[1] in the description of the selection and training of the Rulers, sets out the problem clearly and gives part of the answer. The rulers must be clever and possess other desirable qualities. They must be specially educated for their task. They must stand apart from the other sections of the state, especially the military and commercial, and form their judgments on higher considerations. H. G. Wells, in his *Modern Utopia*, imagines a ruling caste that is in a way self-selected, because the would-be members offer themselves for inclusion; but to qualify, they must have already attained a certain distinction in some profession and possess enough firmness of character and self-sufficiency to pass the test of solitude. No modern state has evolved any real method of selecting its leaders. Too often, there has been no method of finding really able leaders or of dismissing incompetent ones. In the days when men bought their commissions, and promotion was by the strictest seniority, some officers were highly intelligent, others were grossly stupid and incompetent. The typical officer was conservative and made little attempt to understand the needs of his men. A memorial in the chapel of the Guards' depot at Caterham "to those

[1] *Republic*.

recruits who died on parade" is perhaps the best commentary on the officer of the past. We shall discuss the whole question in a later chapter, because it is one of the most pressing of modern problems, both in peace and war.

There is one body in the state that should supply moral leadership—the Church. Its members are recruited at an age when their characters are sufficiently formed to be clear to competent judges. There is a great organization that is strong enough to withstand any attacks likely to be made on it. The interests of the Church are largely international and the truths it preaches transcend race and nation. In times of stress the Church has often stood firm. Einstein refers to the part the German Church played in resisting Hitler.

"Being a lover of freedom, when the revolution came in Germany, I looked to the universities to defend it, knowing that they had always boasted of their devotion to the cause of truth; but, no, the universities were immediately silenced. Then I looked to the great editors of the newspapers whose flaming editorials in days gone by had proclaimed their love of freedom; but they, like the universities, were silenced in a few short weeks. . . .

"Only the Church stood squarely across the path of Hitler's campaign for suppressing truths. I never had any special interest in the Church before, but I now feel a great affection and admiration for it, because the Church alone has had the courage and persistence to stand for intellectual truth and moral freedom. I am forced to confess that what I once despised I now praise unreservedly."[1]

The Church in Norway displayed a most gallant resistance to aggression, as did the teachers. Though Hitler expected subservience from the Polish Church, he did not find all others so accommodating. The difficulty of some organized Churches is that they are part of the economic life of the country and, in essence, members of the capitalist class. This is true of the Church in Spain. It is true, to a much lesser extent, of the endowed Church of England. In the past this has had a political effect. When the Church of England was closely bound up with the landed gentry its political bias was strongly Tory, and, even now, when the main interest has shifted to the towns, certain types of economic change are difficult for the Church to imagine.

The other body that might be thought to have special moral functions is the teachers. The behaviour of the Norwegian teachers, who were undeterred by severe persecution and refused to Nazify the schools, is an example of the part that an organized teaching body can

[1] Einstein, *New York Times*, p. 38, 23rd December, 1940, quoted W. L. Sperry, *Religion in America*, pp. 18, 19.

play in the moral life of the state. All teachers, from nursery school to university, should realize their place in the social scheme. In so far as they are selected and trained they do at least partially apprehend it. But in England there are many schools run exclusively for private profit, and even in others conditions prevail that make education seem anything but moral. The English teacher is under a difficulty. There is a tradition of several hundred years that holds the teacher in contempt. This should not be so. The lower middle class, from which most teachers come, is an extremely moral section of the nation. When students come for training, the great majority have ideals and an eager desire to improve the world and serve their pupils. These good intentions succumb to bad conditions and lack of recognition. Work in schools is hard and monotonous; the training is brief and ends before the student's personality is fully developed. Once teaching has started, there is little opportunity for change, refreshment or new ideas. The community as a whole pays them little respect. There is a curious blindness in appointing committees by which many rather brutal and bullying men reach the top. It is little wonder if on the whole in England the teachers fail to play the part their talents and profession mark them out for, or which they do in fact play in some other nations.

The chaotic conditions of moral leadership among civilized nations to-day can only be a source of wonder to theorists. The best that can be said for it is that it allows diversity and so does not stifle altogether the voice of the reformer.

Books referred to in this chapter:

F. C. S. Schiller, *Social Decay and Eugenical Reform.*
David Masters, *Miracle Drug.*
Robert Browning, *A Grammarian's Funeral.*
Margaret Cole, *Beatrice Webb.*
Francis Bacon, *The New Atlantis.*
G. K. Chesterton, *The Napoleon of Notting Hill.*
G. B. Shaw, *The Apple Cart.*
De Tocqueville, *L'ancien Régime.*
E. Wilson, *To the Finland Station.*
Benda, *The Great Betrayal.*
H. G. Wells, *A Modern Utopia.*
J. P. Veale, *The Man from the Volga.*

Chapter 6

EXTENSION OF THE MORAL FIELD

ETHICS is not a stationary body of precepts for conduct. There may be an ultimate standard against which action can be tested, and this may remain constant, but the field in which this standard is applied is continually being enlarged as human activity comes under more conscious control. It is this need for extension of the ethical field that makes the intelligentsia so important, because it is largely through them that the extension takes place.

When Aristotle wrote, his subject-matter was personal conduct; man is essentially a "political animal", but politics and economics are only studied as the necessary conditions of this "good life" which, in its highest form, is a life of contemplation not actively engaged in practical affairs. For the early Christians their moral life was a private affair, and Caesar was the arbiter of his proper realm. It is only by degrees that moral considerations have extended to one sphere after another, carrying with them each time a readjustment of conduct.

In this chapter we will discuss the general mechanism of such change, taking as example the moral attempt to deal with the destitution caused by the Industrial Revolution, an attempt which is drawing to its conclusion. More briefly we shall discuss the attempt now being made to stop wars, which is far less well advanced.

The essential element in these as in other moral changes is the existence of a suitable belief or myth. So long as one type of belief exists no change can take place; once another is established the alteration is almost bound to follow. When a new situation arises there may, naturally, be no suitable belief in existence. The production of the moral belief may take some considerable time. Some situations can exist for centuries before they become the subject of moral thought. The institution of slavery went without remark for so long that it came to be regarded as part of the order of nature. It hardly disturbed the Greeks of the classical period, even though the slave was often merely the citizen of an adjoining state overcome by the fortune of war. The existence of large-scale agricultural slavery troubled the Romans only as lessening the supply of potential army recruits. There was no application of moral principles. Negro slavery in America had existed for a long time before any one perceived that it was wrong. In this case, it was a long-established condition that at last came within the sphere of ethical criticism.

The destitution caused by the Industrial Revolution was a new problem growing up in a pattern of society that was, by that time, very old. The pattern stretches back, with breaks and variations, to the very dawn of civilized history.

The great agricultural civilizations of Egypt and Mesopotamia developed among their ruder neighbours, who had not adopted so satisfactory a way of life, complete with religion and art, laws and science. We look back to them as the exemplars of our city life, as our forerunners in architecture and all the arts that require a roof over their heads, and as the inventors of the laws of land and trade. There are many great institutions that depend on the riches of a settled community: a monarch and nobility who live in luxury, a priesthood devoted to the service of the gods and not engaged in any productive work, and a professional army, equally supported by the more industrious peasant. We find all of these established in the world of 1400 B.C., in the reign of Thothmes III, and they have remained with us ever since. They are the children of agriculture, as truly as are the paintings on the walls of the king's palace and the gold cup from which he drinks. The nomad races, though they have holy men and armed marauders and chiefs, never developed anything similar till they learnt from the settled lands.

The agricultural pattern of life, with its peasants, craftsmen, priests, soldiers, nobles, came down to us through the ages, sometimes differing in form and sometimes shaken by strange upheavals and disasters, but in many ways essentially the same, till it had acquired an internal stability and acceptance that made it seem comparatively free from problems, and gave those who enjoyed it a sense of rightness and a lack of responsibility towards the other parts of the system. The eighteenth century, just before the French Revolution, was perhaps the perfect moment of that long tradition. The cheerful amorality of such a book as Hickey's *Memoirs* brings out with startling clarity the complete lack of moral sense, and the complete carelessness of others that characterized this age. Hickey moved through the financial and political world of London, or the financial and legal world of Calcutta, admiring the yet more completely amoral Bob Pott, with the same swagger as Casanova pursued his amorous adventures. *Tom Jones* would be ruined as a work of art if any one in it ever stopped to think whether an act was right or wrong; and even the satirists of the day were engaged in scourging the manners, not the morals, of the age. If we come, a few years later, to *Mansfield Park* and the country, where ideas penetrate slowly, there is a formal morality combined with the most complete indifference to everything outside the tiny world of well-to-do idle people. We can see there the isolation in which the rich lived. No one ever does any work, takes any interest in affairs or appears to realize that any one else exists; yet at this very time the manufacturing cities were filling up with destitute wretches whose existence was gradually to transform society.

In addition, the eighteenth century, out of which the Industrial Revolution grew, was the great age of individualism, and also of private property. Someone has characterized the eighteenth century as the period "in which the principles of private property were let loose on the world". The great phrase was "a man may do what he likes with his own"; and "his own", owing to the conditions of work, meant the lives of a particularly helpless section of the population.

In this, the first stage, there is a complete severance between the main body of society and the unfortunate urban proletariat. This severance is one of knowledge, thought and sympathy. The rich, literally, did not know how the poor lived. They held intellectual views that prevented them from understanding what was going on; and they either did not care, or felt that the poor were somehow to blame. Before there could be any change, this intellectual and emotional division must be overcome. We shall show how this was the task of the more emotional writers, as well as those who were commissioned to find out the facts.

Both these elements are necessary in the process of dealing with a social problem. Until the facts are known nothing, naturally, can be done. But a mere knowledge of the facts is hardly enough. They must be emotionally realized, the sufferers must be felt to be akin to the rest of society. There must be a sense that something is wrong, and that there is a problem for society to solve.

When the problem has been realized, some solution must be propounded. In such a case, the solution is generally the work of the intelligentsia, and is not likely to carry immediate conviction. Only after argument and experiment will sufficient facts have been dis-discovered to prove the merits of one type of belief. For this belief to be fully established, both parties must be persuaded, the weak as well as the powerful. They must feel able to demand what the others are already half-convinced they must concede. Thus there will grow up the last essential in the chain of progress, a sense of *right* or *wrong*. "We have a *right* to better treatment", say the workers. "It is a *wrong* that we should be deprived of so-and-so." We have already mentioned the importance of this claim to a *right* and the incorporation of an idea in the social pattern of justice. Once the matter has gone so far, the battle, from the ethical point of view, is won. The translation of the *right* into actual fact may be slow, but it will almost certainly come, even in face of very determined opposition from those who find the existent state of things satisfactory.

In the early nineteenth century the separation between the well-to-do and the urban proletariat was almost complete, both physically and spiritually. The civilization of England was of the country-house type, and *Mansfield Park* represented most of it. When men thought about economics it was as a completely non-moral subject. The sole interest was wealth, and the *Economic Man* the explanatory myth. Nassau Senior is quite explicit. The proposition on which economics depends is "every man desires to obtain additional wealth with as little sacrifice as possible. This is the ultimate fact beyond which reasoning cannot go". In order that he may do his best, wealth must be secure and private enterprise well rewarded.

"Experience shows that the greatest and longest continued sacrifices will be made in those countries in which property is most secure and the road to social eminence most open. The inhabitants of Holland and Great Britain, and of the countries that have derived their institutions from Great Britain, the nations which up to the present time have best enjoyed those advantages, have up to the present time been the most ardent and successful in the pursuit of opulence. But even the Indians of Mexico, though their indolence makes them submit to poverty under which an Englishman would feel life a burden, would willingly be rich if it cost them no trouble."

He imagined a total divorce between the enjoying and the producing classes. His whole theory of wages was based on different types of production for the two classes, and he ends up his discussion of whether wages can be raised by saying that the fund for the support of labour "cannot be increased by the increased production of those commodities which the workers do not use—by the increased production, for instance, of lace or statues".

A speaker in the House of Commons, Sir James Graham, declared that "the life of the poor must necessarily be limited to eating, drinking, sleeping, and working". This attitude was reflected in the outward form of cities. Hammond, in his lecture on the "Growth of Common Enjoyment", says that, about 1830, Preston was the only town in Lancashire with a public park and Liverpool the only one with public baths. The leaders of religion were hardly more sympathetic towards the needs of the workers. Whitefield said there was "scarcely any recreation that could be called innocent," and as late as 1853 Lord Wicklow, opposing the measure to prevent children acting as sweeps, "did not see why children of any age might not act as servants to be introduced into any trade".

There was not even apparently the least imagination of the mental conditions of the poor. When Government help was suggested in emigration, Lord John Russell suggested that "those who were thinking of emigrating should co-operate and make their own arrangements without Government help and supervision". This might be possible for the Mormons—as Dickens pointed out—but for the waifs of London the suggestion was either nonsense or deliberately obstructive.

The boy described by Lord Shaftesbury was hardly likely to take part in such an enterprise. "Curious indeed is their mode of life", he said. "I recollect the case of a boy who, during the inclement season of last winter, passed the greater part of his nights in the iron roller in Regent's Park. He climbed every evening over the railings, and crept to his shelter, where he lay in comparative comfort."[1]

The genuine surprise and horror that thrilled England on the publication in 1842 of the report on the conditions of children in industry showed how unknown the situation was.

This ignorance and carelessness, then, is the normal state before moral sentiment begins to work on a problem. In America the publication of *Uncle Tom's Cabin* was the event which had the utmost emotional significance in turning men's minds to the problem of slavery. In it the slaves were shown to be men like others, and a direct appeal was made to group-sympathy. Dickens, perhaps, played something the same part in England. In his *Uncommercial Traveller* he has sketches describing the life of some of the poorest of London's citizens. He is deliberately arousing emotion and plainly saying that a problem exists. He probably had far more readers than those who gave a more formal account of horrors. Dickens visited the East End, workhouses, schools for paupers' children, and the homes of the workless. One of his most moving passages is about the woman with lead-poisoning:

"It was a dark street, with a dead wall on one side. Nearly all the outer doors of the houses stood open. I took the first entry and knocked at a parlour door. Might I come in? I might if I pleased, sur.

"The woman of the house (Irish) had picked up some long strips of wood about some wharf or barge and they had just now been thrust into the otherwise empty grate to make two iron pots boil. There was some fish in one and there were some potatoes in the other. The flare of the burning wood enabled me to see a table, and a broken chair or

[1] Hammond, *Life of Lord Shaftesbury*.

so, and some old cheap crockery ornaments about the chimney-piece. It was not till I had spoken with this woman a few moments that I saw a horrible brown heap on the floor in a corner, which, but for previous experience in this dismal life, I might not have suspected to be 'the bed'. There was something thrown upon it and I asked what this was.

" ' 'Tis the poor craythur that stays here, sur, and 'tis very bad she is, and 'tis very bad she's been this long time, and 'tis better she'll never be, and 'tis slape she does all day, and 'tis wake she does all night, and 'tis the lead, sur.'

" 'The what?'

" 'The lead, sur. Sure 'tis the lead-mills, where the women gets took on at eighteen-pence a day, sur, when they makes application early enough, and is lucky and wanted, and 'tis lead-poisoned she is, sur, and some of them gets lead-poisoned soon, and some of them gets lead-poisoned later; and some, but not many, niver. And 'tis all accordin' to the constitooshun, sur. Her constitooshun is lead-poisoned bad as can be, sur, and her brain is coming out at her ear, and it hurts her dreadful: and that's how it is, and niver no more, and niver no less, sur."[1]

Then follow a few more horrors concerning the symptoms of the patient and then the young girl from upstairs comes down to announce that she, too, is going to the lead-mills to save her starving babies. It is a fine emotional climax.

In another passage, he describes the dreadful state of the uncared-for pauper children who roamed the streets.

"Walking faster under my shame of this public injury, I overturned a wretched little creature, who, clutching at the rags of a pair of trousers with one of its claws, and at its ragged hair with the other, pattered with bare feet over the muddy stones. I stopped to raise and succour this poor, weeping wretch, and fifty like it, but of both sexes, were about me in a moment, begging, tumbling, fighting, clamouring, yelling, shivering in their nakedness and hunger. The piece of money I had put into the claw of the child I had overturned was clawed out of it, and was again clawed out of that wolfish grip, and again out of that, and soon I had no notion in what part of the obscene scuffle in the mud, of rags and legs and arms and dirt, the money might be. In raising the child, I had drawn it aside out of the main thoroughfare, and this took place among some wooden hoardings and barriers and ruins of demolished buildings hard by Temple Bar."[2]

Writings such as this, by the most popular and respected author of his day, could not be without emotional significance. They must

[1] *The Uncommercial Traveller*, XXX.

[2] Ibid. XXXIV.

have moved men's hearts to pity. But pity is not enough. It is merely the emotion necessary to prepare men's minds for action, and when we look in Dickens for constructive proposals for reform, we find very little. He has clearly not even understood—if we understand to-day—the causes and cure of unemployment. He has not conceived the regulation of unhealthy work. In a later essay, he returns to the lead-mills, and explains how he went over them and how well everything was arranged. There were respirators, a canteen, a wash-room—rather short of towels—and a "female attendant" to warn the women not to eat with dirty hands. There were, of course, some disadvantages.

"Some of the processes are unquestionably inimical to health.... The door of another oven, or stove, about to be cooled and emptied, was opened from above for the uncommercial countenance to peer down into. The uncommercial countenance withdrew itself with expedition and a sense of suffocation from the dull, glaring heat and the overpowering smell. On the whole, perhaps, the going into these stoves to work, when they are freshly opened, may be the worst part of the occupation, and yet, on the whole, there is nothing to complain about.

"But I made it out to be indubitable that the owners of these lead-mills honestly and sedulously try to reduce the danger of the occupation to the lowest point.

"American inventiveness would seem to indicate that, before very long, white-lead may be made entirely by machinery. The sooner the better. In the meantime, I parted from my two frank conductors over the mills, by telling them they had nothing there to be concealed and nothing to be blamed for. As to the rest, the philosophy of the matter of lead-poisoning and workpeople seems to me to have been pretty fairly summed-up by the Irishwoman whom I quoted in my former paper. 'Some of them gets lead-poisoned soon, and some of them gets lead-poisoned later; and some, but not many, niver; and 'tis all accordin' to the constitooshun, sur, and some constitooshuns is strong and some is weak....'"

The only field in which Dickens had some constructive ideas was education. When he saw the untamed and untended child wretches of London, he was filled with a sense of *waste*. With so little care, they could be made of use. In another essay he grows enthusiastic over the pauper schools of the Stepney Union, where for the modest sum of 4½d. per week "which includes all the salaries of the teachers and the rations of the teachers", children are so trained that they make good sailors, good bandsmen for the Army, and the girls good domestic servants. He does not look for any higher achievement; but even this is making good citizens from the otherwise waste products of the London slums.

The official attitude as expressed in the Poor Law of the day was more unsympathetic than Dickens, and as far from any vision of the solution of the problem. This is how Margaret Cole explains it in her life of *Beatrice Webb:*

"The Commissioners of that date, the Whigs, the philosophical and business Radicals who had come to power as the result of the Reform Act, knew exactly what they wanted. They wanted to end the practice now known as the Speenhamland system, under which the Poor Law was, in effect, used to provide subsidies for low wages, and, more important, they wanted to force men, under the threat of heavy penalties, to work in the hungry factories of the North. As justification for their course of action, they had three main beliefs, all of which commanded considerable assent among the "advanced thinkers" of the time. The first, deriving from Puritan and manufacturing sources, was that, save for exceptional acts of God, a man's poverty and destitution were his own fault. Riches, as Nassau Senior and his like taught, were the reward of hard work and abstinence; conversely, poverty must be the result of idleness and indulgence. Anyone, therefore, who attempted to mitigate poverty by any other means than by driving the poor to hard work and abstinence was doing them a great moral disservice, as well as undermining the foundation of the community.

"Secondly, they believed in the 'iron law of wages', and in the existence of a Fixed Wages Fund of limited amount, which belief involved that any money paid out in poor-relief in the rural areas came inevitably out of the pockets of those in work—the more money doled out to paupers, the lower the wages would be for the decent, respectable labourer. And, thirdly, taught by the Reverend Mr. Malthus, they believed in the bogy of population continually tending to outrun the means of subsistence, and that any relief given to destitute persons, by encouraging them to breed, would produce more and more hungry mouths to eat into the livelihood of their sober and industrious neighbours.

"The logical conclusion of the Commissioners' beliefs, of course, would have been a policy of simple starvation, of letting those persons who could not, or would not, find work, die out along with their wives and families. But people are not, in general, quite so cruel as their logical beliefs, and even in 1834 it was too much for public opinion, as well as being contrary to the laws of England, simply to starve the workless. The most that could be done, therefore, was to grant relief, but on such conditions that the hardest and worst-paid job would be a comparatively pleasurable alternative—this was the famous 'principle of less eligibility'. These conclusions the Commissioners forced rapidly upon a not unwilling Parliament and summarized them in the prohibition of 'all relief whatever to able-bodied persons and their families other than in well-conducted workhouses', in which old married couples were separated and children torn from their parents.

Without any general plan having been devised for the cure of these conditions, small improvements were thought of. Hammond puts the turn of the century with the passing of the Ten Hours Bill, a measure which ensured to the worker at least the possibility of a human life. But this Bill was met with the bitterest opposition because, implicit in it, was the destruction of the whole system of unregulated capitalism that had created the condition it sought to alleviate. If a man's right to do as he pleased with his own, to make what contracts he could, and to seek his own profit, receiving from the Government merely protection, were challenged, what were things coming to. On 14th May, 1846, *The Times* reported Mr. Labouchere's speech on the second reading of the Bill.

"Mr. Labouchere maintained that the proposal was neither more nor less than one to diminish by one-sixth the productive power of the great manufacturers of the country. The adoption of such a measure was fraught with danger to the permanent interests of the working classes. If the Law adopted the principle of interfering with labour in our factories, it must adopt it generally in all labour throughout the country, and must turn one-third of the people into commissioners to watch the hours passed in labour by the other two-thirds. A reduction in wages must follow the passing of the Bill, and he was convinced that if the working people could be convinced of the certainty of that result, they would to one man protest against it.

"Mr. Trelawny contended that it was as necessary to resist the demands for this concession as it was when they demanded other foolish and absurd panaceas for their sufferings. They might as well adopt a *minimum* price for labour as a *maximum* for labour itself. If they limited the hours of labour in manufactories, why should they not limit them in the rural districts during the harvest season, and in London when the attractions of the fashionable world pressed so heavily on the rest and comforts of 15,000 milliners."[1]

Emotion won the day with the chimney-sweeps, and it was dis covered that chimneys could be reconstructed and yield to other treatment. Emotion carried the Ten Hours Bill, and it was found that

[1] *The Times*, 14th May, 1846, on the second reading of the Ten Hours Bill; proposed by John Fielden, Oldham cotton trader.

profits were as high as ever. The stage of intellectual demonstration had been reached, and as years went by and more and more controls were enforced, the eighteenth-century claim to complete independence wore thin. The belief that certain conditions were necessary for profits was proved false, and a path of reform opened which has been followed through the years.

But this is hardly enough. There were during the nineteenth century other intellectual contributions. The chief was the Marxian system of economics, which, as we have said, did not seriously disturb the placid content of British thought. More important were the attempts to decrease destitution. Beatrice Webb made it clear, in the teeth of officialdom, that it was the causes of destitution that must be dealt with: sickness, old age, unemployment, the loss of the breadwinner. The social legislation of Lloyd George, the present schemes of social security, are recognitions of the truth of this attitude.

These intellectual demonstrations, this provision of ideas, are very important, but they are not fully effective until they are translated into a *right*—claimed and admitted.

In a book, first published in 1894, *Tales of Mean Streets*, Arthur Morrison gives a very clear picture of the acquiescence in misery that is as seemingly inevitable as the repression of a Chinese girl. The sketch of *Lizerunt* depicts women submitting to the maltreatment of a drunken brute with never a thought of resistance or revolt. That was how their world was, and they made no effort to change it. They do not even demand that it *should* be changed. The Labour leader who holds street-corner meetings is represented as the most dishonest ruffian of the lot, and is denied respect or sympathy. That may be the author's prejudice. For, deeply as he feels the degradation of the people, he has not imagined any real possibility of improvement. He sees no hole in the wall of ignorance or the pressure of economic factors quite beyond their understanding or control.

For a right to be successfully claimed there must first be the idea itself—the imagination of a different way of life. As we have said before, this can come only from outside, with teaching and knowledge. Secondly, when the idea has come there must be some belief in the power to achieve it. If the Ten Hours Bill represented the first stage in the victory of ethical principles, the beginning of the admission *from above* that the workers had *rights*, the great dock strike of 1889 was the first really successful claim *from below* to those rights. No longer would the propertied classes dole out what they thought might be safely granted—the poor would demand what they thought was their due. The series of strikes that came with the new century represented the growing power and demands of the workers. Their *rights*, once they claimed them, were granted in practice, bit by bit; the community as a whole had really granted them in theory already.

We are now seeing the end of this process. The individualistic capitalism of the early nineteenth century is dead. The belief that a man may do as he likes with his own is disposed of. We are nearing the end of another stage of thought. The conditions of life and work in the last century produced, very naturally, the belief that the worker was *by nature* inferior to the employing class. This belief in inferiority was embodied in such phrases as the "lower orders", and was made real by the failure to provide any but the most scanty education. This century has seen the gradual destruction of this educational differentiation. Since 1902 it has been possible for the exceptionally able to rise to high positions: it is now going to be a great deal easier for the average to secure a level of culture that will make nonsense of the claim of superiority. Our peculiar system of education will for a long time prevent the complete destruction of the myth of upper and lower classes, but it will gradually weaken. This improvement in individual human beings, as well as regulation of work and schemes of social security, will practically solve the problem of the Industrial Revolution.

As we have said earlier, beliefs bring with them arrangements which enable them to be translated into actuality. The belief that the state as a whole must control the activities of some of the citizens in the interest of the majority could not be effective till the state had developed the apparatus of control. In the days of Cromwell, the Civil Service seemed to consist of John Milton, the two Frosts and a few clerks. Such an establishment was quite inadequate for any but the simplest purposes. The vast development of the Civil Service in this century has been due to the belief in the necessity of control; it has also made this control possible. Thus when we look back on the failures of the past we must realize that they had not as yet the means to do what we have done. It is, of course, the same in education. The belief in the need for education created the schools, but they could only grow gradually. Nothing is more difficult than a sudden expansion in education. The teachers cannot be trained till there are the schools, and the schools cannot exist till there are the teachers. In countries such as Russia or China, where the state decided that a great increase in literacy was necessary, enormous resources and ingenuity had to be used. In such a state as Nigeria, where there is a less vigorous policy and the difficulties are greater, it seems almost impossible to get the system started. The circularity of the problem keeps it rotating, and nothing happens. Thus the solution of a moral problem in a state is a very complicated matter, and cannot happen quickly.

The problem of the prevention of war has arisen, like the problem of the relief of destitution, as something new in a very old setting. There have always been wars; and from remote times there has been a standing army in stable, rich civilizations; yet it is only fairly recently that war has been felt to be wrong. It has been a disaster, a misery, something to pray against, but the idea that it was *wrong* did not occur. The prevalence of war has been due to an idea as well established as the idea of private property. It is probably the aristocratic military caste that has kept wars alive through the world. In many countries of Europe, the nobles originated as an invading tribe of different race from the peasants. These invaders have maintained their military organization, their tradition of fighting and their pride of skill through generations. The myth of the conqueror, of the warrior, has become part of the formative beliefs in man's history. When this is joined to a belief in the value of the ownership of territory, to a belief in the greater glory the more is owned, the mental drive behind most of the wars of Europe is clear.

In classical times the Spartans were the purest example of this aristocratic military caste and the maintenance of a certain way of life based spiritually on the myth and economically dependent on the labour of serfs. The best modern example is the Junker class of East Prussia. Invaders ruling a conquered territory with inhabitants of a different race, they maintained a specific code and cult. Their position as lords has given great prestige to their ideas, and their connection with the government has enabled them to impose their standards on others.

It has been the custom in recent times to find in economic factors the causes of all wars. The Roman capitalist, say the history books, preceded the armies, and the manufacturers of Lancashire determined the British domination of India. We are now told that Europe will never be at peace till economic causes of unrest are removed. It is important to distinguish between the militarism of the capitalist and the miserable discontent of the peasant. The latter almost certainly does not lead to wars nowadays, at the worst to a little frontier raiding. The starving Balkan peasant has not the energy or the means to do more than steal a few chickens; and the restless tribes of India's North-West Frontier can be kept comparatively quiet by being enlisted in the British Army, or given improved methods of agriculture.

The great capitalist has more influence, but he only works through the myth. He does not say, "I want a market for my steel". If he did, no one would take much notice of him. Instead, he appeals to the leaders of the military caste, if there is one, and says, "Now is your moment of glory", and the military leaders, having spent fifteen years making their preparations, cry, "The country's honour is at stake—join in the glorious fight for freedom". However clearly a war may be based on the desire for economic expansion, the appeal is made to the nation in the name of military glory and the great myth.

It was common in the days of William James and later to talk as if man was so innately pugnacious that wars could never be brought to an end: that men must fight as they must breathe. Among many primitive races war is an amusement; and a year or so of raids is part of the young man's education, as three years at Oxford is with us. Before the white man came to New Guinea, youths in their first manhood spent two years going off to raid and burn neighbouring villages. After that, they married and settled down. The white man has forbidden these wars, and has offered as alternative a year or two's service on shipboard or in Australia. All the youths go off as before, as before they return enriched in experience, with now, in addition, a few trade goods and some knowledge of pidgin English. They marry and settle down. There appears to be no hankering after the old customs. It is probable that healthy youth needs a break between childhood and maturity, and society should provide one. But to claim that this natural restlessness leads inevitably to war is to judge of man's nature wrongly.

When one reflects on what is required for a modern war, or how it is waged, it is clear that it bears no relation to any natural human activity. It is an utterly artificial thing and the only people capable of planning it are a body of experts working continuously over a very long period. Perhaps twenty years is necessary to mount a modern offensive war. It is only in the last few months that the emotions of the nations are involved and worked upon. It is the myth the myth of glory—that motivates this planning and makes the nation respond to the final appeal.

This myth was unchallenged till very recently. Perhaps the first war in which moral standards were brought to bear was the Boer War. At that time a large part of world opinion condemned the British action, as did a small but influential party at home. George Cadbury felt so strongly that he bought the *Daily News* to be the organ of protest. The ground on which this war was condemned seemed to be that it was not *right* for a big power to attack a small farming community, particularly when England's action was really taken in the interests of the gold and diamond miners. Those who opposed the war did not offer a solution of the problem caused by the completely intransigent attitude of Kruger, or explain if the Transvaal was to be permanently barred to the exploring capitalist because Kruger wanted a patriarchial state on Old Testament lines, administered with an austerity that prohibited hymns in church as frivolous. What Kruger wanted and what Rhodes wanted were incompatible, and the critics did not offer a method of composing their differences. Still, to have decided even that *this* war was wrong, was the first step in bringing the activity under ethical control.

We have reached the stage we mentioned earlier in this chapter of realizing that a problem exists. Wars are wrong—how are they to be stopped? We have not advanced very far towards the solution. It is clear that the first thing is the destruction of the myth. The next, quite secondary, thing is to provide some method of settling disputes. This is easy when once the will to war has ceased to operate.

Our first step in the destruction of the myth is the Nuremburg trial. This is the first occasion on which the imagination of an aggressive war has been declared wrong. When Napoleon set out to conquer Europe he was a tyrant, but he has remained in the history books as a hero rather than as a criminal.

The somewhat emotional article that appeared in *The Times* near the beginning of the trial, set out the motives of those who staged it. As time has gone on the defendants have shrunk and shrunk. Honour is gone, glory is gone, truth and dignity are gone. If only the whole myth could be dealt with in the same way and the whole nation convinced, good might result.

" In the prisoners' dock sat twenty broken men . . . merely as individuals their fate was of little importance to the world. But they represented sinister influences that would lurk in the world long after their passing. They were living symbols of racial hatreds, terrorisms and violence, and of the arrogance and cruelty of power; they were symbols of fierce nationalisms and militarism, of intrigues and war-making which had embroiled Europe in generation after generation, crushing its manhood, destroying its homes, and impoverishing its life. Civilization could afford no compromise with social forces which would regain renewed strength if it dealt ambiguously or indecisively with the men in whom those forces precariously survived.

"The crimes the judges sought to condemn and punish had been so calculated, so malignant, and so devastating that civilization could not tolerate their being ignored, because it could not survive their being repeated.

"The trial represented the practical effort of four of the most mighty nations, with the support of fourteen more, to utilize international law to meet the gravest menace of our times—aggressive war. The common sense of mankind demanded that law shall not stop with the punishment of petty crimes of little people, it must also reach men who possessed themselves of great power and made deliberate and concerted use of it to set in motion evils which left no home in the world untouched."[1]

When the will to war has disappeared and the myth has been dissipated the solution of differences becomes easy. The Scandinavian countries have for many years referred important matters of sovereignty to the Hague Court, and have found no impediment in national honour. They suffer no more than the modern litigant who has surrendered the glory and excitement of the duel for the dusty formality of the High Court. The U.S.A. and Canada foresee no occasion on which they are likely to be at war, and take no precautions against each other. More recently, England and Eire have decided that fighting is unbecoming in such near neighbours. War, when viewed at these close quarters, becomes infinitely sordid and depressing. The melancholy history of the Black and Tans, the indignities of sniping and reprisals, have suddenly been understood, and though there is a good deal of provocation there is a determination to endure it and hope that things may change for the better. If this change of mind can come quite suddenly, as it has come in the relations between England and Eire, the problem of the prevention of war resolves itself into the best means of producing the change. Some rely on fear and see in the terrors of the atom-bomb the best argument for peace. Others hope that the pain of defeat in Germany and Japan may bring about a re-arrangement of ideas. Pain does have this effect in certain cases, and this is the main psychological argument for its use as punishment; but, as we have said earlier, it needs a good myth to drive out a bad one, and it is hard to be sure how the myth of national glory and conquest can best be destroyed.

[1] *The Times* report of the Nuremberg trial, 22nd November, 1945.

Books referred to in this chapter:

Aristotle, *Ethics*.
William Hickey, *Memoirs*.
Casanova, *Memoirs*.
Henry Fielding, *Tom Jones*.
Nassau Senior, *Outline of the Science of Political Economy*.
J. L. Hammond, *The Growth of Common Enjoyment; Life of Lord Shaftesbury*.

Harriet Beecher Stowe, *Uncle Tom's Cabin.*
Charles Dickens, *The Uncommercial Traveller.*
Margaret Cole, *Beatrice Webb.*
Karl Marx, *Capital.*
Arthur Morrison, *Tales of Mean Streets.*
William James, *Moral Equivalents of War.*
Margaret Mead, *Growing up in New Guinea.*

Chapter 7

ETHICAL FAILURE

We discussed in the last chapter the way in which a social problem is solved. The mass of misery and degradation which was the immediate result of the Industrial Revolution was at first ignored; then it aroused pity and a faint indignation. Then it became a wrong and an injustice. At last, as knowledge grew and as social arrangements were altered, a solution was invented and is now being put into practice. Thus, in about 150 years, a great alteration has taken place, in both the material conditions of life and the ideas that go with the conditions. When Sir William Beveridge laid down the following simple propositions no one seriously disputed them. Had they been uttered one hundred years earlier a *Times*' leader would surely have found him insane.

"When the fighting in the world ends there are three things which each of us will need. . . . The first is a lasting peace, the second thing is that each man and woman, so long as he or she is able to work and serve and earn, shall have an opportunity of doing so. The third thing we need is that each man and woman shall be assured of an income sufficient for honourable subsistence and maintenance of any dependants when for any reason he or she is unable to work."

This statement is the final expression of the abandonment of the principle of individualism and the acceptance of social responsibility.

We can say that this whole process and the final solution is ethical because it conforms to our principle that the general happiness of the group is an important aim of moral endeavour. Those who contributed to the solution, either by arousing public feeling or by increasing understanding of the economic factors involved or by devising new social and political arrangements, had an outlook that saw beyond themselves to the group of which they were part. In addition to this, the solution they offered was a rational one. It was a deduction from principles that were already in existence, and it could be integrated in the social complex. It could be made to work as part of the social order, and it would, in fact, produce the results that were claimed for it.

This process of benevolent and intelligent change is of the utmost importance in any society. Without it there can be no progress, the nation becomes stagnant, and in the end the internal tensions cause it to break out into revolution. On the other hand, all change is not of this nature. There are movements which are bad in essence. They represent not benevolent thought directed to the public good, but hostile and destructive ideas, the product of unrestrained egoism or of a hatred of one's fellow men. The best recent example is the Nazi party. Originally an alliance of various elements each seeking its own gain, led by men like Goering, whose own craving for power and riches dominated all else, it came, as the end drew near, to see in universal destruction the only end it could achieve, and it welcomed this end with insane joy. "Together with the monuments of culture", cries Radio Werewolf, "there crumble also the last obstacles to the fulfilment of our revolutionary task. Now that everything is in ruins, we are forced to rebuild Europe. In the past, private possessions tied us to a bourgeois restraint. Now the bombs, instead of killing all Europeans, have only smashed the prison wall which held them captive.... In trying to destroy Europe's future, the enemy has only succeeded in smashing its past; and with that everything old and outworn has gone". Hitler, shouting for destruction, crying that "losses can never be too heavy", represents this same destructive aim.

Besides the groups which desire changes of a definitely anti-social kind, there are others whose solution of the problems before them is not wrong but merely unrealistic. Their thought moves in a world of wish-fulfilment divorced from the actual facts of economic and social life. The cause of this may be simple ignorance. The enthusiasts for a scheme may not know enough about such difficult subjects as economics or international finance to be able to invent or criticize a workable scheme. In other cases—and many of these are religious—the believer deliberately turns his back on the unsatisfactory world about him and finds a solution of his difficulties in a heaven of his own devising. Such men are employing, in the social field, the same mechanism of dissociation that operates in individual delusions. Frequently such fantasies make the participants happy, but, because they cannot be integrated with the ordinary course of life, they have a very limited use as a solution of the problems of society. It is behaviour of these two types, the unethical and the non-rational, that we want to discuss in this chapter.

Karl Mannheim has suggested that the great danger of our age is the liability of the industrial population to attacks of violent unreason. The industrial worker, by his circumstances and training, is forced to live a life of the utmost regularity. He is for most of his time under formal control, either by the foreman or the machine. Yet from time to time he breaks out into violent action of an unreasonable or unethical kind. The question for the social moralist is: under what circumstances do these outbreaks occur, and what determines the form they take. For there is all the difference possible between a lynching mob hanging a negro, and a revivalist meeting singing Hallelujah hymns.

It has been said, by those who wish all thought to be rational, that it is essentially the myth that sets free these unreasoning forces in society. There is no doubt, as we have said earlier, that the myth corresponds to a type of thought far more primitive than that used in rational thinking. The myth gives an embodiment to emotions, and, by embodying them, liberates them into action. But, as we have also said, not all myths are bad. The irrational need not lead to evil, though the rationalist tends to think it does. Rather, what matters is the nature of the myth, and this in turn largely depends, as we have already said, on the conditions under which it is formed. The myth that is the product of hate and frustration is the dangerous one, and among an industrial proletariat frustration and hate are the most probable emotions. Yet even among them irrational emotions are by no means always bad. The miners who listened in ecstasy to Wesley preaching salvation could hardly have been worse treated by society. They were liable to riot, to fight or to commit other disorderly acts, but it was the story of God's goodness that really moved them. At all periods of history the good myth has been potent as well as the bad, and men have sought bliss in Heaven as well as hell on earth.

Another attempt to describe the conditions in which unreason becomes dominant is given by an American, Hadley Cantril, in his book *The Psychology of Social Movements*. He finds that these disturbances arise when a man finds himself in a "critical situation". A critical situation exists "when an individual is confronted with a chaotic external situation which he cannot interpret, and which he wants to interpret". He is then highly suggestible, ready to follow any leader. His habits of thought and action, his formed ways of life, are broken through, and his actions are controlled by impulse, or those motives which he usually keeps in check. This concept of the critical situation is certainly valuable, but it fails to explain why so few outbreaks occur, for many men find themselves in such situations and do very little about it. Man is held for most of his time in the grip of an inertia which prevents him from doing anything very different from what he has done before.

Were not this inertia the strongest of forces society would be always flying apart, like a glass bowl, the moment strains and stresses were set up in any part. On the contrary, society holds together, in more or less the same pattern, through all sorts of circumstances. Wars, famines, economic upheavals, pass and leave the cohesion of society little undermined. We may grow more friendly in the black-out, but we are Englishmen still.

There is a tendency, a natural biological tendency, for a person or a group of persons to find a stable, and if possible, fairly satisfactory pattern of life and to keep to it. That is the essential element on which the permanence of society depends. The pattern of life may not be a particularly good one, but, once it is formed, it is re-enacted day by day, year by year, according to the length of the repeat: and it continues till something from outside occurs to disturb it. This formation of a pattern has great psychological as well as social importance. It conserves energy, and for most people the amount of available energy is only sufficient for the purposes of life. When there is surplus energy one gets the restlessness of youth, or, in certain cases, the unexpected outbreaks of middle age. It is, perhaps, only people of the creative, artist or author class who have an excess of energy; and in their case the mental creativeness is generally made possible by a life in which other activities are very unexacting.

From the social point of view this patterning is equally important. Without it co-operation would be almost impossible. Men must go to work at the same time, shops must open. We may laugh at the man who is so regular that you can set your watch by him, but he is only the extreme example of a necessary virtue. In smaller matters regularity gives the same comfort. If Mrs. A is always in the Cadena at eleven o'clock, her friends know where to find her; and if Mr. X ***never*** drinks whisky there is no need to order in a bottle when he is coming to play bridge.

It is only by appreciating this static element in life that we can understand the tolerance that men show of their conditions. When a very reasonable agitation would result in a marked improvement, e.g. in the conditions of school-buildings, teachers, parents, children—all those, in fact, who suffer from official inertia—continue to make the best of things, and do nothing to bring about reform. So, too, workmen will risk mutilation by using machinery that breaks the law by having its moving parts unguarded, when a complaint would compel improvement; and the vast mass of the unemployed during the years of depression did practically nothing to force a really serious consideration of their position.

Professor T. H. Pear, in his paper on *The Study of Society*, mentions investigations that have been made into the effects of mass unemployment. If ever there were a "critical situation", this is one; but the behaviour of the victims shows how much more than a problem is needed before action takes place. A village of Austria, Marienthal, containing about 1,486 inhabitants, was the subject of the study. Every one had been out of work for three years. It was found that in spite of the universal misery very little was done by the sufferers. They had lost interest in politics and all the bigger social questions. All conflicts were played out on the lower plane of individual quarrelsomeness. There were sporadic outbursts of unreason, when the emotions, dammed up by the conditions of life, demanded a symbolic expression. Then allotments were planted entirely with flowers, and luxuries bought instead of necessities. But, in the main, lethargy set in. Hardly any one carried a watch and there was universal unpunctuality even for meals. As was to be expected, low incomes brought with them a great deterioration of health and resistance to disease. The paralysing effect of unemployment was measured by classifying all families, according to a number of observational criteria, into those who were still unbroken and resistant to social degeneration, even if in despair; those who were resigned; and those who were broken, apathetic, or no longer looked after their children or kept up appearances. The figures were, unbroken 23 per cent., resigned 69 per cent., broken 8 per cent. It was clear that the last group would grow steadily as time went on.

There was apparently no tendency to revolutionary behaviour or the development of any new pattern of thought or behaviour; all that happened was that the old lost its grip.

The same lack of initiative appeared among unemployed in America. Investigators questioned unemployed people about their activities. As was to be expected, there was a great falling-off in all those that cost money. Fewer visits were paid to friends, or the cinema, fewer trips to the country or the sea; but there was hardly any compensatory increase in other cheaper occupations. More men just spent their time around home, but hardly any took to serious woodwork or gardening. There seemed no increase in education. The only new activities noted as a result of unemployment were sleeping during the day and standing at street corners and talking. These was no voluntary effort to find a new way of life.

Much the same appears in individual accounts by British unemployed published by the B.B.C.[1] It is an uncritical work, consisting of some sketches of their conditions given in the first person by unemployed men. Its purpose was to arouse sympathy rather than to study the conditions or suggest remedies. It was apparently successful in its

[1] H. L. Beales and Lambert, *Unemployed.*

aim. In many of these statements the writer hints darkly that if something is not done to alleviate his condition he will take to a life of crime. But whatever they may threaten, few starving men do. The man who, when hungry, actually burgles a baker's shop and then pleads in court the bad management of society, is very rare. The criminal is a totally different type, with his own pattern of existence.

A more enlightening American book[1] gives cases of unemployed families and traces the course of their difficulties and their adjustments. In each case a way of life was shattered, and in most of them, after much pain, a new way of life was built up; but it was done with as little disturbance as possible, and the chief desire of the family was to maintain the pattern of their own life, and to keep their place in the pattern of society. In most of the cases studied a profound change occurred in the family. The father ceased to be the breadwinner and the mother or the grown-up children took his place. Sometimes the roles were completely reversed and the father became the housekeeper while the mother earned the living. On other occasions the support of the family fell on the older children. This meant that the family pattern had to be altered now that the father had lost his position of economic leadership.

In other cases there was a change of social grading. A musician earning fifty to sixty dollars a week with a wife who prided herself on her membership of the professional class lost his work and could only get employment as a labourer at twenty to twenty-five dollars a week. His wife had to give up her servant and learn to do housework and cook. She had to readjust her ideas of dignity.

In most cases, after a period of strain and unhappiness, some adjustment was reached. The musician was proud of his success as a labourer, and his wife of her cooking. They had learnt new standards of value, had a new place in the community, had retained their self-respect and developed a new pattern of life not too different from their old. In other cases wives and husbands stood together and found a way to happiness. In some cases the children found the support of their economically useless father irksome and drove him out, splitting up the family. Then a much more radical change of pattern took place before a new stability was achieved.

The important thing to note in all these cases is that the family and its members were trying to keep the pattern of life they knew, or devise another as much like it as possible. They were not anxious for revolution. They did not wish, while in the grip of an incomprehensible

[1] Bakke, *Citizens without Work.*

force, to attempt to break out into any great change. The old standards, the old morality, had them in their grip.

It is clear that it needs a great deal of uncertainty and chaos to cause the average man to desert the pattern of life that he has established. He does not rush to change. He resists it, and in the very great number of cases will even perish rather than vary from what he knows. His experience is limited, and his power of acquiring new ideas, even when they are offered to him, is small. As a starving peasant of a rice-eating country will refuse wheat, so countless people, in difficulties, will absolutely refuse to vary their way of life in the slightest. They would rather make themselves ill by overwork than lower their standard of living, or of housewifely ostentation.

It is in these circumstances that the harmless forms of non-rational behaviour arise. Many of them, as we have said, are religious, for in religion men can find escape and help without disrupting the material pattern of society that exists around them. Moreover, religion is, for most of us, an activity which is included in the normal organization of our lives. In adopting a different form of it, or in intensifying our attendance at its accustomed services, we do nothing that requires explanation or defence. We are still safe in our own world. The Oxford Group, which is typical of a harmless non-rational movement, fulfils these characteristics well for the middle classes. Its members, judging by observation rather than statistics, have all had at least a secondary education, and most of them are comparatively well-to-do. They fall mainly into two groups (excluding, of course, the professionals): the young who are in search of excitement, and the middle-aged who require help in their adjustment to reality. For the young there are the rather boisterous group activities, the meetings that are on the level of a fifth form at school or the first year at an American university, and there are also the intimate talks about their sins, often sexual ones, and the emotional crises of public confession. The older people have got themselves into some difficulty. They are drug-addicts, their source of income has failed, they have quarrelled with their families. They are intelligent enough to feel that they cannot resolve the problem alone, so they seek aid, partly divine, partly human. Very many of them find it, and develop a new adjustment to life—though frequently the change is only very slight.

The other characteristics of the Group accord well with the previous pattern of life of its members. The Group does no social work, being in this quite unlike the Salvation Army. It holds its meetings in the comfortable seclusion of a college or a good hotel. Its financial arrangements have, for its ordinary members, a somewhat adolescent irresponsibility that is only possible for those who have never had to take money very seriously. Here, without surrendering any of the types of behaviour that characterize a social class, the members gain an assurance, an excitement, a new outlook that is of help to many.

A very different type of escape, and one, again, in accordance with myth and tradition, has grown up among the negroes of Harlem. These men from overcrowded tenements, economically exploited, socially inferior, flock to Father Divine's Heaven. Here they can, in the religious manner traditional to American negroes, find safety, comfort, equality, music, love and care. Heaven has its unpleasant side, but only for those who think, and so resist its spell, or for those who are worth exploiting financially. For the lowest and the poorest there is comfort and shelter, and a community so closely shut off from the outside world that the incompatibility of the two does not trouble the members.

The financial crisis and unemployment in California before the second German war produced the Townsend Plan by which every person over sixty should have a pension of about 200 dollars a month, the recipients to promise to spend all of each allotment within the thirty days. This simple scheme fitted in perfectly with the American tendency to base their lives on money values. These pensions gave the elderly security and dignity, and by increasing the spending-money available, and by withdrawing the older man from employment, held out hopes of jobs to the young. It offered people exactly what they wanted, in a way that was exactly in accordance with their preconceived ideas. It is in complete contrast to the religious escape of the Harlem negroes, just as the negro myth differs from that of the white American.

These groups and activities are harmless. The dispirited small-town American sending his contribution to forward the success of the Plan, the poor negro singing:

> Father, I thank thee, Father, I thank thee,
> Father, I thank thee, for what you've done for me.

are misguided, perhaps, but they are not doing any one any harm. On the contrary, they are quite ready to share their felicity, and to help others as well as themselves. The mob of poor whites, exasperated by low cotton prices and the competition of cheaper and more efficient negro labour, embarking on a lynching, are more socially and ethically objectionable. It is important to consider the reasons for the different type of behaviour, since materially the circumstances are very similar. In both cases there is economic difficulty—a chaotic situation which threatens their dearest interests and which they cannot understand or control.

The existence of the lynching mob, as opposed to the revivalist meeting, seems to depend on psychological factors.

There appear to be two elements which determine the type of activity adopted: the myth or pattern of behaviour current in the group, and the extent to which the minds of the actors are stored with hatred and violence. In addition, violence is more likely when restraint is weak. Cowards will be lawless when they think they can do it with impunity. The rioting which characterizes a certain stage of social development illustrates this. In certain communities rioting of a certain kind is traditional. In Ulster in the past Catholics and Orangemen had almost established a ritual. In India, Hindus and Mohammedans, or, failing any Hindus, the two sects of the followers of the Prophet, add variety to their lives in this manner. In none of these cases are the riots really religious. The religious basis is merely a convenient method of defining your opponent, and the flouting of religious customs is simply the way of announcing that the fight is ready to begin. So with lynchings and racial riots generally. The fact is that in a society that lacks harmless means of excitement, among people whose economic and social condition have made them thwarted and hostile, violence and destruction, carried out in the accustomed way, provide a necessary excitement and relief. As resentment of this kind is most likely to exist among the poor, the riotous elements in society are usually the poorest. But other factors may operate. The less intelligent members of the "upper classes", finding themselves outdistanced by cleverer and less-well-born competitors, may try to find in Fascism a way of silencing their opponents. In the same way, the unemployed intelligentsia will devise schemes of bloody revolution. Such a violent movement, wherever it may start, is greatly assisted by the hope of impunity. The actions of a mob are always harder to control because emotion grows in a crowd, and because no one feels fully responsible for his actions. When, in addition, there is a hope, or almost a promise, of legal impunity violence grows. The lynching of negroes could probably be quickly stopped in America if a few leaders of lynching mobs were executed, or even gaoled for a considerable period. Instead, juries regularly acquit, even in cases where there is no defence. Usually, no one bothers to bring the case into court at all. So in any country where there is felt to be a weakening of the central government, rioting will break out because the mob hopes to be undisturbed.

Before such rioting can turn into revolution, other conditions must exist. Chaos must have spread so far that the whole basis of life is felt to be shattered, not merely some personal part. The necessary condition precedent to any revolution is a general break-down of the central government. This usually comes about owing to the bankruptcy of the state. In England it was finance that brought about the outbreak of our comparatively decorous revolution. It was the king who was bankrupt, not the state. Pre-Revolution France was in much worse plight. In Russia by 1917 everything had gone to pieces; and so it had in the Germany of which the Nazis took control.

Nothing could illustrate better the conditions in which reason and morality break down than the Germany of the 1920s. Everything then was in a "critical" condition. The inflation and the economic crisis had brought ruin and utter bewilderment; politically, the country was bankrupt with a government that had, apparently, not the slightest wish to govern; the youth saw no outlet for their talents, and had been taught no stable moral code; the industrialists feared Communism from the East, and the army disarmament from the West. In addition, everything was made worse by contrast. While the *rentier* starved, the dealer in currencies reached fantastic heights of wealth; while the nation smarted from defeat, it looked back on real or fabled glories. No one understood why these things had happened, and all felt that they must end. The nation was completely suggestible, ready to follow any prophet, ready to discharge its accumulated hate in any direction, ready to accept any belief that accorded with the myth that had been established. In addition, when the first Nazi gangs began beating up Jews or other possible opponents of the party, there was no police action. The lawless movement grew. It was exciting to engage in this group activity, and it was safe.

When a nation is in this condition it is not merely an act of chance or of God what leader arises. One has often speculated how it would have been with Germany if Hitler had been a good man, and if the dream he held out to the nation with all the force of his eloquence had been one of peace and love. The answer probably is that Germany would not have responded. At any one time there are many voices calling, and those that are heard fit in with the mood of the nation. In our own recent experience the change from Chamberlain to Churchill was a change in the national consciousness. At a certain moment the chivalric myth of England was predominant and we accepted the voice promising "blood, toil, sweat and tears", and cast from us the out-dated commercial pacifism that we had believed in so long. Thus, had Hitler spoken to distracted Germany with the tongues of angels he would probably have been disregarded. Germany wanted new power and domination; she wanted to forget her own guilt in the imagined guilt of others, and she was eager for the excitement of blood that the myth held out to her as the natural satisfaction.

Once a man has become entangled in some group whose activities are largely anti-social it is very difficult for him to escape. And the worse the activities are, the more difficult. On the one hand his group must triumph or society will exact from him the penalty of his crimes, on the other the group itself will destroy him if he wavers. And his own soul is equally divided. The more he dislikes his activities the more guilt he will feel, the more fear, and the more he will seek safety by brutality.

If it is hard for an individual to reform, it is equally difficult for a nation. Once it has committed itself to a certain course of action, once it has planned its economy, it must continue its course. Fear and guilt drive it on, the inexorable economic pattern, and the myth that has been invented to explain or justify action. "You have an empire to lose, and a danger to face from those who hate you for your empire. To resign it now would be impossible—if at this crisis some timid and inactive spirits are hankering after Righteousness even at that price! For by this time your empire has become a despotism, a thing which it is considered unjust to acquire, but which can never safely be surrendered."

Books mentioned in this chapter:

Sir William Beveridge, *Full Employment in a Free Society.*
Trevor Roper, *The Last Days of Hitler.*
Karl Mannheim, *Man and Society.*
Ginsburg, *Reason and Unreason in Society.*
Hadley Cantril, *Psychology of Social Movements.*
Beales and Lambert, *Unemployed.*
Bakke, *Citizens without Work.*
Rose Macaulay, *Going Abroad.*
Frank Tannenbaum, *Darker Phases of the South.*
Chapel Hill, *Southern Regions of the United States.*

Chapter 8

LEADERSHIP

IN a previous chapter we raised, very briefly, the problem of how to produce in a state the best type of governing class. It is true that then we were discussing the intelligentsia and they seldom if ever actually govern. They supply the ideas which others put into practice. If we ask how to ensure that the best ideas are made available, and that the intelligentsia perform the functions of a moral *élite*, the answer probably is that we do not know. The group from which the intelligentsia takes its rise, the educated middle and professional classes, can be to some extent controlled. It can be educated to an advanced stage, given a sense of importance and obligation and helped to extend its experience. But within that group the children who will form the intelligentsia are the hardest to control. They are the most original, the least suggestible, the most wayward. Often they do not show at school anything like the full measure of their ability. Their place in society is as critics and originators, and in their youth they are very apt to antagonize those teachers who expect to be treated with respectful belief. All that society can do is to offer them freedom of thought, freedom from persecution when they express unpopular views, and trust that by the time they reach maturity and power they will have acquired acceptable moral standards and a sense of public duty.

But the intelligentsia will have influence only so far as their ideas are taken up and put into practice by others. The actual rulers are a governing class, elected representatives, or a bureaucracy. It is these men who consider the ideas of others, translate these ideas into administrative arrangements, and provide the practical realities of government.

These governors seem to require several characteristics: they must be clever, able to understand problems and able to think; they must have knowledge and experience so that their thinking is efficient; they must have high moral characters and be eager for the public good, and they must wish to rule. It is this last characteristic which distinguishes the man born to office from his equally clever and virtuous brother who adopts some other way of life. If to be a successful governor requires all this, it is easy to see why the hereditary principle is successful only when applied within very narrow limits.

The desire for leadership is one of the characteristics that are innate in certain people. A group, particularly the type of group formed by man, requires a leader, but not as many leaders as followers. Nature seems to have realized this, and the impulse to lead is born into only a certain proportion of children. It would be a very interesting experiment to assess in each class of a school the proportion of children who possess this characteristic in a marked degree.[1] One might then develop a fanciful argument on the size of the "natural" human group. It would certainly be found that both girls and boys possessed the quality, and it is clear that the dominance men claimed for so many centuries forced women to satisfy their impulse to lead in a variety of less desirable ways. It would also be found that the impulse and power to lead is not confined to any one social class. It exists in all sections of society. Powers of leadership develop in some very young children. A certain number of three-year-olds show this characteristic in a very marked fashion. Other children do not manifest it till much later.

As with other innate characteristics the power to lead is greatly influenced by training. Suitable conditions will encourage its development, and thus has arisen the belief that it is a quality especially possessed by the public-school boy. It can also, particularly if not very strong, be so discouraged by bad conditions or the lack of any adequate field for its display that it almost dies away. Moreover it depends largely on the treatment a child receives whether he becomes a good or a bad leader. We have said very briefly in the chapters on "Frustration" and "Socialization" how a child is helped to develop a healthy integrated personality in which leadership plays its part, and how he is taught the type of leadership characteristic of his society. If, on the other hand, the child is repressed and prevented from developing in accordance with his nature, he will become warped and hostile—not at all the kind of person to contribute the best leadership to his group. Thus, in considering the methods by which a community may be provided with good leaders, one of the first considerations is to discover the natural leaders early, and give them the sort of education that will help them to develop their gifts and incline them to make the best use of them for the community.

In some people the desire to lead is so strong that they are really unfit to hold any but a dominant position. Napoleon was such a man. Unless he was in command he was useless or dangerous. Others only rise to the height of their powers when they stand supreme. They are but indifferent colleagues, and frequently bad or sulky subordinates; but when at last power has come to them they amaze by their ability

[1] It seems to be about 1 in 25.

and control. On the other hand, many men have no desire to command. They may not, of course, possess the intelligence. A system of society that maintains a ruling class forces many unsuitable people into positions of power. Nicholas II, the last of the Czars, would have been quite well placed as a pious country gentleman; he had neither the ability to understand nor the wish to control the course of events. Charles I of England was little better; his obstinacy was either provided by his ministers, or sprang from an education that had taught him he ought to control events but had not given him the power to do so properly. Even when there is great intellectual ability there may be no wish to dominate, and a man otherwise suited for control may prove unfitted in an emergency. There is a story told of a certain Indian Civil Servant who, on the outbreak of a dangerous riot, sat down calmly in his office to write a full report on the matter, leaving the military outside to fire a volley too many into the crowd. The civilian was not frightened, his prose lost none of its accustomed lucidity, but he had no impulse to be at the head of affairs, taking his part in riding the whirlwind of events. His absence from the scene of action produced a major political crisis as well as a considerable loss of life. It is therefore necessary, if the state is to be well conducted, that those suited to rule should hold the power. To a large extent this happens of itself if there are suitable social arrangements. The natural leader is recognized as such by his fellows, and promoted to positions of authority by popular voice. But this is rendered difficult or even dangerous where a system of political or social or cultural inequality exists. Under those circumstances leadership is forced on a few whether they are really suitable or not, and large sections of the population, which probably contain their full proportion of potential leaders, are deprived of the education and opportunities that would allow their members to take a prominent place in the state.

> Some mute inglorious Milton here may rest,
> Some Cromwell, guiltless of his country's blood.

If the exclusion from political life is only partial, then the uneducated demagogue arises who, spurred on by a sense of his own deprivation, strives to destroy what has been refused him. Where the exclusion from political life is more complete, the leader who must lead whatever befall becomes the revolutionary, and seeks an absolute destruction of the absolute barrier. At the same time, with high office the perquisite of a definite class, the danger to the state from the well-born incompetent becomes greater. There is no competition to spur on effort or eliminate inefficiency. A man has no reason to exert himself, and if he realizes that he cannot do what is asked of him, he simply ceases to try. The campaigns of the Duke of York when commander of the British Army show that realized lack of mental power brings carelessness rather than a will to greater exertion.

Thus a society in which opportunities are unequal suffers a double damage. Men who might have been highly efficient leaders never receive the opportunity of power, and others who are unfitted for it have power thrust upon them. Through the greater part of European history equality of culture and opportunity was unknown. Occasionally, in moments of great national stress, the customary barriers fell, and some peasant rose to be his country's saviour. Many a mercenary soldier of genius usurped the place of his employer and founded a dynasty; but on the whole the system held. Now, for the first time, reasonable political equality has been achieved in many countries, and cultural equality in some. There should thus be great gain: more leaders will be available, they will be better prepared for their task and less liable to be bitter against society; and the incompetent will not occupy high positions to the public danger. The leaders of the future must feel that the state is theirs, that they are its loved children and have a loyalty to it. They must be free of hatred and must understand the pattern of life for which their state exists. They must be able, from youth, to look forward to taking a place such as their desires and talents fit them for, and they must know that their abilities will not be denied scope. This is no small matter, and it presupposes an organization of society that has been achieved in comparatively few countries.

It has frequently been argued that the future rulers should be selected young and educated apart from the mass of the citizens. This was Plato's idea and it has frequently been urged as a reason for maintaining our public schools as class institutions, or for selecting lads for training as, e.g., naval officers at the age of twelve or thirteen and restricting the applicants to the sons of the well-to-do. There is no doubt that certain types of leaders require certain specialized knowledge and must also acquire sets of ideas somewhat different from those held by the bulk of the population, but it is being realized that this knowledge is learned more quickly and the ideas not more slowly if they are presented at a somewhat later age. In practically all cases, it is now considered that the dangers of segregation outweigh the advantages. In fact, the future leaders are only one section of a large group which includes all the professional classes, the higher business executives, managers of industry and, of course, the intelligentsia. This group will all have approximately the same standard of intelligence, and will need, at least to the age of eighteen, much the same type of education. Beyond eighteen the needs of the different sections will differ, the doctor requiring one discipline and the architect another; but for all there will be training till the age of twenty-three or later. It is only by that time that they will have acquired sufficient knowledge and firmness of character to fit them for their part in the state. It is of this group that the future ruler forms a natural part, and there is no reason for segregating the future Home Secretary from the future doctor, architect, army officer or technical expert. There is, rather, every reason for keeping them together as long as possible so that in later life each may understand the way of life of the others.

This group of children, which will, when its time comes, control most of the life of the state, supplying it with its thought-power, its skill, its moral and material leadership, is not very large. Only about 3 per cent. of children of seventeen are still at school, and in 1938 only about 13,000 children all told were receiving full-time education at the age of eighteen. It is thus clear that for the part it has to play, our supply is quite small. If we wish to develop more leaders or more technicians, more doctors or more officers, we must increase the size of this group.

In fact, however, this group has not in the past produced all our leaders. In England we have had a more rapid growth of political than of cultural equality. Thus it has been possible for a man to rise to high office in the state though it was impossible for him to go to a secondary school. Before 1902 the secondary school did not exist, and there was no education available to the poor child, whatever his abilities, after the age of thirteen. This condition of things produced the Labour politician who left school at eleven or thirteen and has fought his way to high office through the trades unions. To take a few examples more or less at random from a record of Members of Parliament. Ernest Bevin started work as a farm-boy at ten, was a page-boy and tram-driver in Bristol, and then became a trades union official. Mr. Thomas Brooks, M.B.E., who has not risen quite so high in his country's service, was educated at the elementary school and night classes and became a miner. Mr. William Dobbie, who sits for Rotherham, became a coach-painter at the L.N.E.R. works when he had left his elementary school. The list is full of such cases. These men must all possess outstanding ability, drive and energy. The comparatively liberal organization of society which, while giving them all the handicaps of poverty and lack of organized education, yet allows them to fight their own way, has made them useful citizens instead of rebels. They are the products of a transitional stage in society. In the future, there will be few or none of our leaders who have had such an upbringing. They will not escape an extended education, for it should be impossible for abilities such as theirs to go unnoticed. At eleven they will be carefully garnered up, placed with their peers and given an education which, through the secondary school and the university, will guide them to the fullest use of their powers. An extended education is a thing whose worth is hard to assess. A man has either had it, or not had it, and it is only a matter of conjecture what he would have been like if his fortune had been different. Yet those who have enjoyed it esteem it one of the blessings of life. A writer of the eighteenth century has given as good an account as any of the advantages we hope to secure for the child through such an education:[1]

"To educate is to form the judgment—that is to give the mind a taste for the good and the power to recognize it; to make the mind keen to recognize false reasoning when it is somewhat obscured; to teach it not to be dazzled by a show of vain talk with no sense in it, or satisfied with mere words or vague principles, and never to be content until one has penetrated to the very bottom of a thing. It is to make the mind quick to seize the point at issue in an involved matter and to recognize what is irrelevant. It is to fill the mind with those principles which help to find the truth of all matters, and especially of those where the truth is most needed."

There are other advantageous elements in an extended education besides these somewhat abstract ones. Mere knowledge plays a considerable part in our powers of judgment. The ability to endure long periods of intellectual work or to pursue a difficult train of thought comes with practice. The attention of the trained man is far steadier, and his sympathies, if he has been properly taught, wider. All this is of importance to the individual and the society of which he forms part. Moreover, an extended education gives the opportunity for the thorough teaching of the moral ideas and principles on which the value of a leader's work so largely depends. These qualities show already in the professional classes, and it is these classes that our politician or other leader will now join.

If one considers the type of the professional man, he has certain characteristics that mark him out from the mass of the community. He is cleverer and better trained than the majority. He does more difficult things. He has also a code of honour and service that distinguishes him. The professional man works hard. He has acquired his skill through many years of effort, and he would think it a wrong to himself and society if he did not use it to the best advantage. He must

[1] Nicole, *Essais de Morale*, 1714.

not refuse his skill to those who need it, and he must not be careless, for his own honour as much as for the sake of the client who has put his life and fortune in his hands. Further, this sense of duty extends to voluntary work in a professional capacity. The voluntary hospitals were largely staffed by doctors who give their services; solicitors are expected to take a certain number of poor persons' cases; teachers are expected to help their children in countless out-of-school activities, and university professors will lecture to any group that cares to invite them. This acceptance of voluntary public work is different from the public work done by other classes. A carpenter may canvass for his political party. He is not often asked to repair the chairs in the debating-hall free of charge. The standard of common honesty is also high in the professional group. One can nearly always obtain credit by giving one's university address; and it is an acknowledged fact that a dentist's bad debts are far less in Oxford than in a non-university town of a similar size. The reason, among others, is that the professional man has too many interests, too keen a sense of his membership of the group, to be greatly concerned about small gains or to wish to achieve them in a way that is socially undesirable.

A professional man's sense of honour is strengthened by being a member of a professional organization. These societies not only look after the interests of their members, they maintain a code of behaviour, the minimum standard that can be accepted if they are to perform their proper function. The disciplinary action of the B.M.A., for example, is real and much dreaded; but for the ordinary doctor who has learned the ideals of his profession it is something he is not likely to encounter. Just as the ordinary citizen avoids conflict with the law, and judges his actions by a standard far above the minimum prescribed, so the good doctor gives to his patients a care and consideration that could never be *demanded* of him.

This is the group that the future politician or leader would join. With it he would receive an extended education and in its company he would learn the standards of behaviour, the public concern and the width of view that such an education should give. History is so plentifully spattered with examples where these virtues did not exist that it is refreshing to contemplate a world in which at least some of the politicians will have all the qualities we could desire. However, for many, politics is still a "dirty game" to be played with relish by those who like it; and the type represented by Laval has not vanished from the assemblies of the world. The only way to destroy it is to make the ruler a better man by giving him the kind of education that in the past has produced a higher type of citizen.

What should happen after the period of formal education is a harder question. Perhaps our rulers, if they have discovered their vocation, should be made to experience as many different ways of life as possible. As every potential officer now enters the Army through the ranks, and as many industrial executives work their way through all the departments of a factory, so our politicians might, having tried the life of a collier or a teacher, experiment with local government or a trades union. As things are at present this is what happens to most Members of Parliament. Most have tried some other profession and come to politics as a development of their interest in life. It is hardly possible to give a complete analysis of the previous occupations of Members of Parliament, they have had experience in so many walks of life, but an analysis of the careers of the first twenty in an alphabetical list will give some idea of the diversity. Among this twenty are three who have been company directors, three clerks, one sailor, one newspaper printer, one boot-operative, two lawyers, one journalist, one auctioneer, two teachers or lecturers, one Civil Servant, two trades union officials, one engineer, and only one who seems to have made his life in politics. This diversity of experience is a very important thing in helping the House of Commons to understand the needs of the nation, and, if this width of knowledge can be retained when a higher standard of learning and culture is available for all, then there is a good prospect of an understanding as well as an efficient government.

It is hard to over-estimate the importance for the governor of a wide experience of life. The extent to which rulers know and share the life of the majority in the state is one of the marks of democracy. In the past the aristocratic tradition was entirely directed to isolating the future ruler from contact with the world about him. Court etiquette, education by tutors, the limitation of friends, the necessity of marrying within a narrow circle, the refusal of experiences that are the common lot, all these cut the prince off from his subjects. He could not rule with the understanding of shared experience.

The effect of this was almost entirely bad. The future ruler grew up ignorant of everything that really mattered; and was so shut in by a wall of prejudice, so convinced that the values he knew were in fact the only ones, that he could not look at facts in an intelligent way. Among the minor tragedies of history there is hardly one more pathetic than the fate of the Habsburg Maximilian who went off, at the suggestion of Napoleon III, to be King of Mexico. This well-meaning and harmless prince had received an education and upbringing that completely severed him from almost all his fellow men. He had been taught to think that the dignity and grandeur of his family were of supreme importance. He had learned nothing but court etiquette, and when he came to Mexico all questions of finance, organization or war were completely beyond him. He could, and did, write a treatise on court ceremonial, but the reasons why the Mexicans should revolt against him were completely incomprehensible. On his way across the Atlantic to his new realm he set down his meditations on his undertaking. Nothing could more clearly indicate the curious blindness with which he set out, or his high and unselfish ideals.

"In all the circumstances of my life I shall be only too happy to make every sacrifice, no matter how great, for Austria and the future of my house. Throughout the centuries every dynasty in Europe has adhered to the prudent practice of appointing cadet sons to conspicuous posts where they might further the interests of their mother country. This policy has been known to embrace diplomatic and political as well as commercial fields. . . . Owing to the accelerated pressure of modern conditions the power of our family has diminished, while the adaptable Coburgs take throne after throne. Our ancient house has in recent years forfeited two sovereignties. In view of these misfortunes I have become convinced that it is the sacred duty and desire of every Habsburg to wipe out this stain. I cannot but believe that a good impression will be made upon the world, and especially upon our weakened Austria, if the Mexican enterprise attains success."

Arriving in this frame of mind among a people to savage that they did not understand the importance of the resurgence of the house of Habsburg, it is little wonder if all Maximilian could accomplish was to face a firing-squad with undiminished dignity.

Political leaders, then, require ability, experience and the goodwill that comes from stable and contented minds that have absorbed the best traditions of their age. They are not the only leaders needed by the state. We are growing more and more conscious of the need of leaders in all departments of the state, and nowhere has this need been more acutely realized than in the Army. In consequence, the technique of training and selection has been the subject of much thought and care. In so far as the modern officer is superior to his predecessor, this has been successful.

It is comparatively easy for society to impress on its professional classes certain standards. The group is selected early, receives special training, and, when trained, the members respect themselves and their profession and are anxious to live up to the best in it. They are all volunteers. No one survives the long training for a profession unless he is not only willing but anxious to do so. The Army receives unselected men, often unwilling recruits. It must make them first soldiers, and then officers. It is very interesting from the social point of view to consider the aims and techniques of the Army; for if the Army can achieve so much, other groups within the state, if they knew what they wanted and would take the trouble, could accomplish as much.

We have spoken before, and shall speak again, of the importance of indirect methods in moral training. In the Army the larger part of the training is given by these indirect methods. The man is introduced to a certain way of life, he is taught certain relevant knowledge, and is introduced to certain ideals and attitudes. It is these that have the main effect upon him. Army discipline has largely ceased to be punitive. Punishment exists, and is kept in fairly close reserve, but for most of the time it is the more positive incentives that are relied on to produce results that could not, in any case, be achieved by punishment.

Before training can start with any hope of success, it is necessary to know what is wanted. The three chief qualities that the Army wants in its soldiers are discipline, skill at arms, and courage—and each of these is complex. Discipline is not merely obedience to orders; it is this, but it is also a way of life, a feeling of self-respect, a willingness to give of one's best; personal smartness, pride in a job, and intelligent co-operation. Skill at arms is not only understanding the very complex pieces of machinery that a soldier has to handle, it is also an intelligent understanding of methods of fighting, the lie of the ground, minor tactics. It is more than skill: it is readiness, initiative, dash, self-reliance. Courage is the physical courage of the man who does not know what fear is, and also the more laborious struggle of a man who is frightened, but for certain reasons tries to overcome his fear. Further, the man who has truly entered into his task will show firmness of purpose and steadiness under such conditions as captivity, when the courage called for is of a very different type from that needed on the battlefield.

Moral courage is a quality different from physical courage, and it becomes increasingly necessary the higher a man rises in command. It is the power to take responsibility, to assert oneself when reason says that self-assertion is necessary. The newly-appointed corporal who decides that he must report a man, the junior officer who feels he must correct his general on an important question of fact, are exercising in varying degrees the same quality as Eisenhower giving the order for the armada to sail for D-day. The leader, to be successful, needs this quality, and any system of training must be so arranged that each man is encouraged to show it.

The Army tries to produce this good soldier by a number of

means. It was imagined in the past that a soldier's life was parade-ground drill, continued till the man had lost all power of thought and initiative. Undoubtedly drill plays a part in Army training, and in some corps, notably the Guards, it forms a large part of the preliminary training. Yet even the recruit, suffering on the parade-ground, knows that it brings certain advantages—quickness, self-respect, achievement. The commandos, most individual of troops, adopted the Guards drill; the officer cadet at Sandhurst springs to attention at the voice of a Grenadier sergeant: it is clear that the system must have other than repressive functions.

But more than drill, the whole ordering of a soldier's life is intended to promote the wider discipline. Regular hours of work and inspection to see that the work is properly done; personal cleanliness; an orderly life with proper periods for work and recreation, and, above all, a sense that the work done is of benefit to the community. On the intellectual side, an attempt is made to increase a man's self-respect by making him feel a member of an important unit. This is done by the teaching of Empire and regimental history—every recruit learns at least the battle-honours and outstanding achievements of his regiment. Then there is the commemoration of heroes, and the development of *esprit de corps* by suitable competition with other units. All this should have a real effect on a man's outlook on life.

Skill at arms is largely taught directly. The mysteries of the Bren gun or the proper method of throwing a grenade can be learned in class. The other qualities of judgment and dash have to be learned by wider activities. Each man, as he develops, must be allowed to use his judgment, and take responsibility. He must be encouraged to desire the end and plan the means, and he must be in the highest state of physical fitness.

In this and the development of some kinds of physical courage, physical training plays an important part. It is possible to watch a rather timid lad develop self-confidence and a much greater degree of courage under a systematic course of physical training. The proof to himself that he can co-ordinate his muscles exactly, that he can time a jump, and that when faced with what appears a difficult task he will not fail, has a profound effect both mentally and morally. Games such as football have the same effect. A lad learns to take the initiative, to act quickly, to think rapidly and in co-ordination with others. The feeling of success that these qualities brings heightens his self-respect and self-confidence. Boxing, too, is valuable, for in addition to the other results it teaches the boxer to tolerate a moderate degree of pain. The excessive fear of pain is one of the elements of cowardice, and if the youth can learn that a certain amount of pain does him no real harm, then his confidence is again strengthened.

The same theory underlies the battle-school and the use of live ammunition in training. The crack of machine-gun bullets is very unnerving at first, but when they do not hit you a certain confidence develops. The rattle and bang of battle are apt to make one start and to inhibit thought till one learns, like a cavalry horse, to disregard them. It is thus possible to train men into at least a semblance of calm when they are first confronted with the reality.

Apart from this, the simpler kinds of physical courage can hardly be taught. It seems a part of a man's natural make-up. This physical courage is not always associated with the highest mental development. Probably the bravest fighting-men in the Army of the British Commonwealth are the Gurkhas, and, though they are unsurpassed in their own sphere, they have not the complicated mentality of some other troops. This is not to say that valour is an affair of stupidity, but it is frequently connected with a certain lack of imagination. As one woman said about her experiences in the blitz: "Of course, I never for a moment thought it would hit *me*." That is the happy state of many a hero. The imaginative man who lacks this natural insensibility to danger often displays an equal courage. But he does it at the cost of more effort, and for different reasons. It has been pointed out that the real antidote to fear is another equally strong emotion. Love will make men or animals careless of their own peril, and many a man will die rather than surrender his self-respect. So, too, a man convinced of the justice of his cause, or the supreme importance of holding his post, will hold out to the last. The defenders of Calais in 1940 knew that it largely depended on them whether the rest of the army got away at Dunkirk; and the Guards who charged at Cassino had their regimental pride to support their other resolution. The whole complex of Army training is devoted to producing this latter type of courage. In not all men is it successful. Some cannot stand long-continued strain. Among those escaping by boat from Marseilles after the fall of France and making the nerve-wracking journey along the coast to Gibraltar, a few went mad. But a very large proportion of men can be induced to relegate the impulse of self-preservation to a secondary place, and behave as if their own survival were comparatively unimportant.

If the general Army training is successful, if a man becomes orderly, skilful, self-respecting, resourceful, brave, then a number of other virtues follow. He is good-tempered, and has a sense of humour. He is just and reasonable, honest and friendly. He is, in fact, on an adult plane, what the well-brought-up child is, and he is free from the vices that follow frustration. This Army aim is not, of course, always achieved. There are bad soldiers as there are bad teachers, shopkeepers or navvies, but, considering the fact that the Army swallowed at one bite a large part of the nation, there have been comparatively few of them. The war-time Army, naturally, appeared to many men very frustrating. They were taken away from their wives and families, and from their civilian occupations or peace-time ambitions. Yet even these men learned something. The regular soldier, married on the strength, has a very different position and outlook.

We have discussed at some length the training of the soldier because it illustrates clearly what happens when one section of society has an idea of what it wants in the moral sphere, and sets out to get it. It is a favourite belief of some civilians that there is nothing to be said for the Army, that those soldiers who are not hopelessly stupid are bloodthirsty jingoes. This is strangely untrue of the Army to-day. With doubtless many failures, with probably many officers unequal to their task, the Army has set itself the largest task of moral training that any body has ever undertaken. If the Army can make conscription play the part in the national life that the idealists hope it will, then a great step will have been taken in developing certain characteristics in the nation.

Many people, however, find it difficult to share these idealistic hopes. They point out that the nation has accepted the discipline imposed on it by conscription in order to attain a single, clearly-envisaged common aim—victory in war; and the training can therefore provide no guarantee against the kind of reaction with which—to compare small things with great—visitors to Eights Week at Oxford are used to. After the last race the crews plunge into the river, swim across to their college barge, and, sitting on deck, enjoy their first cigarette after the long weeks of strict training during which smoking was denied them. Moreover the training fails to instil moral standards in conduct not directly connected with military efficiency, so that in time of war the young soldier is not adequately equipped to withstand the temptations which will face him in the situations in which he will find himself—as is indicated by the grave concern about the behaviour of our army of occupation in Germany—while in peace-time it introduces grave moral dangers of its own. The pacifist would of course go further still and claim that any training which habituates young men to contemplate taking part in the activities of modern warfare, with its wholesale destruction and slaughter, is in itself a moral wrong.

The army officer should first have learnt the qualities of the good soldier, but to fit him as a leader he needs additional qualities. Nowadays he rises from the ranks and has had at least a short period as a lance-corporal. He then learns the rudiments of the art of command. The youth of nineteen in charge of a hut of twenty-five men must study the individualities of those he has to control, and devise means of getting his orders obeyed. It requires thought and practice. It is the kind of training a school prefect gets, but it is more serious. A boy who has refused to play the school game, who has found school regulations silly and the whole set of ideas puerile, will be intensely serious and responsible when he is really part of a machine ordered for the most important tasks. For those who think in abstract terms the philosophy of leadership is presented clearly by the first experience of command.

For an officer, probably the most essential qualities beyond those needed by the good soldier are natural leadership, power of man-management and decision, and ability to teach. For all of these knowledge is essential. Unless an officer knows his job, and the men know that this is so, when the real test comes they will not readily follow him. Natural leadership, the desire to lead, we have discussed.

Man-management consists in studying the mental, moral and physical welfare of all one's subordinates, and doing everything practicable to promote their well-being. Good man-management raises the morale and efficiency of the troops; and, in addition, troops will move readily and follow a leader who they know has their welfare at heart. The commander in Africa in 1941 who got hold of a plane and flew all his troops' letters home and brought back a full load of "home mail" raised the morale of his men and his own prestige with them out of all proportion to the actual act concerned. Approachability, fairness and understanding are other qualities that go to make a good manager of men. Man-management is taught in the Army by direct example, by lectures and by practice.

Taking command of a situation can be learned, and every wise officer regularly practises himself and his subordinates in this. A confident manner, and, where necessary, a loud voice, when backed by knowledge, go far to making a competent leader. The power to make fairly rapid and at the same time sound decisions on the best information available, however meagre this may be, becomes more important the higher an officer rises in the service. It cannot be taught except by example and practice. Many men lose their power of decision with advancing age, and when this happens the officer must be retired immediately to avoid the direct results of his indecision and the bad example he sets to others.

The ability to teach is an essential part of an officer's equipment in peace or war—except for the senior ranks where most of what teaching they do is done by example. The officer is taught how to teach by courses of instruction, by example and by practice. Before one can teach, one must, naturally, know one's subject.

It is clear that a prime factor in all the four qualities—leadership, man-management, power of decision and teaching ability—is knowledge, knowledge of every type. From a knowledge of the regulation which authorizes the special issue of an extra pair of boots to the knowledge of the integral calculus which enables an officer to lecture with confidence on the theoretic range of a searchlight. In fact, we can only say with Gilbert:

> And so in matters animal and vegetable and mineral
> I am the perfect pattern of a modern major-general....

It was this discovery of the importance of knowledge that led to the revolution in training and discipline that has taken place in the Army.

Plato's definition of a general, quoted with approval by Lord Wavell, as a man who sees that his troops get their rations, sums up concisely many of the qualities of leadership that we have considered. To understand what the troops need and to be competent in supplying it needs great ability, knowledge and a sense of group duty. Wellington, in his first campaign in India, demonstrated that winning a war depends on adequate preparation quite as much or even more than on ability in the field. It was harder to collect bullocks and rice, required greater expenditure of brain-power and moral effort, than to outmanoeuvre the enemy commander. The greatest generals, such as Marlborough, Wellington or Montgomery, have always been distinguished for this very quality.

The question of how best to choose and train officers first became really acute during the 1914–18 war. Previously the officer had been part of the social system rather than a man chosen for special qualities. The days when Arthur Wellesley rose in six years to the rank of lieutenant-colonel and command of the 33rd by purchase and influence gave the Army many less able officers than he. Even until 1914 birth and the ability to pay large bills at Sandhurst or Woolwich were the chief qualifications necessary in the would-be officer. Even to-day a senior officer who really started his career in the ranks is spoken of as a "ranker", with mingled doubt as to his social qualities and respect for his undoubted abilities. But amid the expanding armies and the destruction of the first German war officers were vital. In England we continued to kill off our public-school boys as second-lieutenants.

The Americans, with more time and fewer social distinctions, were driven to thought.

When they came to analyse out the qualities needed by the prospective officer, the first that appeared vital was general intelligence. The American Army tests, which were the prototypes of many of our present intelligence tests, gave satisfactory results in that they greatly reduced the number of men who failed to complete their training successfully. They got rid of the stupid. These tests were mainly linguistic, and they have been followed by others that try to use non-linguistic material for assessing general intelligence. These tests are important, for intelligence is certainly one of the indispensable qualities of the leader. But with more experience it has been realized that more than intelligence is needed. A man may want to lead, be intelligent and yet fail in sympathy, in group feeling or the power to persuade others to follow. The technique that has been used recently in the W.O.S.B.s has taken into account other qualities in addition to intelligence. These qualities, moral and emotional, are very hard to assess in any abstract way. It is not much use asking a man questions or even interviewing him. The modern technique has various elements. Perhaps the chief method is to put the man in a variety of situations and observe his behaviour. These situations may call for organizing ability, agility, courage, a sense of the group. If he can devise a solution, get the others to follow him, show the forceful patience that brings an enterprise to success, then he is felt to have a fair number of the desired qualities.

The highest types of leadership carry with them something more. There is about the great man a romantic emotional quality that gives him something extra in power. A soldier will respect and follow an officer who has the technical virtues, but for the greatest effort there must be a dramatic appeal. Most of the great leaders have had it. Men like John Nicholson of Delhi have become the centres of legend. We have mentioned Henry V. It is perhaps worth quoting Alan Morehead's very vivid account of Montgomery's speeches before D-day, because Montgomery believed in this aspect of leadership. His speeches follow in modern form the great tradition of the leader.[1]

"I travelled with him one week. Each morning, we got out of the train about nine o'clock and drove to a village green, or a clearing in a forest, where a brigade of soldiers would be drawn up in a hollow square, with their senior officers standing out in front. Montgomery talked to each of the officers. Then, with a band playing, and the troops turned inwards so that they could see him closely, he walked

[1] Alan Morehead, *Eclipse*, p. 80.

between the ranks. He walked slowly, peering at the men face to face. Except for the distant band, there was complete silence, an atmosphere of theatrical tension. At the end of the inspection Montgomery would get on to a jeep in front of a loud-speaker and tell the soldiers to break ranks and gather round him.

"That was always an astonishing moment. Five thousand men in heavy boots would charge together towards the jeep like stampeding buffaloes. It caused a heavy rumble of the earth, and often the jeep would be nearly overwhelmed. And then Montgomery's speech would go like this: 'I wanted to come here to-day so that we could get to know one another, so that I could have a look at you and you could have a look at me—If you think that's worth doing. We have got to go off and do a job together very soon now, you and I, and we must have confidence in each other. And now that I have seen you I have complete confidence . . . complete confidence . . . absolutely complete confidence. And you must have confidence in me.'

"That was the beginning. For a hundred yards all round him row after row of young upturned faces, an atmosphere of adolescent innocence and simplicity. They sat on the grass, keeping utterly still lest they should lose a word. . . .

" 'We have been fighting the Germans a long time now,' Montgomery went on. 'A very long time . . . a good deal too long. I expect, like me, you are beginning to get a bit tired of it . . . beginning to feel it is time we finished the thing off . . . and we can do it. . . .'

"And so on, with a little joke now and then.

" 'Now I can't stay any longer. I expect some of you have come a long way to get here this morning and you want to get back (some of them had been travelling since 4 a.m.). I just want to say good-bye and good luck to each one of you.' "

"That was the speech, followed by three cheers for the general, and Montgomery made it four or five times a day for much more than a week.

"The speech spoken by Montgomery had magic. No mention of God or Divine Assistance. No mention of England. Not a single Eternal Verity. No hate, no revenge. But I doubt if the soldiers remembered the words. The whole performance succeeded because it was the expression of a wanted emotion. Without their consciously knowing it, the speech adopted an attitude which the soldiers wanted to have. At the end they felt they knew Montgomery, as he believed he knew them. They felt they were thinking on the same plane as he was, that they would indeed go into the assault together."

It is true that Shakespeare did not write Montgomery's lines for him, as we have said before, but with that difference the descent of this type of leadership from the days of chivalry is clear.

We have considered leadership in the Army at some length because it is one of the fields in which leadership has been most closely studied, and thus it illustrates the social problems and the ways in which they can be solved.

It is not possible to consider all types of leadership in the State. We have taken two, political and military, but there are many others: in industry, the Civil Service, and education. In each case the leader requires general intelligence, general culture, the moral stability and excellence that comes from a well integrated personality; and, in addition, special qualities and knowledge. One of the tasks of the coming years is to do what the Army is already doing and the Civil Service just starting to do—analyse the qualities required in the different types of leaders, and, having performed this analysis, to find the means of selecting and training the leaders we so badly need.

Books referred to in this chapter:

Thucydides, *Histories*, bk. III: Career of Cleon.
Plato, *Republic*.
Bertita Harding, *Phantom Crown*.
Carol Bunker, *Who's Who in Parliament* (1946).
Lord Moran, *The Anatomy of Courage*.
Lord Wavell, *Generals and Generalship*.
Alan Morehead, *Eclipse*.

Chapter 9

LAW AND RELIGION

Any established society requires methods of teaching its ethics, and sanctions to see that the code is obeyed. Education of the young, which will be discussed later, is the great means of perpetuating the ideas and customs of the state, but the adult also needs instruction and guidance. There are two main departments of state which are concerned with this: law and religion. Law is directly concerned with the state, with men's relations to each other, and their moral conduct. It has a number of separate functions, but, taken together, these functions have the comparatively simple task of regulating human relationships in fields that can become matters of public concern, e.g. law does not prescribe the love of husband and wife, unless the loss of love leads to brawling or a desire to be permanently separated. Religion on the other hand has a much more complex function. It is always looking beyond man to God, and direct's man's behaviour not only in relation to his fellow men, but also in its relation to some deity who takes a benevolent or hostile attitude towards him. Furthermore, religion looks to the heart of man and concerns itself with his secret thoughts. It claims a much closer scrutiny than law, and can condemn not an action but the flickering of an intention.

At various times in man's history, law and religion have been very closely connected. The same person, often the king, has been both priest and lawgiver. At other times the priests have become the real rulers. Yet even under those circumstances law and religion have different spheres. The two may inculcate roughly the same behaviour, they may certainly influence each other, but they are not the same, and in practice no one has any real doubt about what kind of actions are the province of one or the other. Moreover, the sanctions that they use to enforce their orders are different. If at one time the Church could call in the "secular arm" to enforce its commands, it has long lost the right, and perhaps even the wish to do so.

Sir Frederick Pollock begins his essay on *Lay Fallacies in the Law* with the statement: "The most fruitful of these fallacies, if indeed it be not the common root of them all, is the assumption that the law of the land purports to be a general guide for the conduct of life." Law is not such a guide because it covers only a small part of man's activities;

because in actual fact if we all insisted on our legal rights, life would be almost intolerable; and because, even in those fields in which law does speak, the infinite variety of circumstances makes any general pronouncement largely inapplicable. Yet, granted that law is not ethics, it still plays a large part in causing ethical behaviour in the community.

Law can be thought of as having two main sources. There are the customs of a group, which we regard as very important, because on their exact observation the safety of the group depends. The elaborate ceremonies of primitive people, the exact arrangements for marriage, the proper ways of hunting or dates for sowing corn are connected with religious beliefs and cannot be varied without danger. If anyone is so hardy as to disregard these customs, the group must be safeguarded against the punishment that might fall on them all. In this lies perhaps one of the most definite origins of law and punishment.

But within the group itself there is also a code of behaviour that does not claim divine sanction. These are the manners and modes of thought and behaviour that are common to the members, and, unanalysed and undefined, they form the cement of the group. In Greece, when the blind adherence to custom grew weak, when the gods became less terrible and less exacting, this code of behaviour grew more conscious and important as individuality developed.

Among primitive people such a code hardly needs to become fully conscious. The psychological forces of cohesion are so much stronger. Pitt-Rivers has pointed out that the characteristic of the morality of primitive people is the unification of the individual with the group.

"In all his activity the primitive savage merges his self-regarding sentiments in—or rather projects them on—to his social group, either his clan or his tribe. He feels himself an inseparable part of that group and identifies his passions with those of other members of the group. There is no sense of sacrificing himself for the good of the whole group. There is merely a spontaneous and unrationalized feeling of identity with the group. . . ."[1]

In classical Greece such a group feeling had quite disappeared. The individual and the group were as separate as they are to-day. There thus arose the problems that we still have of the relationship of the individual to society, especially when the interest of the individual and the group appear to conflict. Philosophic thought, considering the facts of communal life, was struck by the fact that, after all, certain modes of behaviour were necessary if states were to exist. This idea developed among the later Greek and Roman thinkers till it emerged as a principle of law. This they named "natural law",

[1] Pitt-Rivers, *Clash of Culture*, p. 215.

which, man being by nature a member of a community, "is the sum of the principles, founded on human nature, which determine the conduct befitting him in his rational and social quality". Natural law was, roughly, the principles of conduct found in all civilized societies, and, though societies differed in many matters which were really indifferent, there were certain fundamental principles that obtained everywhere. These principles guided the good man's actions, and, if the bad man failed to observe them, there were powers and penalties that could be used to coerce him. This idea of natural law is of course merely the formulation of ethical feeling, but it is ethical feeling that seems to arise normally in a society organized in a certain way. It is possible that in classical times natural law appeared rather different to the Greeks and the Persians, but the natural law of countries with the same basic civilization is largely similar, and the concept has persisted to our day and can be successfully appealed to in cases which no statutory law covers: for example, the club bore who, being expelled without notice or explanation, confounded the committee by claiming that their action was contrary to natural justice. He retained his membership, and died their most valued member.

These two elements—custom that is peculiar to the group and often felt to be necessary for its safety, and the general body of behaviour demanded by social living—are organized into the body of positive law in any state. The degree of organization and the field that it covers depend of course on the nature of the state. Trading communities early develop laws of contract and sale, agricultural peoples laws of land-tenure or water-rights. Western civilization has seen the development of more than one code of law, and these codes differ in various ways. They all, however, possess the underlying principles that derive from the law of nature.

Once a law has developed, or even when it is in process of development, it becomes one of the chief agents for controlling action. It does this in many ways. We are apt to think of law as mainly criminal and its function as essentially punitive; but, for the greater part, the law is regulative and only punitive in a secondary way. In some cases there is no punitive element at all, and the law merely prescribes the most convenient way of carrying out certain necessary actions, such as the transfer of property. Moreover, for most people, the value of the punitive part of the law is educative. It points out clearly, by its threat, that certain actions are not to be done, and it turns men's minds away from doing them. So successful is it that far the greater part of the population do not *wish* to steal or commit murder; and many carry their ideas of honesty and mercy far beyond what any law enjoins.

In essence, perhaps, criminal law grows from a conviction that members of the same group must not engage in private quarrels. If they do the group is split up and peaceful co-operative action becomes difficult or impossible. Thus, early societies reach out for agreed "penalties" for certain acts which will satisfy the aggrieved party and prevent him wreaking his own vengeance. What the penalty shall be depends on the power and prestige of the sufferer, and is as little as he might be expected to accept quietly. Sometimes it is the offender's all.

One example of this is the attempts made in different societies at different times to get a money payment accepted as compensation for a killing instead of the prosecution of a family feud. The most famous example of this is probably the trial scene on the shield of Achilles, but the Laxdael saga, some 2,000 years later, shows just the same process at work in Iceland. There, unfortunately, Gudrun, acting against the good sense of the men and the advice of the king, insisted on continuing the feud. The story shows the critical point when public order was struggling with conservatism and passion. At a later date other ideas intervened. A killing was not just an unhappy event that must be compensated for by an agreed payment to preserve public order; it became a moral crime, and various confused ideas of expiation and deterrence came in to make nations impose death on the slayer.

Other examples of the same kind are the fines or compensations decreed in the early laws of the British Isles for various types of damage. If a man killed the king's cat, who was warden of the king's granary, he must hold the cat up by the tip of the tail and pour grain round it till the cat was quite covered; and this, apparently, whether the killing was accidental or deliberate. Such a payment saved the offender from any further vengeance from the king, who, being powerful, could demand a higher price for clemency than an unimportant person.

The law of debt and theft was similar. The creditor, being the more powerful, could make his debtor a slave under Roman law: up till comparatively recent times in England he could imprison him, although the imprisoned man was unable to take any active steps to gain enough to meet his debt. The thief or the poacher again was in a weak position compared with the owner, and the law was correspondingly severe. A man who killed the king's deer died, as did the man who stole a sheep. If the law did not impose these severe penalties, the lord or the farmer might keep his private army and impose them himself. They remained a part of the law long after their cause was forgotten.

Thus the criminal law was part of the mechanism for ensuring peace within the group, and we have curious laws to-day which are concerned with the same purpose. We do not fear organized defensive violence, but we do still fear brawling, and the law of assault is comparatively strict. A bus-conductor who *looks* as if he might throw you off the bus has committed an assault though he does not touch you, and the passenger must be correspondingly discreet. So too in criminal libel (the libel which might prove a provocation to assault); the greater the truth of the statement, the greater the libel, because the greater the provocation. In other cases of libel truth can be pleaded as a defence.

Punishment of this type was intended to serve its modern purpose as a deterrent, rather than its earlier one of preventing worse consequences. There is no doubt that punishment has a very useful deterrent effect. There are people who have failed to develop the sense of the community and a feeling for its welfare, and only egoistic motives influence their conduct. Punishment in these cases can act in one of two ways. It can operate through fear, or by making the criminal activity unprofitable. The tough, the footpad, the bag-snatcher are often characters that can only be restrained through fear, and on more than one occasion the judges have completely stopped a certain type of violent crime by ordering flogging as a punishment. The other type is represented by the black marketeer, the smuggler, the dope-pedlar. If they are convinced that the chances of detection are high and that the fines will be big enough to take all profit from the business, they will employ their capital in some less lucrative but more secure direction. Further, as has been said above, the punishment lends point to the prohibition. As the schoolboy says, "We know that you really mean it", and the waverer comes down on the side of law with scarce a struggle in his conscience.

The weakness in this theory of deterrence is that there are large categories of offenders who cannot be deterred because their mental processes are not normal. There are the very stupid who cannot do the sum in moral arithmetic which compares present pleasures with future pain, and there are the neurotics, driven on by an internal compulsion that exterior threats cannot touch. It is because these classes are really so numerous that many magistrates believe that when a man has been found guilty his future treatment should not be a matter of law but referred to an advisory board and settled mainly by psychological considerations.

One of the weaknesses of punishment as a deterrent is that it is hard to assess the severity of the punishment required. For an intelligent man who really approves of the law a very small punishment may prove efficacious. The householder fined ten shillings for failing to draw his blackout curtains was more careful afterwards. In many cases a visit from a policeman and a subsequent note saying that "on this occasion" no further action would be taken, was equally effective. On the other hand a severe punishment may be necessary with the stupid or where the regulation is not approved of. When once there starts, as there well may, a competition between hardihood and punishment, hardihood will generally win till life itself be lost. This became clear in both the Army and Navy. For centuries flogging was the standard punishment, and, however severe it might be, men would always risk it to satisfy certain desires. The flogging was a challenge, and it was accepted. The more modern punishment of "C.B." which is tiresome, undignified, and in which a man may find himself scrubbing out the cookhouse at 10 p.m., is no challenge, holds no glory and is, on the whole, a more effective deterrent.

One modern development is interesting. In many cases where positive good conduct is required rather than the mere avoidance of specific errors, punishment is being supplanted by persuasion. The "courtesy cops" of pre-war Bank Holidays have their successors, and *The Times* gives this pleasant picture of gentler London:

"Scotland Yard is so keen to secure the goodwill of motorists that a new kind of "courtesy cop", who will take every possible step to steer the motorist clear of trouble, is appearing. Practically every street in the West End is used for parking purposes, but Scotland Yard, with the job of controlling traffic, has the important task of keeping in motion the flow of vehicles. Traffic experts at the Yard have agreed that if motorists show more consideration in parking their cars some of the congestion in the West End can be avoided and prosecutions against motorists will be kept at a minimum. The new-type policeman, specially selected for his "kerbside manner", will charm the motorist into considerate behaviour. Not to persecute, nor to prosecute, but to persuade, is the job of the 1946 traffic policeman. For the present, a small number of constables supervised by a traffic inspector constitute the force of "courtesy cops", but if the experiment is successful the force will be enlarged."

In a similar way the Japanese in the Pacific were anxious to produce a change in behaviour, but they lacked the power to enforce it. They saw that the islands could be made far richer and more productive if agriculture were improved. They established research stations and employed experts to advise. They were met by a determined conservatism.

"All the wealth of good things developed by the trial farms is available to the native. Some rather elaborate persuasion is employed at times to get him to make use of his privileges. Village delegations are brought to the farm, instructed and supplied with seeds. An expert goes back with them to see the seeds are used. Every school has a farm, equipped from the trial farm, and agriculture is the chief subject in the curriculum. Selected graduates are sent for a post-graduate year on the trial farm, free of cost, and go back to their villages with the necessary seeds and tools.

"But the chief professor of agriculture in the South Seas is the policeman. . . .

"The policeman at one village was a pleasant fellow with none of the officiousness that sometimes characterizes police in Japan. In fact his training and his duties were not at all those of the ordinary police. He was a graduate of an agricultural college, and his chief task was not to apprehend criminals but to teach agriculture. The mild-mannered Kanakas commit few offences. They do not need punishment as much as guidance. Therefore the South Sea policeman is trained in first aid, treatment of simple diseases, sanitation, the construction of better houses, road building, educational methods, Shinto principles of morality, and chiefly farming.

" 'A good man', acknowledged the king. 'But we do not need him. We know more than any stranger about cultivating the soil of these islands. And our gods will be angry if we follow new ways.'

"Polite and obstinate natives opposed to a tactful and firm policeman. But it is the policeman who always wins. Not by force, for he is only one against many if the scene be a remote island. He must forget his authority and go to work on his own garden. He must travel from one native's farm to another carrying seeds and tools, and he is always ready to bend his own back and get his own hands into the soil to demonstrate his teachings. Of couse, any young natives who have had agricultural training at the school trial farm are his able supporters. And as the irreconcilables have died and the youngsters have come on, the revolution has gathered speed".[1]

If punishment is diminished in severity, modern times have also decreased the occasions when it may be necessary. The Victorian idea of thinking what a man may want to do and then forbidding it, has been followed by an effort to make arrangements for it to be done without public harm. One very small and awkward fountain at a cross-roads in Birmingham bears the legend, "This water to be used for drinking only". The picture doubtless in the scribe's mind was of a

[1] Willard Price, *Japan's Islands of Mystery.*

tramp standing tip-toe to wash. How different the considerate city of Rouen which has set aside a stretch of the quays as a tramps' refuge with water laid on just where it might reasonably be needed, and steps down to the river for washing shirts. Again, a very successful Army experiment allowed each man to get a pint of beer in the dry canteen with his tea. The men wanted the beer, but taken in those surroundings there was no tendency to drunkenness. Looked at in these ways the punitive aspect of law continually decreases.

Regulative law probably dates from a time earlier than criminal. No group of people can co-operate without some regulations, and the group of elders or the chief, in council or by his unaided thought, must have made rules. Some of these rules must have been of the utmost importance, some trivial but requiring regulation. Of the first class must always have been water-rights in a dry land, periods of hunting, proper worship of the gods, the order of battle and the payment of dues. It is possible that these were enforced by penalties for breach of the regulations, but these penalties were of quite secondary importance; the important thing was the rule, and if it was in accordance with general sentiment it would be generally obeyed. Rules such as these represent a mean among egoisms. They are less than most men would like, and more than some would get if there were unrestricted competition. It is the situation of the nursery and the bag of sweets or the favourite toy; and just as children quickly learn to appreciate a solution, so men cherish regulations that give reasonable satisfaction to all.

In comparatively modern times a similar instance is the growth of the Law Merchant which has been built up by the combined efforts of traders and judges. This law codifies the practices of trade. It is in accordance with the best general feeling of merchants and if it enforces practices superior to those that the worst would like, it does not make demands in excess of what the better can give. A very similar development is the law relating to trustees. These laws again are largely educative. They teach the well-intentioned what is expected, and they guide them in their decisions in particular cases. At the same time they prevent the minds of the less good imagining practices to the hurt of their clients.

There are many matters, indifferent in themselves, that have to be regulated, such as the rule of the road, the colour of lights in buoys and the letters on the number-plates of cars. This too must have been a function of law from very early times and perhaps it was felt to be more important in the past, when ceremonies, having a religious sanction, might, if incorrectly performed, bring down the wrath of God as well as man.

Lastly, perhaps, is the section of the law that deals with the best way of doing things and has no punitive side. If we wish to sell a house, draw up a will or a marriage settlement, arrange a trust fund for our own children's education or get a divorce, we employ our solicitor to do it because he knows the best way to see that we get what we want. If we draw up our own will the only penalty we suffer is that our ghost uneasily attends a court of law and sees its provisions set aside as void for uncertainty. It must be very provoking, if ghosts feel such things, to hear our legatee deprived of the £200 we left him because we called the investment "stocks and shares" which have no legal existence.

This regulative aspect of the law has a moral side far greater than might appear. Through his dealings with such matters the lawyer impresses on his clients, or often on the community, his ideas of what is right and just. Many people forced into moral problems, which are often financial, take them to their solicitors for solution. "What would be a fair distribution of my property between my children?" "What sort of reward would be suitable for my faithful housekeeper?" "Ought I to disinherit my wife, after the way she has behaved?" On the answers given by the solicitor, on his conception of natural justice, much happiness or misery may depend, and through a series of decisions a general moral idea will be diffused through the community. As an example, by a recent law, passed under the prompting solely of this idea of reasonable right, a man cannot leave his wife totally unprovided for, though previously he could. The financial helplessness of the wife when compared with the husband had been adjudged wrong by general feeling, and that general feeling passed into law. Now the law will coerce those people who lag behind the moral standards of their age.

On the other hand a law passed against the great body of public opinion will fail. The most notorious example is the failure of the prohibition law in America, which not only failed to stop drinking, but gave a tolerated position to a large number of criminals and others engaged in breaking the law.

From early times one of the important functions of the law-giver, of the person rather than the law, has been the settlement of disputes. The ordering of society has so far advanced that the settlement by personal violence has been abandoned and both parties are willing to submit quietly to the decree of the judge. Sometimes an actual wrong has been done, sometimes it is a question merely of convenience. The most famous settler of disputes, because the most cunning in detecting falsehood, was Solomon; and his judgments, enshrined in literature, remain as models to this day. His mantle descended on others. More recently Paul Kruger, who ruled the Transvaal in most patriarchal style, gave judgments that for shrewdness and ingenuity would not have disgraced the Great King. Nowadays these matters come, more impersonally, before a court, but the judge still has to exercise personal discretion. Sometimes the legal aspect of the dispute is clear, it is only a question of which party is telling the truth, as in many compensation cases. In others the problem admits of no really satisfactory solution and can be dealt with only on general moral grounds. At a time like the present a continual stream of cases comes into court concerning occupation of houses. The parties ask for guidance as to who shall have the house. The law is vague; they do not wish to seize it from each other by violence; they ask for a decision. The courts apply the principle of "greatest hardship" and try to decide whether a man with five children will suffer more by being denied it, or a man with a job on the spot and an invalid mother. The cases are almost insoluble, but the fact that the criterion is a moral one, and that the judge is regarded as strictly impartial, renders the decision as little obnoxious as possible. In cases of this kind the ethical aspect of law is most prominent and its educative value highest.

One of the services that law renders to society is that it decreases the chances of disputes by letting it be known in advance what the decision will be. Each case is decided in accordance with precedent and as Pollock says, "If it be known that a decision once arrived at will be followed in the like case in time to come, the uncertainty of men's affairs is diminished and occasion of dispute taken away. The same question must not be agitated for ever. . . ."

This establishment of uniformity is most helpful. A thing once done creates an expectation of its repetition, and the child in the nursery and the Council of Elders alike invoke precedent for their acts. Nothing so tends to lessen frustration and ill-will as a comfortable experience of the uniformity of nature and men's actions: and the growth of social virtues is much easier under this shelter of predictability.

The lawyers are conscious of the part they play in the ordering of society, and the training a lawyer receives sharpens his perception of right and his sense of duty to the community. They have been anxious to bring the law into conformity with the general moral sentiment of the community, and from early times there has been the idea of equity which was designed to supplement the law where, owing to exceptional circumstances, its rigorous application would have brought it into conflict with the vague concept of natural justice.

The oldest idea was of "equitable power, exercised by some one, usually the king or great officer of state, who can dispense with rules according to his discretion, conceived as a reasonable discretion, but not defined beforehand. The other and modern form is the rational interpretation and qualification of the rules themselves by a dialectic or scientific process."[1] It has now of course been the subject of so much thought that it has passed from individual discretion to a body of precedents not very different from Common Law; but its existence and the precedence it takes over other forms of law show that there is ever present in the legal mind the belief in justice as something preceding and guiding all law.

This being roughly the function of the law it is easy to see that it holds a high place among the social mechanisms that make for morality. And with the law, so the lawyer. The lawyers of England are among the professional groups that impose on themselves standards of professional conduct. Lawyers are exposed to temptations, and these temptations are likely to be the more enticing because no man uninterested in money becomes a solicitor. Therefore the Law Society has no small task in maintaining a level of honesty that will give the public confidence, and set a standard of behaviour. In the same way barristers and judges are the guardians of national rights and honour. A good barrister will not take a case that he considers bad, and will not encourage his client to proceed with a claim he feels to be unfounded. A judge will decide a case on grounds of "public policy" or criticize a Government minister whom he thinks is exceeding his powers. This is made possible for British judges by their appointment "for life, so long as they are of good conduct".

When Hitler seized power in Germany one of his earliest acts was virtually to destroy the legal system. For some considerable time previously the judges had failed to administer the law fully, and had more and more allowed themselves to be persuaded or intimidated; but when he was fully established even this remnant of legality vanished, and the whole legal structure passed into Nazi keeping. It was clearly too strong an element in the moral stability in the state to be left alone.

To turn to the second of these two organizations for maintaining the moral code of communities—religion.

The range of beliefs and practices which we group together under the name of religion is very wide: from the Aztec priest holding up before his god the still quivering heart of his human victim, to the mild Jain who refuses to take any life, however low in the scale of

[1] Pollock.

existence; from the conception of the holiness of God as so utterly alien to all normal life and interests, so "wholly other" in modern phrase, that the mountain made holy by His presence must be fenced off lest even a beast tread on it, to the conception of that same holiness as the basis of an intimate communion with God as constituting the blessedness of even the humblest among mankind, "For thus saith the high and lofty one that inhabiteth eternity, whose name is Holy: I dwell in the high and holy place, with him also that is of a contrite and humble spirit".

When this variety is contemplated together with the great diversity of moral codes in different social and economic conditions, which we have already considered, it will be no matter for surprise that there is no fixed and constant relationship between religion and morals. This may seem somewhat of a paradox to those who have accepted Arnold's definition of religion as "morality touched with emotion", and been brought up in the belief that the demands of religion and the moral law are identical; but as Canon Quick has finely said, those who are called to the glorious liberty of the children of God, which consists in the knowledge that worshipping God and doing good to one's neighbour are two inseparable aspects of one ideal life, "will never understand their liberty, nor use it aright, unless they have learned something also of that great sum of toil and tears whereby the human race has purchased what their Church preserves for them as a birthright".

Some types of religion are not directly concerned with moral conduct: their aim is to secure by the correct performance of the appropriate ritual some particular blessing. The character of the worshipper may merely be indifferent, as when Amos complains that the wine for the religious feasts has been paid for out of fines unjustly inflicted in the law courts; or the rites themselves may actually be of a silly, degrading, or immoral character. This is particularly likely to be the case when the object of the rite is to secure the beneficent working of the great forces of nature on which the life af an agricultural community depends. Such rites are closely akin to what is called sympathetic or imitative magic, the attempt to bring about some desired result by imitating it as closely as possible. Anthropologists are familiar with it all over the world; a modern example is thus described by Sir James Frazer:

"In times of drought the Servians strip a girl to her skin and clothe her from head to foot in grass, herbs and flowers, even her face being hidden behind a veil of living green. Thus disguised she is called the Dodola and goes through the village with a troupe of girls. They stop before every house and the Dodola keeps turning herself around and dancing, while the other girls form a ring around her and sing one of the Dodola songs, and the housewife pours pails of water over her. One of the songs they sing runs thus:

We go through the village;
The clouds go in the sky;
We go faster, faster go the clouds;
They have overtaken us,
And wetted the corn and the vine.

"In such custom the leaf-clad girl appears to personify vegetation, and the drenching of her with water is certainly an imitation of rain."[1] The most obvious way of imitating and so stimulating the generative and reproductive forces of nature on which the fertility of the soil and the increase of flocks and herds depended, was by ritual prostitution. This institution, which still exists in India where "devadasis", or dancing girls, are attached to some of the temples, is best known to the general reader through the licentious worship of the Canaanite Baals at the local high places. In denouncing such worship the early prophetic writers who collected the stories in Genesis entirely repudiated the idea that the supply of corn and wine and oil was in any way dependent on the performance of any ritual acts whatever :

While the earth remaineth,
seed time and harvest
and cold and heat
and summer and winter
and day and night
shall not cease.

Such natural forces are not necessarily fully personified and conceived as gods with individuality and character. Many of them are purely functional beings, as can be seen in Varro's list of some of the ancient Roman nature spirits, over which St. Augustine makes so merry. "They made Proserpina goddess of the corn's first leaves and buds; the knots Nodotus looked into, Volutina to the blades, and when the ear began to look out it was Patilena's charge; when the ear began to be even-bearded (because *hostire* was taken of old for *to make even*) Hostilina's work came in; when the flowers bloomed, Flora was called forth; when they grew white, Lacturtia; being ripe, Matuca; being cut down, Runcina".[2] But the tendency is always to increasing

[1] Sir James Frazer, *The Golden Bough*, Vol. I.

[2] St. Augustine, *The City of God*.

individualization, and so animism shades off imperceptibly into polytheism. China holds a unique position in that her popular religion (apart from Buddhism which was imported from India) is an animism, which in essentials is on much the same level of development as that of the African native, but which is combined with the very high moral standard and noble ideals of conduct embodied in the Confucian code. As the ethical element in animism is very slight, this example tends to show that the ethical and religious development may go on side by side in almost complete independence of one another.

Religiously, the conception of gods may represent an advance, but it is apt to introduce an increased tension between religion and morality, because if we attribute the working of natural forces to the personal action of a morally responsible being, such a being can only be conceived as capricious at best, and may be cruel.

> Streams will not curb their pride
> The just man not to entomb,
> Nor lightnings go aside
> To leave his virtues room.

This is well brought out by J. N. Farquhar in his well-known *Primer of Hinduism*: "The main idea which the Hindu has with regard to worship is that every god must be worshipped according to his own wishes. The command of a god must be honoured no matter what it may be. The Hindu mind possessed no settled conception as to the moral or religious character of the gods, and consequently no man could tell beforehand what might be demanded by any god or goddess in the way of worship.

"The origin of the great gods of the Aryan peoples will make this idea still clearer. They were originally powers of nature, and therefore had *natural* attributes. Sun and wind, fire and rain, had no necessary connection with morality But they were powers, and therefore to be honoured and pacified by men. One could not guess beforehand what their wishes might be, but it was man's interest to gratify them, whatever form of worship they desired, whatever kind of action they ordained.... We are now in a position to realize how it has been possible for the Hindu to admit such things as the following into his worship: unlimited idolatry, human sacrifice, cruel torture, temple prostitution and obscene sculpture. The same idea explains how the Hindu did not regard it as unbecoming that Kali should be the patron divinity of robbers and murderers."

On the other hand, when the god is conceived as the patron of a group, whether tribal or political unit, he is bound to take an interest in securing among the members of that group the kind of behaviour which makes for the stability and success of the group. Thus, among relatively primitive, as well as advanced, peoples the god is the sanction behind the group morality. Among the Bechuana, for instance, thunder was regarded as the accusing voice of god, and, hearing it, the tribesmen cried: "I have not stolen, I have not stolen. Who amongst us has taken the goods of another?", while the code of Hammurabi (roughly contemporary with Abraham), in whose day Ur was already an ancient, even a decaying civilization, begins by stating that Ilu, the supreme, and Bel the Lord of Heaven and Earth, ruler of the destiny of the world, called Hammurabi, the godfearer, "to create justice in the land, to destroy the wicked and evil, that the strong oppress not the weak". It is, however, an undue simplification to say, with a modern professor:

"It is a generally recognized fact that men just work out among themselves their own rule of procedure and ideals of conduct, and then find sanction for them in the mind of God. Men first become ethical themselves and then they cease to think unethically about God. Consequently, to discover the accepted standards of an age or group, we need only examine its conception of God, for it will reflect most of the higher ideals of his worshippers."[1]

This presentation ignores the fact that religion is often a drag on moral progress, and the gods, even where they are conceived as the guardians of morality, often fall far below the standards demanded of their worshippers. The amorous intrigues of Zeus were made the subject of scathing criticism alike by Plato and by the early Fathers of the Christian Church. Where these tendencies are exploited by a corrupt priesthood or other vested interests, the demoralizing influence may be very great, and perhaps all the greater because it can assume very sophisticated and subtle forms in the hands of unscrupulous men. Pascal exposes the ingenuity with which the Jesuits discovered ethical sanction for any course of action towards which their penitents felt an inclination. Assassination is proved morally superior to duelling; for while *I* may be able to kill my opponent without sin by the simple expedient of directing my intention on some morally unexceptionable or even commendable, motive, such as the defence of honour, I can by no means be sure that if *he* kills me *he* will be free from the sin of intended murder. By privately killing my enemy, therefore, I not only avoid the necessity of risking my own life, but I escape the charge of being accessory to the sin which my enemy would have

[1] J. M. Powis Smith, *The Moral Life of the Hebrews.*

incurred by fighting a duel with intention to kill me. "A most pious assassination", is Pascal's caustic comment[1].

Secondly, this paragraph fails to recognize the influence which religion may have on the actual concrete detail of moral behaviour. When the historian H. A. L. Fisher includes among the "prospects flattering to human pride which it is a pleasure to recall to memory" on the long journey from Neolithic man to Hitler, "the cleansing tide of Christian ethics", he certainly means that Christianity introduced new ways of behaving, new standards of value to be embodied in human conduct and character. In particular, the paragraph obscures the difference between religion as the sanction of the tradition, customs and behaviour of a particular group, and religion as the critic of those very standards, and the champion of the claims of those who lie outside it. When religion appears in the first role it is often impossible (or undesirable) to leave different customs existing side by side as merely different, without any attempt to appraise one as better than another. This might appear at first sight to be the line taken by Sir Charles Napier, when Brahmins protested against his prohibition of widow-burning on the grounds that it was a *religious* custom. He readily conceded that in that case it must be allowed to continue, "but", he added, "my nation also has a custom. When men burn women alive, we hang them and confiscate all their property. My carpenters shall therefore erect gibbets on which to hang all concerned when the widow is consumed. Let us all act according to national customs."

It is however clear that what he believed himself to be maintaining is not merely a different national or even religious custom but a superior morality. This claim was explicitly asserted in 1891 during the agitation against the bill for raising the "age of consent" in girls to twelve. When it took the form of an appeal to Queen Victoria's pledge to "pay due regard to the ancient rights, usages and customs of India" the Viceroy (Lord Lansdowne) himself "replied by saying that the Queen's Proclamation had always been understood by every reasonable man and woman and interpreted by the Government as subject to the reservation that

> in all cases where demands preferred in the name of religion would lead to practices inconsistent with individual safety and the public peace and condemned by every system of law and morality in the world, it is religion and not morality which must give way."[2]

[1] A. A. Bowman, *Studies in the Philosophy of Religion.*

[2] Eleanor Rathbone, *Child Marriage.*

On the other hand, it is often the function of religion to correct the narrowness and exclusiveness of the group morality by extending its protection to those whom the group does not include, and making claims on their behalf which have no basis in their status within the group. Among the ancient Greeks, strangers were under the protection of Zeus Xenios. It seems an extension of the same idea when, in a well-known passage in the Iliad, the duty of forgiveness to those who have injured us is made to rest on the allegorizing picture of penetential prayers as his daughters:

For Prayers are daughters of great Jove, lame, wrinkled, ruddy-eyed,
And ever follow Injury, who, strong and sound of feet
Flies through the world, afflicting men. Believing Prayers yet
To all that love the Seed of Jove, the certain blessing get
To have Jove hear and help them too; but if he shall refuse
And stand inflexible to them, they fly to Jove and use
Their powers against him, that the wrongs he doth to them may fall
On his own head, and pay those pains whose cure he fails to call.

And when the great Stoic, Marcus Aurelius, makes his famous appeal for a world loyalty that shall transcend and consecrate all particular loyalties:

" The poet says ' Dear city of Cecrops ' [i.e. Athens];
cans't not thou say ' Dear city of Zeus?' "

there is no doubt that, whatever the rational and philosophical basis of his universalism, whatever Zeus may stand for as a cosmological symbol, the sentiment owes its emotional appeal to the religious associations clustering around the name of Homer's "Father of Gods and Men".

But it was to the Hebrew prophets of the eighth century B.C. that we owe the revolutionary insight that religion and morality are indissolubly bound together, and that for this very reason, both are universal in their scope. If Jehovah cares for justice and mercy more than anything else, more than sacrifice and ritual, more even than the victory and prosperity of one particular group, then he must love justice and mercy wherever and by whomsoever they are shown, and hate their opposites, even in his own chosen people. Amos, in the earliest of the books of the Old Testament, represents him as saying, "Are ye not as the children of the Ethiopians to me?"

These are not matters of ancient history: all these tendencies are alive and active among us to-day. It is only too common for people to regard God as an arbitrary and capricious tyrant, whose decrees, having no logical connection with anything else in our experience, can be disregarded with impunity unless he should suddenly take it into his head to exact summary vengeance. The writer was travelling one Sunday evening in a crowded train, when a little girl came into the compartment to take a seat just vacated. She whispered something to her mother, who produced out of her bag a tablecloth to be embroidered in coarse, bright-coloured wools. Thinking "What a sensible woman!", the writer smiled at the child and said: "I think it's very clever of you to be able to sew in the train: I can knit, but I do not think I could sew." The mother beamed at this and remarked complacently, "I tell her it's *wicked* to do it on a Sunday", leaving the writer gasping in the attempt to envisage what her conception of God and His demands must be like.

Again, the urgencies of a world war have made us only too familiar with the idea of God as the champion of a particular culture and the peoples who act as its bearers. The Jewish scholar Klausner sees clearly that the teaching of Jesus is not fitted to be the religious sanction of a group loyalty and the distinctive content of a group culture. As that was what religion meant to the Jews of his day, Klausner holds that they were quite right in rejecting him.

"In the self-same moment he (Jesus) both annulled *Judaism*, as the *life force* of the nation, and also the nation itself as a nation. For a religion which possesses only a certain conception of God, and a morality acceptable to *all* mankind, does not belong to any special nation and consciously or unconsciously breaks down the barriers of nationality. This inevitably brought it to pass that his people, Israel, rejected him."

In the course of history Christianity, as the bearer of the whole cultural and political content of an advanced civilization, came to play in the communities which inherited the traditions of the Roman Empire the very same role as Klausner here assigns to Judaism. There are indeed thinkers who, following the French sociologist Emile Durkheim, hold that religion is essentially a function of society, and the god worshipped is "a collective representation" of the sentiments which his worshippers have in common, a symbol of the social consciousness which unifies any given human society, and the means by which it is impressed on all the members, thus ensuring the continued existence of the group. To thinkers of this school it is natural and inevitable that Christianity should become identified with the whole manner of life of the Saxons fighting against the Danes. When Asser writes of the battle of Ashdown:

> "There was also in that place a stunted thorn-tree, which we have seen with our own eyes. Round this tree the opposing sides came together with loud shouts from all sides, the one party to pursue

their wicked course, the other to fight for their lives, their wives and their conutry."

he is using almost the very words with which the Greeks at Salamis exhort one another in the *Persae* of Aeschylus, words which are almost equally appropriate to any nation on the eve of an important battle. But most people will feel an irony in the turning of Christ into a warrior captain, as in the words of the Wext Saxon King Ethelwulf before the battle of Englefield:

" They be more than we, yet fear them not. Our captain Christ is braver than they."

At a time like the present, when national unity is strained to breaking-point and there seems no common loyalty to ideas and ideals to hold society together, there is a tendency to exploit religion to supply this same need. The required focus may be sought in Christianity, in itself as we have said a fellowship which transcends national frontiers. Thus a recent letter to the *Modern Churchman* says:

"It is most important, however, *in the interests of the nation's social and cultural unity* that we should have a national system of religious education in the undenominational schools of the nation, and that now our nation itself should undertake the teaching of the Christian religion in all these schools so as to secure Christian unity and to advance Christian culture." (Italics mine.)

More logical would seem to be the erection of the national unity with the ideals and values of which it is the vehicle into a new religion. It is widely recognized to-day that Communism and Nazism are both really new religions, rivals self-confessed to Christianity. This search for a new religion to fulfil the social functions of religion in the community need not take such violent and revolutionary forms. An American educationalist has just published a book based on "functional analysis" of religion, what it actually *does* in the life of a community. He claims to show that it consists of ten basic experiences, essential to a stable and mature life as a member of society:

(1) sense of worth
(2) social sensitivity
(3) appreciation of the universe
(4) discrimination of values
(5) sense of responsibility
(6) social fellowship
(7) quest for truth
(8) integration of experience into a working philosophy
(9) appreciation of historical community
(10) sharing in group celebrations.

The author accepts the philosophy of naturalism: to talk of a personal god is to him no longer possible for an intelligent man. But religion is still necessary to encourage and develop these essential experiences. Even if the teaching of the Churches were acceptable, they cannot, operating one day a week as they do, touch the fringe of religious education in this sense. It must concern itself with, and express itself in, all human activities and relationships. Hence we have a confession that a purely secular education breaks down, and the demand that the state education shall be religious, but the religion it is to teach is to be a new religion, radically incompatible with the claims of the Christian Church, and not a neutral basis or a common core to which Christian teaching can be added at the option of the parents.[1]

[1] E. J. Chave, *A Functional Approach to Religious Education* (University of Chicago Press; Cambridge University Press).

Books referred to in this chapter:

Sir Frederick Pollock, *Essays in the Law; Essays in Jurisprudence and Ethics.*
Pitt-Rivers, *Clash of Culture.*
Laxdael Saga.
Willard Price, *Japan's Islands of Mystery.*
I Kings.
Edmond Fleg, *Life of Solomon.*
William E. Hocking, *The Meaning of God in Human Experience.*
J. M. Powis Smith, *The Moral Life of the Hebrews.*
J. Frazer, *The Golden Bough*, Vol. I.
A. A. Bowman, *Studies in the Philosophy of Religion.*
H. A. L. Fisher, *History of Europe.*
Joseph Klausner, *Jesus of Nazareth.*
J. N. Farquhar, *Primer of Hinduism.*
Eleanor E. Rathbone, *Child Marriage, the Indian Minotaur.*
E. J. Chave, *Functional approach to Religious Education.*
St. Augustine, *The City of God.*

PART III

THE EDUCATION OF CHILDREN IN VIRTUE

INTRODUCTION

IN this the third section of our book we wish to show how the principles we have already discussed apply to the more formal education of children. It is a very definite stage in the evolution of society when the education of children becomes part of the duty of the community as a whole, and ceases to be the concern of the individual family or of some special group. In simple societies, children are left to grow up, learning from their environment what they need to know for life. Some of the activities may require great skill, and therefore take long to learn, but their utility is obvious, and the child practises with full knowledge of what he is doing. The ideas that are characteristic of the group are learned equally simply. They are not very complicated in most cases, and as there is no competing system the child cannot go far wrong. The more complicated a society is, the harder it is for children to learn what it is necessary for them to know before they can take their full part in the group. Reading and writing have not, for children, the obvious utility of paddling a canoe, and even the practice of an art may, among us, involve a long period of practice when the value of the activity is very far from clear. This is particularly so in music, where real pleasure may be long delayed. Moreover, some of our myths and social ideas are very complex and hard to understand. Nor does any one system hold complete dominance. There are always competing systems, either in the group itself, or in other groups in close relationship. The child therefore has a choice, and at many periods in history there has been eager competition for control of the child's mind.

Formal education in Europe has had different origins and different aims. The education given by the Church has been partly technical in that it was intended to train priests; partly moral, because it taught religious beliefs and practices, and these were felt to be the best means of producing virtue. The education given by a craftsman to his apprentices covered the trade and an outlook on life. From the Renaissance the nobles and the rich developed an education, of varying content and merit, which aimed at making a man a citizen of his world and equipping him for his part in the state. It was not till the middle of the nineteenth century, when the industrialization of Europe was making ever bigger demands on the workers, that universal education began to seem desirable. The realization that alike in war and peace, in politics and commerce, knowledge was a valuable thing came rather slowly; but, when it did come, one European country after another collected its children in schools and proceeded to teach them those things that it thought the community needed.

At first this education had really little concern with morals. Home and Church were supposed to deal with them, it was the duty of the schools to impart useful information. In some countries, the idea that instruction is the real purpose of the schools still remains; in England there have always been schools that placed moral training high. Their methods were frequently, in the past, most unlikely to have any good effect, but that did not alter their good intentions.

The moral education of children, as soon as it was seriously undertaken, whether at home or school, has always presented theoretic as well as practical problems. What was man's nature? Were the impulses of a child inherently bad, so that they must be eradicated, or might they be left to develop in the confidence that in time they would conform to the general pattern desired? Need parents or society really do anything about it, and, if they did, was a gentle guidance all that was necessary, or must pain be the sharp spur to virtue? Fichte's assertion, "Education must ensure that it totally destroys the freedom of the will", can be set against George Bernard Shaw's declaration, "The vilest abortionist is he who attempts to mould a child's character".

On the whole, education, at least till the present century, inclined more to Fichte than to Shaw; and too often brutality was the only method of influencing a child that his parents or teachers imagined. It was not the poor and ignorant who treated their children worst; the rich, thoughtful and religious were often the cruellest. Even a happy child, like Mrs. Sherwood, had to find her happiness under strange conditions.

"Lady Jane Grey speaks of the severities to which she was subjected by her noble parents. I had neither nips nor bobs, nor pinches; but I experienced what I thought far worse. It was the fashion then for children to wear iron collars round the neck with a backboard strapped over the shoulders; to one of these I was subjected from my sixth to my thirteenth year. It was put on in the morning and seldom taken off till late in the evening; and I generally did all my lessons standing in stocks, with this stiff collar round my neck. At the same time I had the plainest possible food, such as dry bread and cold milk. I never sat on a chair in my mother's presence. Yet I was a very happy child; and when relieved from my collar, I not unseldom manifested my delight by starting from our hall door, and taking a run for at least half a mile through the woods which adjoined our pleasure grounds. . . . I still was carrying on in my Latin studies and even before I was twelve I was obliged to translate fifty lines of Virgil every morning, standing in these same stocks with the iron collar pressing on my throat."

The contrast that Samuel Butler draws between the treatment of poor Ernest and of the countrywoman's child, points the harm done by those who thought most. At the start of this century the idea was beginning to grow that children would develop well if left alone. Shaw's protest was reasonable enough, but the advocates of freedom had soon gone farther, and, instead of merely refraining, negatively, from "moulding," had developed a theory that denied a child all the guidance that his immaturity requires. It was not that a child was really good if he was encouraged to develop in the right way, but that he was really bad—and better so.

Once education in school had become universal it was but natural that the teaching of morals should seem a desirable addition to the imparting of information. This was the more necessary as the conditions of industrial life made it difficult for the family to give a training in virtue. But, if moral training was to be given in school, methods had to be devised which fitted the instruction of a classroom full of children, and which could be learned and applied by all teachers. In the arid immobility of school, eighty children at once must learn virtue together. The method must not rely on the lively interchange of daily life, it must be suited to an unvarying environment and the domination of a single individual. It is not surprising that the moral lesson, with its adjunct the moral story, became the most favoured method.

These matters were much in men's minds at the beginning of the century, and in 1909 two fat volumes were published, one by Longmans and the other by Watts, surveying the practices of different countries. One was the work of an International Committee presided over by Professor M. E. Sadler, in the other Gustav Spiller collected and published "*in full* most of the Moral Instruction Syllabuses and other ethical matter of the schools of France, Italy, Japan, Belgium, Germany, Switzerland, the British Empire (including England, Scotland, Ireland, India and the Colonies) the U.S.A. and of many other countries".

It would be almost impossible to read all this matter, but it is clear that the countries surveyed believed firmly that moral exhortation, apt story and pregnant picture were the methods by which children might be made good; and Mr. Spiller, discussing it all with firm ethical zeal, shares the view. He notes for instance with particular satisfaction the progress Japan has made in inculcating virtue and peace among her children.[1]

"The Japanese Board of Education has published two rolls of some sixty large pictures to illustrate the moral lessons given in Japanese schools. In them the mother plays the chief and central part, and sympathy forms the keynote. The political and warlike elements are almost wholly lacking. . . . Finally the walls of the school, as in France, would be decorated to some extent with ethical pictures and maxims."

Pictures were not the only valuable material. The subjects of the curriculum themselves could have their effect, though perhaps not so direct and simple a one.[2]

"Of the more common subjects there remains now only mathematics, at first sight a most unpromising subject to deal with ethically. Yet Mrs. Boole has shown that not only can arithmetic be taught thus, but that it cannot very well be taught effectively in any other way. In general the selected examples might largely be of an ethical character, and the need for precision in all thought might be illustrated by applying mathematical methods to what are ordinarily considered non-mathematical subjects."

From such material he hoped to see evolved a complete formal scheme of lessons displaying step by step a theoretic plan of morality. "The teacher would, consequently, have to treat of the twelve relationships of life and the four cardinal virtues, and these together properly developed and unified, would yield the complete material for the moral instruction lesson."

The belief in this type of moral training was very widely spread. It was, in fact, the one positive type of moral training imagined. In the form of the moral story it held sway for long periods in the literature of the nursery as well as outside. Its defect was its lack of reality or aesthetic merit. Three years after Mr. Spiller's vast work, Felix Adler was presenting to the world a system of teaching morality spiced with anecdote and example. In all solemnity and for many pages he discusses the moral lessons that can be extracted from the Iliad and Odyssey, and when literature fails to give all he needs, turns to real

[1] *Moral Education in Eighteen Countries*, p. 107.

[2] Op. cit., p. 33.

life. The life story of Jacob and William Grimm illumines a lesson on fraternal love. When he has traced their undivided career from birth to death he ends with the solemn words, ". . . and as they slept together in the same bed when they were children, so they now slept side by side in the grave."[1]

None of these authors faces the problem of what happened supposing the child failed to respond properly to these edifying talks. If the contemplation of Jacob and William Grimm failed to produce fraternal love, what then? The answer lies in *The Fairchild Family* where Mr. Fairchild, finding the children quarrelling, first canes them smartly and then takes them to look at the mouldering bones of a man who had slain his brother. Behind the moral tale lay punishment, and the age which thought most highly of the one practised the other with peculiar severity. As Keate at Eton is reputed to have said in Chapel: "Boys, you've got to be pure in heart; and if you aren't pure in heart, *I'll flog you till you are.*"

Professor Sadler's committee saw a little farther into the art of moral training. In a very excellent passage he points out a second type of moral training, which was very new then, and even to-day is little more than an aspiration in many schools. There are, he explains in an introductory essay, two views of moral training. "The first view finds its most characteristic expression in the teacher skilfully stimulating and directing from his desk the intelligence and the aspirations of the diligent and well-disciplined pupils who sit before him in the class-room. The second presents us with a very different picture of the most characteristic form of effective school life: that of a more or less self-governing community occupied with vital movement of all kinds, full of freedom and initiative in a great variety of tasks, getting experience of the labours and relationships which lie at the foundation of all society, dynamic, self-expressive, educatively practical, busy with the effort to accomplish (under due but unobtrusive guidance) certain things which its individual members wish to accomplish, and in which therefore they find a strong motive to effort."

This account of the school and its moral training based on activity is important because it stresses, as we have done earlier in this book, the value of purpose, of actual experience and of the group. It is the insistence on these that marks the difference between our views and those of the individualistic morality which holds out the promise of separate rewards or punishments and says nothing of purpose or the bonds that hold man to man. But excellent as Professor Sadler's

[1] Felix Adler, *Moral Instruction of Children.*

description of one part of moral training is, it does not cover sufficient ground. Before the school can be organized as self-governing and active, the children must be ready. They must be willing to co-operate, and have some idea of group activities and obligations. It is not at all difficult to imagine children with whom all such activities would be impossible because they were too rowdy, too egoistic, too hostile. The first stage, in school as in life, is to produce benevolence and a friendly attitude towards others. The child in a school where he is happy and well done by rapidly develops this attitude, but because it comes easily it is none the less of the utmost importance, and the school must deliberately set itself to cultivate it. Where the majority of children are rowdy and difficult to manage it is almost certainly the fault of the school, and it is the management that should be blamed, and not the children.

There is also a further training needed. Purposive group activities are very important in training the child, but the teacher should not leave them to take any convenient form. He should have in mind some scheme of qualities that he wishes the activities to develop. The Army, as we have said, relies largely on indirect moral training of this kind to produce a definite set of qualities that it does not seem possible to produce by other means. The public schools rely on games or the prefect system to teach boys habits of decision and command. The leaders of our schools must have a wider vision by which they decide what types of activity the children shall adopt.

But this indirect training, though of the utmost value, does not do all that is required. Each child must learn the set of ideas current in his group, and he must absorb the myth. In this, teaching is mainly direct. The moral story is the crudest and least effective means of imparting these ideas. The child learns them from literature, talk, the cinema. They are presented to him continually. It is most important that the school should realize what it is doing and make the presentation of the myth as convincing as possible. This teaching of ideas we have called direct positive training and it is the partner, as we have said, of the direct negative training of punishment.

All children should then experience as part of their formal education the good treatment that produces benevolence, the indirect training in methods of living, of group activity and purpose that develops desirable characteristics, and the direct training, both positive and negative, that teaches them the special ideas and forms of activity of their group. In addition some children will go further and receive a training that will fit them to be leaders in their society whether their eminence comes mainly from technical skill or the possession of certain generalized powers of mind. This special training will be for the exceptional children and the discovery and training of them will be one of the most serious tasks that a nation undertakes.

Books referred to in this Introduction:

Mrs. Sherwood, *Autobiography*.
Naomi Royde-Smith, *The Mind of Mrs. Sherwood*.
Moral Instruction and Training in Schools, ed. M. E. Sadler.
Gustav Spiller, *Moral Education in Eighteen Countries*.
Felix Adler, *Moral Instruction of Children*.

Chapter 1

INDIRECT TRAINING

We have said that part of the duty of the special moral training given in formal education, as well as of the informal teaching of home and environment, is to produce benevolence, independence and the sense of the group. Now these virtues are largely produced by suitable arrangements made for the child and do not depend so much on his own exertions. He experiences certain things, and this outside experience has a direct effect on him and is reflected back in his pattern of behaviour.

For benevolence, as we have said, a child requires to experience affection, to feel himself safe, well done by, to be free from vexatious and frustrating experiences. It is only recently that this has been in any way possible in schools. If we take the schools that received 90 per cent. of the population, we can say that, in the past, in fact until well into this century, they were not fit places for children either in buildings, curriculum, or methods. In the schools for the other 10 per cent., while some had adequate buildings and playing-fields, the curriculum was generally unsuitable, and the discipline based on flogging had a damaging effect on most children. To achieve independence a child needs a sense of power and mastery, and an environment in which he is free from humiliating restraint, whether imposed by unwise love or eager domination, and for the sense of the group something very different from the individual and individualistic methods so usual in schools.

The simplest element is the physical environment. In 1845 *The Times* was reporting an inquiry into an extension of the London and South-Western Railway. Mr. Tite, the architect and surveyor, was giving evidence. "Mr. Tite endeavoured to console [opponents] by . . . referring to the Greenwich and Blackwall Railways, the property close to which had increased in value. . . . The arches were let out for infant-schools, and no annoyance was felt . . . from the passing of trains over their heads; no doubt the South-Western Company would be able to appropriate their viaducts to similar purposes, and in course of time the direct promotion of 'national education' might be numbered among the great results of railway establishments."

It is a far cry from railway arches to the best of modern primary schools, but every step of progress has been won in the face of opposition, and is justified by a change in the relationship between child and school. The experience of teaching in a school that still exists with the premises of even 1900 is very enlightening. Many subjects, and those the most enjoyable, cannot be taken at all; every type of activity is handicapped, if not by the lack of space, then by the noise of the class next door. Children who have come from another school speak longingly of dancing and music, of space for handicrafts or games. There is an attitude of hostility produced simply by bad conditions, lack of fresh air, and the knowledge that what is offered is inferior. The basis on which any system of education that aims at benevolence must be built is the provision of physical conditions that facilitate the children's activities, are healthful, and give an impression of dignity and care.

Reforms in educational method have come largely from the bottom. It is the infant-schools that have been the most progressive in the last twenty-five years. The schools where children worked on backless benches placed one above another in the step-like classrooms, when for long periods it was a sin to move, and where an hour might be spent learning A B C, have vanished, and their place has been taken by open-air classrooms. The great discovery, imagined by Froebel, made into a system by Maria Montessori, and popularized into a method by modern training colleges, is that children are themselves really best able to judge what they need. If they are put into a suitable environment and then largely left alone, they will occupy themselves profitably with what is provided, choose activities that suit their individual stage of development, and pursue them as long as their very unstable attention allows. Children thus start their period of formal education without strain or frustration. From two years old, if a child is lucky enough to go to a nursery school, he finds interest, freedom, kindness. He is not forced to do what he does not care for and feels he is incapable of. He is not rebuked, or exposed to any but the mildest restraints. The result is a school of good, happy children who are at peace with the school and the world.

When D. H. Lawrence in *The Rainbow* described a school and life there, he was, obviously, writing from experience: and he has given his account the particular vividness and sense of reality of which he was master when writing of the actual. The chapter is too long to quote, but the dark malevolence of the school, the unnatural lessons, the continual bullying of the headmaster, the hostile wretched children, all lead up naturally to the moment when the girl teacher loses all control and flogs a boy. It was not only that physical conditions were bad; the material taught and the manner of teaching were both such as to render friendly co-operation between teacher and children impossible.

The two principles of school work that bear directly on this part of morality, are that the material should be felt *by the children* to have interest and value, and that, as far as possible, the work should be arranged individually so that no child is made to feel inferior by inability to progress as fast as the others, and no child should be exasperatingly retarded by having to wait for the less able. The moral effect of this individual work is perhaps as great as its intellectual. It is certain that by its introduction in such a subject as arithmetic a previously unruly class will become good, and children who previously resisted instruction become quite eager to receive it.

Improved relations between children and teacher follow at once. So long as the child feels he is being forced to learn the useless in the most unpleasant way, he will resist; and discipline inevitably becomes a matter for the rod. When a child is asked to acquire various skills, whether in dancing or arithmetic, and feels that the teacher is helping him, himself, as a valuable individual, then practically all the resistance vanishes, and control can depend on the spiritual prestige of the teacher.

When education is thought of in this way, when the teacher is willing to put himself in the child's place, and considers both what he teaches and how he teaches it from that side, the frustration and hostilities that used to be considered an inevitable part of education almost all disappear, and with them the resistance of the child.

Independence follows also from these methods. If children work individually they cannot all be doing the same thing at the same time, and schools cease to feel that complete uniformity is the ideal. Moreover, if relations between teachers and children are friendly and natural, the child will give a lead as well as follow one. There are many occasions in school life when the intelligence of the class, as a whole, is superior to the teacher's individual thought; and every encouragement should be given to the children to make such contributions. True, valuable independence comes with self-respect, and this arises when a child feels competent and well esteemed, and this is much more likely to happen under a system of individual work. Thus we can see the production of a type of school in which the children are contented, enjoying a sense of well-being, of dignity, and of balanced self-importance.

There is no need to fear that they will usurp the functions of the adults and that freedom will turn into contempt. The natural basis of discipline is very strong. The child needs the adult. The adult knows how things should be done. He can show the child how to set about doing what he wants to do. Moreover, the adult has persistence and power of attention. He will carry through to a successful conclusion things that the child would like to achieve and cannot. The child knows that he is dependent. He will come to a kindly adult and ask to be helped and taught. He realizes that decisions as to what is right or proper can be made most surely with adult help; and when disputes arise they are often spontaneously referred to the teacher or parent. Quite apart from the adult's power to coerce or punish, strength, size, skill, knowledge and determination make the adult naturally dominant. The child who has not been rendered hostile by over-control or domination accords willingly and gladly this position to adults. The children who make this natural discipline difficult are those who, having already been mis-managed, are anxious, above all, to reassert their own personality against a hostile environment. When once they feel safe from further attack they will adopt a very different attitude.

In Home Office approved schools for children who have become delinquent, the first sign that a child is responding to the new environment is that he becomes friendly to the adults in charge. This friendliness means that the child is once more prepared to allow the natural relationship between himself and the world, particularly adults, to be re-established. He has become delinquent because he has refused to obey the regulations laid down by adults. He has turned on them and all their works in the bitterness of his heart. Now, as that bitterness decreases, he is prepared to accept from them the help he knows they can give, and to allow the relationship to assume its natural quality of friendliness.

In no department of life have schools of the older type managed worse than in relation to group activities. There has been an emphasis, amounting to an insanity on "doing your own work", on individual progress, on competitive success. The child who does his sums with his arm across the paper, lest his neighbour should see the answers, is as great an offender against co-operation as the child who copies them.

Group work is not the denial of individual responsibility; it is an attempt to raise it to a higher level. The child who is responsible only to and for himself can be as idle and unprofitable as he dare, and he, in his own person, suffers the full consequences. The active member of the group has a far wider and more compelling responsibility.

This is not an argument for the methods of "teams" that schools so love. These teams retain the traditional element of individual work (all the same, and all done independently) and of competition. The method makes what it feels is a concession to modernity by lumping several children's marks together and rewards the team with the highest score. If this system is taken seriously, it leads to bullying and victimization of the less able children; and there have been schools where this element was accepted and even encouraged. In one, where a child was seriously hurt, the parents prosecuted the school and won their case.

What is needed is a totally new conception of group work. Dewey writes bitterly of the ordinary school attitude, and indicates what he thinks is needed. "A society is a number of people held together because they are working along common lines, in a common spirit and with reference to common aims. . . . There is something to do, some activity to be carried on, requiring natural division of labour, selection of leaders and followers, mutual co-operation and emulation. In the schoolroom the motive and the cement of social organization are alike wanting. Upon the ethical side the tragic weakness of the present school is that it endeavours to prepare future members of the social order in a medium in which the conditions of the social spirit are eminently wanting." At Oundle under Sanderson much practical work was done on the group principle. A group of boys was responsible collectively for a certain repair or the making of a piece of apparatus. The work was sub-divided and carried out. This involved in the first place the real delegation of responsibility by the teacher, and in the second an arrangement among the children so that each did something different. There was thus no possibility of individual assessment or of competition.

Teachers of a certain type find it very hard to accept these qualities of work, and are very unwilling to desert the customary methods; but as long ago as 1908 an American school was trying an experiment which deserves to be repeated.[1] Children of eight and nine were told that if they wished they would be allowed to work together in groups on some subject that they themselves chose. They must prepare and plan and say how much time they wanted for their activity. At first there was little response, but gradually first one little group and then another presented plans. Some three or four boys wanted to cook so that they could eat the results, another group wanted to do photography and so on. They were allowed time, makeshift facilities were arranged. They had to do everything themselves and they were not always very successful, but, bit by bit, they learnt both how to manage

[1] C. A. Scott, *Social Education*, 1908.

group co-operation and also to do special things. There were problems about unreliable children who failed to bring the things they promised; about bossy children who would not let the others share in the management, and doubtless there were occasions when the toffee burned or the fixative spilled. But the children showed great common sense and made on the whole a success of their ventures. They certainly learnt a great deal of social value.

If a school feels unequal to this rather heroic work, it can obtain a similar result by allowing groups of children to produce and act plays, prepare maps and reports, make models, and a number of other things more usually considered school subjects. By giving the management of the groups to the children themselves and not leaving it in the hands of teachers, they are given a sense of importance, of independence, and taught what are the problems of co-operation. The natural leader takes his place, and the child who is cursed with both incompetence and the desire to dominate can learn early what place he can achieve in the pattern of society.

To accept this idea in education it is necessary to think of school as being much more than a place of instruction. Probably the first schools of modern times to realize the importance of this type of education were the English public schools. They started with certain advantages and to these they added others—such as the house system, which, starting from necessity, became one of the main elements in their organization. Of their methods three may be singled out because they invented them, and once invented they have been imitated in an increasing degree.

The first is the athletics. As has been said in the discussion of army training, athletics, games, rowing, P.T., have a very considerable psychological effect as well as a physical one; and steadiness, endurance, discipline, self-respect, team-work, are all connected with moral virtues. The boy who has been a success on the athletic side at school is likely to have acquired characteristics that will be very valuable to him in certain occupations in life.

The second is the house system which makes a boy a member of a conveniently small group. This group has direct adult supervision, and the boy can be watched and helped individually. Moreover, he is encouraged to identify himself with the group, and seek its advantage in various ways. Membership of a group is the basis of morality, and for satisfactory results, especially with children, the group must be small enough for him to comprehend it, and to feel a responsible fraction of the whole. A school is often too big a group for this purpose and the house is more convenient. Moreover, in a school divided into houses there are many opportunities for internal competition which is an incentive dear to the English educational system, and reasonably potent in a certain field with children.

The third is the experience of responsibility given by the prefect system. At seventeen or so a boy is beginning to feel his growing manhood, and is anxious to try his power over others. As a prefect he has certain mild privileges, definite duties, a status and a chance of showing what he can do in comparative independence. This experience is much appreciated by most boys; and sends them out from school having learned certain of the arts of command, and having increased their self-respect and self-confidence.

Of these methods of education the first is very expensive. Quite apart from the initial cost of the land the maintenance of large playing-fields costs a great deal. In schools in which the railway-arch tradition was hardly dead such extravagant provision was hardly to be thought of. The house system in its proper form is also costly. The most essential part of it is the close association between a member of the staff and a comparatively small group of boys. The boys must be few if the relationship is to be sufficiently close and intimate to make the house master a real factor in the boy's development. In many schools that have partially imitated the house system economy has caused this particular aspect to be ignored; and the house has come to serve other and less valuable purposes. It is considerations of expense that have delayed the use of these, and other methods of indirect training in the education of normal children. They have been used, of necessity, in attempts to reform small groups of children who have been corrupted by ordinary methods and need to be re-educated.

In addition to these advantages the public schools have handsome or venerable buildings, large, highly-selected staffs, and an organization that allows of meeting-places and opportunities for discussion. They have also the advantages of being boarding-schools, and this, as we shall see presently, gives them far greater opportunities for teaching the myths they wish to impart. Most important of all, they keep their boys till they are seventeen or eighteen. At this age a boy is really able to understand the ideas that are being taught him, and to enter fully into the life that is offered. It is in his last year or two that the boy, now a prefect, captain of football, in the school eight and the classical VI, really understands what it is all about, and reaps the benefit of an organization of which, previously, he was largely the victim. Schools which are still labouring under the curse of past meanness, and which lose their children at fourteen, cannot offer anything really comparable. They have done extremely well with what they have.

A system has just been developed in France to meet the needs of orphaned children of the Maquis, which has some of the features of the public-school system—residence in small groups under special adult supervision—but contains also much greater flexibility and much more home life. The following is an account of the experiment, from *The Times Educational Supplement*, 1st June, 1946:

"In July, 1944, the Germans, fearing an allied landing on the south coast of France, attacked and destroyed a company of nearly 4,000 Maquis in the Haute Savoie. Following the liberation of France, members of the Resistance of Lyons, headed by the Prefect and the newly appointed Inspecteur de l'Academie, requisitioned empty chalets in a number of Alpine villages as homes for the orphaned children of these members of the Maquis.

"The organizers placed in charge of these chalets young married men teachers. Those selected were men who could, in the opinion of the organizers, find happiness and satisfaction in the creation of communities of children, and who were willing to regard their own children as members of the community of which they were to become the heads. It is not, therefore, surprising that the most important member of the staff is the schoolmaster's wife, who is dignified with the title of *Chef du Centre,* since she is foster-mother, plans menus, supervises health, and sets the tone of the community.

"The husband, known as the *Directeur-Instituteur,* is responsible for the education of the children which, so far as scholastic work is concerned, is carried on in the chalet itself on individual lines, not unlike those of the Dalton Plan. Other members of the staff include either a young woman teacher known as a *monitrice* or a nurse (there is one nurse for every two chalets), a cook, a *lingère* who is responsible for all laundry work and for the repair of clothing, a *ménagère* who assists in all the work of the house, and a non-resident part-time handyman. All members of the household, including the *ménagère* and the *lingère,* eat together in a common dining-room at small tables accommodating six or eight children and one adult.

"There are at present no children over the age of fourteen in any of these houses, and the problems of adolescence have not so far had to be faced. On the other hand, the presence of tiny toddlers, from the age of four, raises other problems. These have been met by placing the very young children under the care of a girl aged twelve to fourteen, who is regarded as the "*petite maman*" and works under the supervision of the *Chef du Centre.*

"For minor ailments the children are nursed in the sick-room of the chalet, and are not, in consequence, subjected to the mental disturbance of leaving their little community. For other complaints there is a special village hospital staffed by two highly qualified nurses, and presided over by a doctor who gives full-time attention to the physical rehabilitation of the children.

"The villages, of which there are at present four, each accommodate about 700 children, and are situated in some of the famous Alpine health resorts—Mégève, Combloux, Villard de Lans, Morzine, and Dieu-le-Fit. There are, in addition, a number of small centres which raise the total number of children to over 3,000.

"Any adult visitor is received by the toddlers, in the words of a British Consul, 'as a new sort of teddy bear, to be cuddled and played with'. There is a similar, but more restrained, welcome from the older children. This complete friendly trust could not be extended by children who had been regimented and had come to regard teachers as anything but personal friends. Corporal punishment is forbidden, and, by the organization of every house into teams, rivalry in well-doing is almost the only form of competition encouraged. The teams are often members of the same dormitory in which there may be pinned such a notice as: 'The Panthers are always the first ready; Panthers are never late'. In houses where the majority of children are over ten years of age, the teams play a part in some of the minor duties of the home. The children also work together as teams in their studies, older children helping and encouraging the younger. It is as teams that they take part every evening before bedtime in the '*veillée*'—something best described as an indoor camp-fire gathering.

"Physical development is fostered by a short spell of morning exercises out of doors, and by two hours in the fresh air every afternoon. During the winter most of the children learned to ski, and on some days, especially Thursdays when there is no academic schooling, the older children make long ski excursions.

"There is a wide range of handicraft studies—model-making of all kinds, puppetry, clay-modelling, etc., and most of the work is performed in teams. In addition the children in all the villages have a large repertoire of songs which have been chosen for their musical or poetical quality. With the aim of making song a normal part of every child's life, there are songs before and after meals, and songs while work is done. Indeed, the children clamour for song, and are always ready to sing on the slightest excuse.

"In order that uniform standards of teaching may be observed in all the chalets, and all the subjects correlated as far as possible, there are regular meetings of the staff—something unknown in any other part of the French educational system. Specially recruited for the work, they are given an opportunity of thinking out their subject in the light of the special needs of the children. The result has been freshness of approach, and a high degree of enthusiasm. Although schooling is at present limited to the primary stage—the work of the *Ecole Maternelle* and the *Ecole Primaire*—the development of individual methods of study has resulted not only in every child being able to move at his own speed, but also in giving scope for the older children in the team to assist some of the younger ones occasionally.

"The essential advantage in this combination of foster-parent and teacher in one person is, however, that the child is seen as a whole. The teacher who is trying to build up health and education, and who supervises both, has a much clearer understanding of the child's needs in scholastic work than one who sees the child only in the classroom. Pale and depressed urchins have, under this care, developed within a few months into bright-eyed, keen, happy, and industrious children. Anyone who inspects their work, whether handicraft or exercise books, cannot but be impressed by the fact that the majority are now well on the way to becoming able and constructive citizens.

"It is hoped that these villages will become the prototype of a wide development of child-care throughout France for all children who need this special treatment—not merely for casualties of war. Great developments are possible, provided of course, that the French Government continues its financial support and provided that the young teachers, their wives and assistants, are physically capable of continuing the great task they have undertaken."

French education in the past has been far more rigid, far more an affair of instruction, far more individualistic and examination-ridden than English. It has never made games an important part nor concerned itself with those aspects of life that England includes under character training. Thus it is very instructive to see how complete a change conditions have imposed, and how well the children respond to educational methods that make a way of life the chief element in training. For what we have to say later the prohibition of corporal punishment is particularly important.

In England, if we wish to see the principles of indirect discipline carried out logically we must turn to schools that deal with children spoilt by ordinary systems. During the war and evacuation a number of children proved to be "unbilletable". This word covered children who were unmanageable, unruly, who stole, who wet their beds, and did other things of the same kind. One lad, sent to Evesham, developed a compulsion for cutting down fruit-trees. These children, rejected by foster-parents, were collected in hostels and offered an ordered life, under expert care. With more serious offenders, the children who appear in the courts, very much the same policy is adopted.

In recent years the approved schools—intended for delinquent children—have greatly improved and are developing a new type of education. The children with whom they deal have been failures under ordinary conditions of education. They have not learned enough virtue to live as inoffensive members of the group, and the task of the school is largely to repair the damage that society has already done to these children.

Here again there are various principles that are generally applied and they are not such very different principles from those of the public schools. Perhaps only the first is different.

The schools try to provide an environment in which a child is free, as far as possible, from those restraints and frustration which lead to bad behaviour. The child in a good approved school is far freer than a corresponding child in a public school. He is allowed out, allowed to go to the cinema on Saturday afternoon, allowed larger choice in his occupations. He also lives under freer rules and is in danger of much less punishment. If he runs away he will not be brought back and beaten; he will be transferred to another school that the authorities hope will be more to his taste. He may in this way change his school several times, and a benevolent Home Office will continue to try and suit him.

An exactly similar attitude has been adopted in hostels for difficult children and there too, this freedom has been found to have the greatest effect on children's behaviour. To quote a recent report[1]: "The previous regime in this hostel was very much restricted and the policy of segregation which was pursued had disastrous effects on the boys' discipline, and consequently on their reputation in the village. The present warden has given them as much freedom as possible believing it essential for the growth of self-respect and responsibility. The result of this is seen in the good contacts which they now make in the village. They belong to the choir, attend Sunday School and are members of the local Scout Troop. They attend local concerts and plays and are sometimes asked to take part. One of their most popular activities is to act as casualties for the local Red Cross and A.R.P. practices. They know they may ask permission to go anywhere they wish in the neighbourhood, and that, if it is a reasonable request, it is likely to be granted. This trust has seldom if ever, been abused, and the boys are all taught to use a

[1] H.M. Stationery Office, *Hostels for Difficult Children.*

telephone so that if they miss a bus they can report. It is noticeable that they take comparatively little advantage of their freedom, for there are many occupations in the hostel of absorbing interest, and even in the lunch-hour they usually return to them.

"The most striking thing about the hostel is the relationship between the boys and the matron and warden. It is noticeably informal and this is probably due particularly to the matron. They seem to look on her as a mother and she in her turn never seems to find them a nuisance, busy though she is. When she is working in the kitchen the boys are often around her. . . ."

The second, and here the schools resemble the normal school, in that the child is invited to become a member of the community; but more fortunate than in some schools, he is offered full membership from the earliest stages, and is not left, comparatively unregarded, till he is about to leave.

"In their approach to the individual child the approved schools recognize that his need to be a respected member of the community is as important as his need to be his individual self—with the children who come to them, perhaps more important. Their individualities have frequently had very full if dangerous play; what they lack is the development of a sense of community. Many of them are here, as has been shown, as the result of a conflict, and that conflict has either been external (themselves against society) or internal (themselves against themselves). It is the latter only who are clinical cases for a psychiatrist, and the approved schools consider that only about three per cent. of the boys and girls who come to them are of this kind.

"Recognizing that they have not many children of this kind, the approved schools do not regard the boys and girls sent to them as 'cases'; to them they are Nick or Jenny, or Jim. With them their methods are not those of the hospital (giving treatment suitable for each particular case) but rather those of a club (providing interesting activities for those willing to join in). The child is sent to a particular approved school because, on the information available, that one seems likely to offer the kind of things that would interest and help him. At first he may not recognize them—what happens when he arrives at the school is that he finds there a busy community which he is invited to join. He may refuse or run away; or remain, and still keep outside the community. Because, however, the school is ordered in a way which has a strong attraction for young people, especially for those who have never known order in their lives before, and because it is engaged in many of the things a child enjoys doing, with most boys and girls a time comes when the pull of the community proves stronger than the promptings of selfish desires. The child, expressly or implicitly, asks to join and, once accepted, is often happy for the first time in his life. All approved schools make these two assumptions—that their function is to do the best they can for each child, and that what is best for most children is to become members of a purposeful and ordered community."[1]

It is not enough to offer a child membership of a group, he must also be persuaded that he himself is of value to the group, and that without him the group would be worse off. Even a very small child desires to feel important and to have the impression that he is making a contribution. A bigger child feels the need even more keenly, and is unlikely to develop fully in an environment that does not give him this experience. A very important part of the work of the schools consists in devising situations in which this feeling of importance can thrive. This is largely done through the practical work.

"Before they arrive at their approved schools most boys and girls have come to regard themselves as inferior to others and frequently the first time in their lives they really approve of themselves is the moment they realize they have actually made something by the skill of their hands.

"What gives the teaching in the craft departments its unique value and makes such a strong appeal to so many children is that it is based not on formal exercises but on actual work needed for the upkeep of the school. The cow will not wait to be milked nor the burst pipe sealed till the class have learned the theory of the operation.

"What gives the training lasting value is not only that boys and girls are doing work that has to be done, but also that this work could not be done at all without them.

"A strike of the milk boys or the gardeners or the plumbers would seriously disorganize the school. In addition there is work requiring every degree of skill, or none at all, so that even the youngest or dullest child can feel that there is something he can do which is necessary for the well-being of the community."[2]

This description makes it clear that there are two main aims in this craft-work; giving the child a sense that he is an honoured and necessary part of the community, and increasing his self-respect by this sense and by the experience of achievement. For most children this experience must come through manual work and not through intellectual tasks. It is much harder to devise intellectual tasks that

[1] *Making Citizens*, p. 27. (H.M. Stationery Office: 1s.)

[2] Op. cit., p. 41.

are equally necessary, though this can be done, but most of the boys would never be able to carry them out. There are few intellectual tasks on the level of making a bad rabbit-hutch.

It is in this matter of devising necessary tasks that the ordinary school fails. It is true that, under Sanderson, Oundle had craft shops that would mend ploughs and wagons and do work of public utility in the district. On the intellectual side Caldwell Cook at the Perse School produced plays that the boys felt they had to write. A school magazine, school accounts, vegetables for the school canteen and so on do not fill the range of activities desired, and the failure of the elementary school is largely due, as is most of its failures, to lack of money and suitably trained staff. The ordinary public school which could have what it felt necessary has not realized this need; and has tried to substitute for it inter-house matches and imaginary incentives. When, during the war, schoolboys were employed for a period or so a week making parts for, e.g., aeroplanes, the work chosen was of the simplest repetition kind and called for no thought. It was done faithfully but without pleasure.

Besides promoting this sense of membership of the community and encouraging craft work, the approved schools must concern themselves with the children's health to a degree not really necessary in schools for the more wealthy.

Very revealing is the following account of a boy in a hostel for evacuated children unsuited for ordinary billets: "His I.Q. is 93. Not very low perhaps, but his educational attainment was much lower. He was chronically and (apparently) incurably lethargic and tardy; his nose was always running; he wet his bed nearly every night, and he fouled his trousers regularly. Altogether a nasty, dull, dirty oaf, you might think. But he was never happier than when listening to music or poetry. . . . Incidentally, if I may digress, Leslie serves as a warning against becoming too involved in highfalutin psychology. We discussed his lumpishness, his nose-running, and all the other unpleasant things at staff meetings, and made impressive-sounding notes about "weeping diathesis", but he got no better. Indeed, he got worse. Then one day I was giving the boys their weekly hygiene lesson, when it suddenly occurred to me that we might do worse than try the effect of a nasal douche on Leslie. I will not say it cured him the next day. But after a couple of months of douching, Leslie was a different character; his nasty symptoms near to disappearing-point, and he skittish to the point of being a positive nuisance."[1]

[1] W. David Wills, *The Barnes Experiment*.

Nearly all approved-school children put on weight at such a rate that it is clear they were previously undernourished. This improvement in physical condition has an effect on their nervous stability and general behaviour. It is closely associated with games and athletics. If these are beneficial to the normal boy they are particularly advantageous to a lad who has probably never had the opportunity of proper games before. He can now learn all the virtues connected with athletics and in most cases he does it readily. The great advantage of games over other forms of moral training is that they are really enjoyed and are done with eagerness. Therefore to be prevented from playing for some misdemeanour is quite sufficient punishment.

It is in the matter of supplying the children with occupations that these schools are noteworthy. There is a clear connection between virtue and a power to find amusement. At Cambridge, during an investigation to compare a group of normal evacuee children with those who had been referred to the Child Guidance Clinic, it was found that: "There was a notable difference between these children [normal children] and the problem children referred to the Clinic in their activities and amusements. The problem children seemed usually at a loose end, not knowing what to do with their spare time: they 'play about' or 'there's nothing to do', they 'don't like going to Scouts', they are 'bored', they 'like to go into the town', they 'ask soldiers for money to go to the cinema', whereas the other children seemed to be always occupied with activities, hobbies and games which absorb their energies fully and happily. It is not without significance that many of the problem children seem to find their chief amusement and relaxation at the cinema, which plays a comparatively small part in the lives of the normal children. Only about 40 per cent. of the latter visit the cinema at all regularly, the visits varying from once a month to once a week."[1]

This fact has been understood by the approved schools, and one of the first remedial measures is to try and fill the children's lives with interest. This is the account of one hostel for difficult children:

"This hostel has such a wealth of occupations that all of them cannot be enumerated. The key-note is freedom, and the children are encouraged to find their own occupations and amusements. There is a great variety of indoor games, ping-pong, bagatelle, etc., which can be had on request. There is a small library and the quiet playroom can be used for reading. On a winter visit boys were found playing in the dining-room with a "meccano" set of which not one screw had been lost since the hostel opened. The warden says that, in the main,

[1] Bamster Ravden, *B.J. Psychology*. General Section, May, 1945.

play is invented by the children, realistic games of Red Indians, cowboys, commandos. They make their own play materials, e.g. guns, tomahawks, knives, etc. The boys get clay out of the grounds and do modelling and carve out of chalk from neighbouring pits. The only occupation which is supervised (but not taught) is painting. Individual children have gardens and keep rabbits, or mice if they want to. There is a wireless and boys like listening to certain programmes, 'Into Battle', etc. The children are allowed to leave the grounds by permission, usually in twos and threes. They may go to the pictures, shop or go swimming, etc. Three hostel bicycles have recently been provided through the American War Relief Funds. They are very popular. The warden often takes a few boys with her into the town. They go to the museums, go on the river with her, etc., and in the holidays there may be more organized expeditions. Treats are common but hardly ever all the boys together. A great deal of kindness and interest is shown by local residents. One lady invites eight or ten boys to tea at her house every week of the year. Another has started asking little groups on Sunday."

The approved schools try to do the same things but they are not always quite so successful, usually because there is a shortage of staff to act as supervisors and helpers. Still, the principle is accepted and also the principle of choice, so that children are often divided up into clubs such as young farmers' clubs or dramatic societies each with their appropriate activities. When in addition there is camping, hiking and other diversions in the summer, athletics or games, books, pets and a school cinema there is a wide range of amusements many of which can have a bearing on the child's future life.

In conjunction with this, there is, of course, a certain amount of direct reward or punishment, but on the whole the rewards definitely overweigh the punishment which is kept to a minimum and varied to suit the offender. "Sometimes", it is recognized, "a child who misbehaves may need to be given not punishment but extra affection or perhaps some new privilege or responsibility" and the warden is taxed to explain to others why John is treated in this unusual way. David Wills has described how when he was Camp Chief at the Hawkspur Camp for maladjusted and delinquent boys, a special Camp Council was called to deal with his favouritism towards a certain boy. The boy who called it "demanded, in the name of justice, equal treatment for all". "That", I replied, "within the limits of my human frailty, is exactly what I try to provide." Scornful laughter from Sidney, and several others. "Not uniformity", I said, "but equality. If several of you were ill and the doctor came, you would be justified in demanding that you were all given a bottle of medicine, or whatever it was that your condition called for. But you would look pretty silly if you started kicking up a fuss because Raymond, who has one complaint, is given some physic which tastes sweet and you, with another, were given something that tastes like prussic acid. And you look pretty silly now, only you don't know it. I don't give everybody the same physic. I try, so far as I can, to give them the kind of physic they need. Needs vary a good deal, but my attitude to them all is the same—so far as I can I try to supply them."[1]

Direct punishment is as far as possible avoided.

"Many wardens and matrons realize that where punishment is necessary it should be given in such a way that the child accepts it as the natural consequence of his misdemeanour, and should be fitted not only to the offence but also to the temperament of the individual child. Children are most usually punished by being deprived of special treats, privileges or responsibility, or occasionally of sweets, or they may be sent to bed or to sit quietly for some time. Corporal punishment is generally avoided though in some hostels a smacking on rare occasions is considered useful.

"Sometimes the older children formulate their own rules and punishment either by means of a 'hostel council' or in some less formal way. It is found that they adhere rigidly to their own rules and tend to treat offenders over-severely unless they are given some guidance. Small rewards are occasionally given for good behaviour or for a real effort towards improvement; examples are for younger children—extra treacle for tea, blowing bubbles at bath-time (as a reward for a dry bed the previous night) and for older ones special privileges such as going to bed later, going to the cinema in the evening, or for an outing alone with a member of the staff. In one hostel a special scented soap is given as a reward for personal cleanliness."

The experiences of these hostels where by far the greater number of the children can be said to have been cured of their difficulties is deeply encouraging, the more so because the methods used are those which kind parents could at any time apply; and involve little more than providing an environment in which children can happily pursue interests that seem quite natural to them.

Very much the same principles apply in the approved schools and here there is a statutory regulation to protect the children. "The punishments allowed in these rules are few: corporal punishment (except for girls over fifteen) to the degree used in ordinary primary

[1] W. David Wills, *The Hawkspur Experiment.*

schools; duller meals; isolation for short periods. On the other hand the privileges that can be given or taken away are many. They include unsupervised walks, visits to the local village or nearby places of interest, and to the cinema; use of special rooms and apparatus, smoking for the older boys, expeditions, week-end camps. There is three weeks' home leave a year, pocket-money, payments for proficiency in a craft and so on."

The child is never faced with a blank wall of punishment and is able to see at all times the goal of virtue.

These systems have been markedly successful especially with the younger children in hostels. By far the greater number of these improved greatly. Out of 486 children recorded only thirty-two were classed as not improved, and in most of these cases there was some complicating factor such as mental deficiency.

Of the approved schools much the same can be said. The majority of the children conform and are successful, and this is more astonishing when we consider the material with which they have to work.

"To achieve anything at all with such selfish, vicious, unruly and generally unprepossessing children might seem to be too much to ask of any school. When, however, it is remembered that the need in all children for social approval and participation is as strong as their need for self-assertion, it may be admitted that the school which finds itself too frequently upset by aggressive or rebellious individuals is probably failing to provide for these social needs. . ." But in most cases all goes well.

"The children stay in the schools to which they are sent, and fall in with the life of the community there, not because of a carefully worked out system of rewards and punishments but because they find in the school people and things they like; above all because they feel that here, often for the first time in their lives, they are being given a fair deal."

Books referred to in this chapter:

D. H. Lawrence, *The Rainbow.*
John Dewey, *The School and Society; Schools of To-morrow.*
C. A. Scott, *Social Education.*
H.M. Stationery Office, *Hostels for Difficult Children.*
H.M. Stationery Office, *Making Citizens.*
W. David Wills, *The Barnes Experiment; The Hawkespur Experiment.*

Chapter 2

DIRECT TRAINING

IN the last chapter we outlined the modern technique by which a carefully devised environment allows the child to develop benevolence and the group spirit. He is not exposed to frustration or a sense of restriction, and the path of virtue is made smooth and straight. In explaining the method we did not intend to suggest that a child so treated never gives any trouble to his guardians. No natural child grows up without, at some time or other, resisting his environment. It may merely be energy and a desire to explore; it may be a quite proper desire to do something, which, unfortunately, runs counter to a similar purpose held by an adult; it may be that the irritations of even a carefully controlled environment suddenly become too much for a child, and he relieves his feelings by a fit of temper. Such events are inevitable. But the important thing is that with a suitable environment the period at which they occur soon passes; and the storms that have agitated extreme youth leave behind no lasting hostility or bad feeling. In this chapter we wish to contrast briefly the training by the rod that was so important in the past, and still, unfortunately, exists in some schools. In recent years many countries have forbidden the use of corporal punishment in schools, e.g. Holland, Russia, U.S.A., Scandinavia. England still maintains it, and those who have suffered from it most often speak in its favour. After all it has made them the men they are. There is also a curious impression that the public schools are the last stronghold of the cane, and that to have been beaten two or three times in youth establishes the fact that one had the education of a gentleman. The advocates of flogging overlook the far more dangerous method where each master has a strap in his desk, and uses it with monotonous and unintelligent regularity on all occasions.

The treatment of European children by their elders is a strange and sad story. There can have been few cultures in which the maltreatment of children has been so prevalent, and regarded for the most part with such sanctimonious approval. Among many primitive peoples children are treated with the greatest mildness. The West African child is most conspicuously unpunished, and is charmingly friendly and well-behaved. Margaret Mead records for New Guinea that, apart from a passing smack, no child is punished except for some act that endangers the safety of the group or damages property. A child who loses home care is always adopted by some other family, and no child is ever made to work beyond his strength. It would be hard to find any other major culture, except perhaps the Chinese, that has treated children with the neglect and systematic brutality that has characterized Europeans.

This European treatment of children is probably due to various causes. One is the private ownership of children, another is the competitive nature of private property. If a child is the property of his parents, as he is in law and popular belief, then the parents may, if they wish, maltreat or overwork him. A man may do as he likes with his own. Moreover, should the parents die, no one else is directly responsible for the child's care. This is one of the results that follow from the maintenance of the excessively individualized family. On the other hand, generosity is limited by the need to use all available wealth for competitive ostentation or survival, and another mouth to feed decreases the amount of surplus. The process of thought by which the state has come to accept its part as guardian of children, both to defend them against their own parents, and to guard them when they have none, has been very slow; and is still far from complete, as the Curtis Report shows.

The European child's position has been made worse by the doctrine, held at times by some Christians, that he is born in sin, and that he can only with great difficulty be made to abandon his sin. When dealing with this perverse creature, the greatest strictness in training and the greatest severity in punishment are necessary, if he is to be kept on the slippery path of righteousness. Never did a stone roll down the hill with greater acceleration, than a child, given the chance, plunges into vice. This explains why children were wont to suffer most from those who, themselves, set the highest standard of virtue.

Once cruel treatment is established, it perpetuates itself. The adult corrupted in his youth, satisfies his perverted feelings by cruelty to others, and the Victorian schoolmaster, like many another ruffian, enjoyed the sufferings he felt impelled to inflict.

Quite apart from these reasons, the maintenance of flogging as a method of discipline made a natural appeal to the stupid and the mean. To understand children and to enter into their minds sufficiently to provide an environment suitable for their development required powers denied to many pedagogues. To provide a materially suitable environment cost money, especially in towns. It was a great simplification of the teachers' task to beat children till they endured boring lessons in bad surroundings with passivity. When *The Times* advocated flogging as a method, and railway arches as premises, its policy was all of a piece.

The traditional punishment was flogging, because that is the simplest method of causing the maximum temporary physical pain with the minimum permanent injury. The educationalist therefore spent considerable ingenuity in devising the instrument most fitted for inflicting the exact degree of pain and damage required.

The thought of those who even to-day advocate corporal punishment is not of a very high theoretic order. Pain, and the fear which it causes, are one of the greatest deterrents to action, and as deterrents have played a large part in society and individual life. It can be effective when, for various reasons, no other method of stopping a certain action can be found; and in a very few cases its use is fully justified. But these cases are not many, and in most of them forethought would have rendered punishment unnecessary. A child who persists in doing a very dangerous action in an environment which *cannot* be made safe, may be beaten for his own preservation. But it is far better, if in any way possible, to change the environment. As a cure for anti-social acts a beating is generally only really effective if followed immediately by efforts to help the child to find his proper place in the social group. For example, Vernon Bartlett describes his youthful misdemeanours which clearly arose from boredom and a family who did not spend enough thought on his amusement. The beating that followed his sins seems to have closely preceded his removal to a school where he was introduced to more profitable occupations.

"Robbery had no particular attraction for me, but I was eight or nine years old and rather bored. Why, when the sea and Branksome Woods were within a mile of our house, I do not know. . . .

He tried throwing peas on passers-by, squirting water up through the planks of a bridge, setting the moor on fire and pouring camphorated oil on puddles to see it burn. All had to be discontinued.

"There was I think no solution but to form a little gang of evildoers.

"At first we were content with ringing bells and running away, ragging various people in authority from a safe distance, firing our catapults at street-lamps, or exploring at night back passages or alleys and other people's gardens. But finally all these amusements lost their attractions. Adventure we discovered must be taken in larger and yet larger doses. The obvious next step was to try our skill in competition with that of local shopkeepers.

"There was in the first place, the fun of inventing an excuse for entering the shop at all, for we seldom had the money with which to buy anything. The excuse had to appear so reasonable that it aroused no suspicion, and it was not easy for four or five grubby little boys to make it so. Then we had to decide who was to do the taking and who to distract the shopkeeper's attention. Hesitation or clumsiness on the part of one might give us all away.

"Of course we were caught, caught while stealing an entirely unwanted bottle of gold paint. . . ."

He was beaten by his father.

"I believe all the surviving members of the gang except one have since become excessively respectable and law-abiding citizens. I daresay their respective punishments knocked the desire for adventure out of them. With me it conveniently altered the course of that desire into less objectionable channels. It is more respectable and less risky to pick people's brains than their pockets."[1]

Those who believe in the rod make much larger claims for its powers. It is not, in their view, able merely to stop a certain definite action, it is also a spur to general virtue and more diligent mental effort. Besides, a boy well beaten in youth has acquired a valuable toughness—a beating "makes a man of him", "does him good", teaches him respect for lawful authority, and qualifies him as an empire-builder. There are still a few authorities who have not advanced beyond *The Times*, which, a hundred years ago, scented a weakening of tradition, and thundered on the side of the rod.

"The good old times are gone by when our forefathers were unanimous in adopting the scriptural warning, "spare the rod and spoil the child". In those days there was but one system of education, and that was one of discipline, but now systems are like Falstaff's reasons—"as plenty as blackberries"—modern wisdom has discovered that the passions of youth do not want restraint or correction, but merely guidance: schoolmasters therefore are not to compel the idle or to punish the unruly, but simply to persuade, to show the right way, and to coax their pupils to follow it. There is a significant word of modern coinage that describes this sort of instruction perfectly well—humbug. It is just that—a compound of meddling folly and maudlin sentimentality designed for a market of fond mothers and

[1] Vernon Bartlett, *This is my Life*, p. 12.

illiterate fathers who want to have their sons prodigies of learning, but will not have them whipped."[1]

The history of corporal punishment in schools would be a fascinating study for those who enjoy the contemplation of the follies and cruelties of mankind, and it will be dealt with here briefly. It is treated thus fully only because the battle against it is not won, nor will it be won till all corporal punishment in schools is made illegal, till prefects in the public schools cease to cane boys who walk across a cricket pitch, till no master in a state school has a strap in his desk, and till advertisements such as the following cease to appear in the *Teachers' World.*

"Punishment canes 39 in. × $\frac{3}{4}$ in. excellent quality, pliable, with grip: 3, 2/9d; 6, 5/6d; dozens, 10/–(including postage). Also birches to order."

We are not then attacking a wholly defunct tradition when we give some examples of the brutality of the past. The ideas that inspired it still linger in many minds at present. A teacher's work is trying, and if once he allows his emotions to be engaged, if once he suffers himself to become angry with his class, then blows naturally follow. It is only when a teacher comes to believe fundamentally that such punishment is impossible, that the old tradition will end. For reasons which we have given before, both child and teacher will be happier. The decrease in punishment has already had its effect.

The flogging tradition in education goes back at least to Roman days and the schoolmaster and rod were already in partnership. Horace refers to *plagosus Orbilius* and paintings give us a lively image of the methods used. In Christian times the tendency was reinforced by Solomon's unhappy dictum; and the influence of the Church was added to the natural irascibility of parents and schoolmasters. In the monasteries the flogging of boys was a regular part of the system and it needed the sagacity of Anselm to see where the method erred.

"An abbot, says Eadmer, who was looked upon as a very religious man, was one day deploring to Anselm the impossibility of making any impression on the boys who were brought up in his monastery. 'What are we to do with them?' he asked in despair, 'Do what we will, they are perverse and incorrigible; we do not cease beating them day and night and they only get worse.' 'And you don't cease beating them?' said Anselm. 'What do they turn into when they grow up?' 'They turn only dull and brutal,' was the answer. 'Well, you have bad luck in the pains you spend on their training', said Anselm,'if you only turn men into beasts'. 'But what are we to do then?' said the

[1] *The Times*, 20th March, 1846.

abbot, 'In every kind of way we constrain them to improve, and it is no use.' '*Constrain them!* Tell me, my lord abbot, if you planted a tree in your garden and tied it up on all sides so that it could not stretch forth its branches, what sort of a tree would it turn out when, after some years, you gave it room to spread? Would it not be good for nothing, full of tangled and crooked boughs? And whose fault would this be but yours, who had put such constant constraint upon it? And this is just what you do with your boys. You plant them in the garden of the Church that they may grow and bear fruit to God. But you cramp them round to such a degree with terrors and threats and blows that they are utterly debarred from the enjoyment of any freedom, and thus, injudiciously kept down, they collect in their minds evil thoughts, tangled like thorns. They cherish and feed them and with dogged temper elude all that might help to correct them. And hence it comes that they see nothing in you of love or kindness or goodness or tenderness towards them, they cannot believe that you mean any good by them, and put down all you do to dislike and ill-nature. Hatred and mistrust grow with them as they grow, and they go about with downcast eyes and cannot look you in the face."

This very pregnant passage sums up clearly the argument of the chapter, and shows that the truth has been known many times—and often forgotten; for it was the practice of the abbot, not the precepts of the Saint, that continued effective in education. So much so indeed that when Radley was founded in the first half of the nineteenth century it must be provided with its "whipping room", as much a necessity as any other domestic convenience.

The picture of Eton in the great flogging days shows clearly some of the consequences of this method. From the side of the masters having a boy flogged and flogging him themselves became so much a matter of course that the punishment was inflicted without any concern for its justice. Swinburne, trying to write an unfamiliar metre in Latin verse, was promptly flogged because his master did not know as much about it as he did.

"I wrote", he says in a letter, "a poem on Boadicea, it is in Galliambics, a metre in which there is only one other poem extant, the *Atys* of Catullus: the rules are too long and intricate to give here. I tried to do my week's verses in it once, and my tutor said it was no metre at all and he wouldn't take them, because it was an impertinence to show such a set-up; so it counted as if I had done nothing, and the consequences were tragic."

And in another letter:

"I need not say that I have not the pluck to try my hand again at Galliambics. I should feel at every line as if I were writing my own name in the bill.[1] I showed my verses indignantly (after the catastrophe) to another master and he said they were very good, and I told my master with impudent triumph (knowing he had done his worst) and he was shut up, I can tell you. But that did not heal the cuts or close the scars which had been imprinted on my mind and body."[2]

The effect on the boys was that they learnt little and passed on the brutality to each other. Small boys were unmercifully tormented till they grew too big. Their books were torn up, their clothes rent, any sign of industry was made the cause of renewed persecution. This was the age of fist-fights and in one of these a boy was killed—either by the blows or the brandy that his seconds gave him—but no one seemed to think any action was called for to prevent such a thing happening again.

Usually the boys endured what came to them with the stoicism of those who can imagine nothing better. Occasionally revolt broke out, either of individuals or whole communities. One headmaster who retired not very long ago described conditions even in his young days, as "despotism tempered by assassination" and said he had seen a master's arm broken in a struggle with his class. Men still alive, who started as pupil teachers aged twelve, had to fight all the bigger boys in their class before they could obtain a hearing; and a friend of the authors' took her first scripture lessons in school—on the Good Samaritan—with the Bible in one hand and a cane in the other, because, unarmed, she faced a furious mob.

These are but mild survivals of the disorders of the past; mild because the conditions in schools have lost so much of their severity.

Edward John Trelawney, sent to a small boarding-school at the age of nine or ten, about 1800, records briefly his experiences:

"The headmaster was one of those pedagogues of what is called the old school. He had implicit faith in his divining-rod, which he kept in continual exercise, applying it on all doubtful occasions.

"As my school life was one scene of suffering I am impelled to hasten over it as briefly as possible: more particularly as the abuses of which I complain are, if not altogether remedied, at least mitigated. I was flogged seldom more than once a day, or caned more than once an hour. After I had become inured to it I was callous. Every variation of punishment was inflicted on me, without effect."

He was not one to suffer unavenged. When on a walk, the boys,

[1] List of those to be flogged.

[2] Mrs. Disney Leith, *The Boyhood of Algernon Charles Swinburne.*

by arrangement, attacked the usher in charge, half strangled him, and beat him with hazel-sticks cut in the wood. On their return to school Trelawney overthrew the headmaster by a direct charge, and set fire to the room in which he was confined. He was then sent to sea, where his treatment was not much better, and his reaction almost equally violent.

Of the greater mutinies, the one at Armagh College in Ireland is perhaps typical. Disorder, including placing "bombs", made of gunpowder and brown paper, at the back of the fire, had increased to such an extent that even the headmaster was not exempt. A violent attack was made on one of the assistant masters, who had his head enveloped in a bag and was then pelted with boots, books, dictionaries and any missiles the schoolroom held. When, as punishment, the boys were deprived of their weekly holiday they barricaded themselves in a dormitory, and for two or three days stood a siege, driving off their attackers with shotguns and small shot. They capitulated when they ran out of water. It is interesting that they submitted meekly enough to the consequent floggings; the mutiny was directly caused by the loss of the holiday that they regarded as a *right.* A recent unhappy event in an approved school suggests that even to-day similar provocation may result in violent action.

A case observed some years ago by one of the authors shows how closely methods of discipline are connected with children's behaviour. In a certain school, in a mining district, terror reigned. The first sound heard on entering the school was the slap of someone being hit with a strap; and one could not be in the building for a minute without hearing it. When students were sent to the school for practice, the children siezed their opportunity and rose in riot. The classes became shrieking mobs, totally uncontrollable. Shortly afterwards the old headmaster retired; his successor collected all the straps, and forbade the teachers to use corporal punishment. Within six months the school was quiet and orderly, and any one could teach in it. In a wide experience of schools it has always been found by the authors that the order of the school is best where there is least formal punishment of any kind.

The tradition of brutality, well-established in English education, acquired a definitely religious element in the nineteenth century. The theory of original sin entered on a new era of influence with certain sections of society, and the increased tendency to patriarchy strengthened the father's claim to absolute obedience. The complete absence of any idea of democracy in family life is one of the outstanding social features of the early part of the century. *The*

Fairchild Family, as usual, shows with remarkable clearness how all these ideas worked together. In the chapter significantly headed "Story of the Absence of God" we get an account of how Mr. Fairchild tried to make Henry learn Latin. Henry was about seven years old.

"About this time Mr. Fairchild thought it proper to begin to teach Henry Latin. Latin is a very difficult language and requires many hours of hard labour before any little boy can master it; but it is necessary for every boy to learn it who is to become a clergyman. Mr. Fairchild wished to bring Henry up to be a clergyman, and it was Henry's own wish also. When Henry got his new grammar and dictionary and Latin exercise-book he was much pleased, but he was not so well pleased when he found that he could not learn his Latin Granmar and play with the hare too half the morning as he used to do when he had only spelling and a verse from the Bible to learn every day."

Trouble began at once. Henry proved inattentive and wouldn't learn. After three refusals Mr. Fairchild grew wroth.

" 'This won't do, Henry, you shall say that lesson before dinner-time or have only bread and water for dinner.' Henry made no answer."

At dinner-time it was not known and bread and water followed, and Henry was left in the study till tea-time. At tea-time Henry was equally stubborn. Mr. Fairchild then took a small horsewhip and making John hold him he flogged him well and sent him to bed, telling him he must say the lesson before breakfast.

At breakfast-time it was no better and Mr. Fairchild at last made an enquiry.

" 'Henry, I know that you could have learned the lesson with ease yesterday morning before eleven o'clock. Tell me now wherefore you would not.'

" 'I don't want to learn Latin', said Henry.

" 'It is my pleasure that you should', said Mr. Fairchild. 'Tell me now at once, will you learn this lesson or not?'

"Henry made no answer.

"Mr. Fairchild got up and walked up and down the room in great trouble.

" 'Henry, listen to me. When wicked men obstinately defy and oppose the power of God, he gives them up to their own bad hearts. I stand in the place of God to you, whilst you are a child, and as long as I do not ask you to do anything wrong, you must obey me. . . I will have nothing more to do with you, go out of my study immediately' "; and poor Henry found himself cut off from all intercourse with the family and, alone and outcast, had nothing to do but wander round the countryside oppressed by the sense of sin. So heavy did this become that after two days, urged on by the dying Charles, he made his peace with his father and submitted to learning Latin.

To all these attitudes, convention, domination, religious sentiment, we can add the pleasure which the executioner received from the flogging. In a strangely unbalanced biography of Dr. Sewell, the rubicund, self-advertising Puseyite, we find this account of his experiences. Sewell was largely responsible for the foundation of Radley and became its headmaster in 1853. He is proud of his flogging and enjoys it. He likes to boast about it to visitors and to explain his system:

"The corporal punishment I adopted after some consultation with physicians was this. I wished to keep the birch for the higher and more disgraceful forms. I was afraid of using the cane across the bare loins knowing it might be dangerous. I used the cane therefore where they would certainly feel pain but would, to borrow Dr. Johnson's euphemism, be 'fundamentally sensible'. A boy was not caned until his name was three times down in the black book, so he had three warnings, and at the end of the week the score was cleared. I kept behind and beyond this a flogging with a regular birch, reserving it for very disgraceful offences. Once I gave a public flogging, I will not say before the whole school, but in the ante-room where the boys could hear but not see. I shall not forget the boys' faces, 150 of them, when I came back into the school. It was a very bad case of running away. I was obliged to treat it as I did. . . ."

Or in a speech: "It was in such trials of extreme severity that I learnt how deeply there was implanted in all your minds the habit of confiding, affectionate obedience. Again and again, in solitary hours, in the silence of the night, I go back in thought to some of those scenes, full of bitter pain and anxiety at the time but now treasured among my most blessed recollections, and I can understand not only how fervent love may submit to the necessity of inflicting chastisement, but also how submission to chastisement will quicken and perpetuate that love. To this hour, some of the most delightful, touching, blessed associations I have are connected with the whipping room at Radley."

What wonder that some Cambridge wag devised the couplet:

> I love her, though she's petulant and cruel;
> As Radley boys adore the Reverend Sewell.

Unfortunately these ideas have not really died out. There is a type of school disciplinarian who, instead of focusing attention on positive good, and forgetting as nearly as possible the ill, is always devising school courts, councils, parliaments and what-nots, so that as many boys as possible may be involved in this work of punishment, and the idea of sin kept as firmly as possible before the children's minds. In the newspapers recently was a case as revolting as Sewell's—worse perhaps, because the victim was only six years old. The following is the newspaper report of the case which the parents brought against the school. The name of the headmaster has been deleted.

"Peter —— six-year-old son of wealthy London parents, was summoned to the Fifth-Form classroom of —— Preparatory School, on the orders of the 'President of the Court' to face justice from his schoolmates.

"Accused of hitting a young boy called John across the nose with a steel door-handle, Peter had offended against the code of school rules.

"Sitting in judgment on him were the 'President of the Court'—the headmaster, bespectacled Mr. —— and two boys, captains of school houses.

"Peter was defended by 'counsel', and a jury of boys from the Middle School heard the evidence. After a two-hour hearing he was found guilty—a verdict that meant a beating from the headmaster.

"So Peter was taken to a dormitory, ordered to bend and touch his knees, and given six strokes with a strip of rubber 17½ inches long, 2½ inches wide and a quarter of an inch thick.

"Mr. —— had never liked the use of canes or birches on the bare skins of his pupils—"so I decided on this modified form of whacking rod", he told me yesterday.

"In the High Court Peter's parents, who had brought an action against the headmaster claiming damages for excessive punishment of their child, lost their case. Mr. Justice Wrottesley gave judgment for the headmaster with costs.

"Pending a possible appeal a stay of execution was granted. Mr. —— walked away to tell me the story of his school's 'self-government'.

" 'You might call —— House a small State with its own rules', he said. 'It is our tradition and there is no one-man rule at the school.

" 'The boys have helped over years to draft our code, which lists behaviour and offences. It sets up three school 'Courts'. We have a court which deals with trivial offences, and the most important of all three is the Central Court which deals with offences such as theft or bullying. Each house sends two advocates for the courts. When a boy is judged by the Central Court he has the right of appeal to a Court of Captains—schoolboys from each house.

" 'We see to it that the juries have no prior knowledge of a boy's alleged offence. There is nothing injudicial about our system.

" 'Officiating officers of the courts wear special sashes across their chests'."

When one reflects that all this parody of legal forms was to punish a child of six who, having picked up a piece of a broken door-handle, ran about waving it till he bumped into another child, it is clear that its author must have had a mind naturally warped and working in a tradition of brutality very deeply established.

It is unnecessary to labour, at this point, the damage such treatment does to all types of children; it is more important to point out its ineffectiveness, and the totally thoughtless way in which it is often administered. The following reminiscence of Sir Ian Hamilton, a man of the highest ability and character, shows how blind and senseless the pedagogue who relies on the cane can be. This was his experience at Wellington, a school at which he learned nothing. The teaching was as careless and unintelligent as the discipline.

"Why Benson (the headmaster[1]) should have pitched on me I cannot say. Our relations had been intimate in a sense. That term he had set before himself the high design of making me punctual—so to his study every morning at 9.30 I brought a little note from my form masters. On reading this he rushed about searching for a cane which was never by any chance to hand. A good deal depended upon the cane. There was one special one which grew larger and heavier towards the business end. The moment he found one he laid on to my back till all was blue, especially my back. When I went to the bathing-lake and stripped I became the cynosure and stupor of the crowd. The blues of the previous week had changed to green and yellow, whilst along the ribs under my arms, where the point of the cane curled, the stripes were dark purple—and yet my dislike to early rising was to remain for many years a constitutional infirmity."[2]

When considering this type of treatment of children it is foolish to say that Swinburne and Sir Ian Hamilton survived it to make their specific contribution to society. The best will survive almost anything, though they may suffer lifelong difficulties which they

[1] Later Archbishop of Canterbury.

[2] Sir Ian Hamilton, *When I was a Boy.*

would otherwise have avoided. It is the average on whom most damage is inflicted. The boy who loses confidence in himself, who is made to feel a slave, may never dare to take the position in life for which his abilities would otherwise have qualified him. Others who have suffered in the same way may try in later life to restore their self-respect by rowdiness, or cruelty, or the domination of others. *The Way of all Flesh* is a most interesting study of these two reactions to tyranny.

In society, as in school, the most noticeable result of the greater mildness with which children are now treated is the improvement in orderliness and general humanity. Boys have ceased to torment cats or throw stones at frogs. In one otherwise rough and unruly school a cat and kitten held undisputed sway in playground and classroom, and walked abroad followed by urchins only craving the honour of stroking their heads. The poem recited with such satisfaction by Mr. Leo Hunter just over a hundred years ago seems to belong to a very different world.

Can I view thee panting, lying
On thy stomach, without sighing;
Can I unmoved see thee dying
On a log
Expiring frog.

Say have fiends in shape of boys,
With wild halloo, and brutal noise,
Hunted thee from marshy joys,
With a dog,
Expiring frog!

To-day any hunting is done carefully with a net for the nature-study aquarium.

The same increase in mildness appears in the undergraduate. Very few drunken youths have rat-hunts in St. Giles or knock policemen's helmets over their eyes. All the jolly revels of the Mohawks, or even of the Boat-Race nights at the beginning of the century, are beginning to look silly or disgusting. The child in the home is cheerfully and actively co-operative instead of passively obedient or "rude and wild". Democracy in the home, and an environment in school that avoids punishment, have made a very great contribution to this pleasant change.

Books referred to in this chapter:

Vernon Bartlett, *This is my Life.*
Eadmer, quoted in Dean Church's *Life of St. Anselm.*
Disney Leith, *Boyhood of Algernon Charles Swinburne.*
Edward John Trelawney, *Adventures of a Younger Son.*
William Trench, *Realities of Irish Life.*
Lionel James, *A Forgotten Genius.*
Sir Ian Hamilton, *When I was a Boy.*
Charles Dickens, *Pickwick Papers.*

Chapter 3

TEACHING THE MYTH

THE external control of behaviour can be achieved most effectively, as we have said, by making a child a member of an ordered community, through membership of which he can develop in what seems to him comparative freedom. It can be achieved less successfully by constraint and fear. But external control is only the first step on the road to virtue, and all later and more valuable control must come from the man himself, and this is achieved by influencing his ideas and beliefs.

We have spoken earlier in this book about the part the myth plays in society. It plays that part by the hold is has over the mind of each member of society, and the teaching of the myth is one of the most important parts of education.

This teaching of the myth is a very complex process, and only a small part is done in school. By the age of five a child has learnt much about such matters as private property, the structure of the family, social status, the ordinary conventions of behaviour. This knowledge is not formalized and explicit; it is implicit in his way of life, and has the influence over his conduct that such implicit beliefs have. This effect of ordinary social life continues, and as the child grows to a man he continues to absorb the beliefs and practices of the world about him, and to reflect in his behaviour the ideas of his age. In addition the literate member of society reads and acquires ideas from books and newspapers. He goes to the theatre or the films; he attends church; he is a member of a political party or even a trades union. From all of these he receives ideas and a point of view. Deliberate, direct moral teaching seems to be a small part of all this, but it is none the less important.

We have said earlier that a set of ideas has most effect on the mind when it exists in isolation, and that the closed society is often considered necessary for perfect political indoctrination. This is true of the education of children. Ideas are learnt, and, if there is nothing to challenge them, they will take complete control of the mind. At the same time modern societies for the most part contain various types of ideas, and thus no set is without a rival. It is for this reason that so many places of education try as far as possible to gain complete control of their pupils, and to keep them in an environment which excludes all antagonistic ideas.

Especially is this so with religious or semi-religious institutions, convents or public schools, which maintain that they are superior in religious knowledge or in gentlemanliness to the surrounding population, or the homes from which the children come. This claim to isolation is made by whole states, as we have seen, when they wish to teach something that is at variance with world opinion. The schools do on a small scale what certain states have tried on a large one. The energy of this demand for segregation varies with the body demanding it. The Catholic Church, always very conscious of its corporate existence and superiority to the non-Catholic world, has generally emphasized it most keenly.

"The Catholic school exists primarily in order that children born of Catholic parents may be prepared for life—and through life for death also—according to the Catholic religion. Because this religion is indivisible it can have no compromise with any hypothetical 'basic' or 'Biblical' Christianity. Because it is an integrating, all-pervading factor, it cannot be taught solely as one of a number of autonomous, uncorrelated classroom subjects. Because it is indissolubly bound up with the organic life of the Catholic Church its teaching cannot stop short at classroom exposition but must find vital expression in an experience of the Church's sacramental and corporate life lived in common and shared in common by pupils and teachers alike. It means an habituation into the regular practice of Catholicism, and a fostering of the full Christian view of life, a process which cannot be allowed to stand divorced from the purely intellectual side of education if the tragedy of divided minds and divided personalities is to be avoided."[1]

Such a policy was carried to its full conclusion in the schools of Port Royal which flourished during the reign of Louis XIV in France. They were based on the moral claim to superiority by the Church, and this claim was reinforced by the firm belief in original sin. The world was wicked, the hearts of children prone to sin; only by a most rigid system of exclusion could the children be kept from outside contamination or from corrupting each other.

"To remedy such great abuses we were trying in the Little Schools to remove from our pupils everything that might harm them. We took care that they never heard or saw anything which might wound the modesty and purity which are so delicate at that age. We tried to have them in blissful ignorance of all things the knowledge

[1] H. O. Evennett, *The Catholic Schools of England and Wales.*

of which might hurt them, and to keep their eyes always closed lest they should ever see anything, the mere sight of which can deal the soul a mortal wound. But while it is good that children should ever have this happy ignorance which preserves in them Christian innocence, it is desirable that they should grow in knowledge and wisdom, that they should not be blind to the good nor careless when evil is to be avoided. . . . On the other hand, since they were busy, to the full extent of their capacity in the study and practice of piety, all leisure to busy themselves with evil things was taken away from them. For this reason the teachers kept them always in their hands."

How close this is to the practice of the Fairchild family is clear, and the same belief in original sin and the same conviction that children have no power of judgment and can only be trained by habituation lie behind both systems—however shocked Mrs. Sherwood would have been to find herself in such ideological company. In the Catholic schools the boys were kept under continual observation and instruction, and the girls were even more closely watched.

"The little girls will never be allowed in the reception room alone neither when they are very young nor when they are older, unless with their father and mother if they wish it—and that only for a very short time. The little girls must never be allowed out of sight lest they fall and hurt themselves. They will not be allowed to play together in a remote part of their room, but they will be watched incessantly so that their little faults may be corrected. The older girls will not be in any way exempt from this supervision: on the contrary, since the ill effects that result may be greater, we shall exercise an equal or greater care that they may never be left without some one to look after them. They must never be allowed to whisper together even for a moment. One of the mistresses should sleep in their room and when walking about the convent to the choir or refectory they must always be accompanied, care being taken that they do not keep close together. In short, continual vigilance should be exercised so that, as far as possible, they may not lead each other astray: for that as a rule is what most corrupts young people."

The slightest sign of independent thought or action involved dismissal. While at school the intellectual food was of the meagrest. Boys learnt Latin and, if they stayed long enough, studied Aristotle according to the established interpretation. The girls read lives of the saints and were prevented as far as possible from doing even the mildest literary exploring, and "this is what almost always results when they have books in their private possession and at their own disposal".

"They are never allowed to open a book that does not belong to them, or to borrow them one from another without the mistresses' permission."

It is no wonder that Racine remarks complacently: "There never was a refuge where innocence and purity were more secure from the contaminating influence of the world".

This was the theory. It is worth noticing, as a proof of the vitality and longevity of this policy of segregation that in 1867—after Oxford had opened its doors to Catholic students, Manning and Ward obtained from the Pope an *Admonition* forbidding Catholic youths to attend.

"What was, in fact, feared, and not unnaturally, was the danger to faith and morals which not only the formal academic teaching, but also the whole intellectual and social atmosphere might bring to Catholic youth fresh from Catholic colleges."

It was not till 1896 that Catholic youth was considered more hardy, or Oxford less dangerous, and the *Admonition* withdrawn.

It is not the Catholic Church only that believes in this policy of the closed society. Nearly all the advocates of boarding-schools have something of the sort in mind, but they do not all give the same reason. Mr. Donald Hughes is concerned at the "sub-Christian" state of the general population and desires a higher moral standard.

"We have noted and deplored the incoherence and instability of the world in which we live. We have agreed that we are aiming at producing, by our system of education, Christians, who recognize an ultimate pattern and are seeking to live their lives in accordance with a recognized ethos. Now that ethos is not the ethos of the general society. I believe therefore that our system must be able to claim as far as possible the undivided attention of the boys; the voice of the sub-Christian society must not be allowed to keep breaking in on him. . . . If we are trying to teach Christian truth as the final authority is it fair to the boy that he should every day have to go through the struggle of divided loyalties by being reminded that his parents have rejected this very authority?[1]

We have quoted earlier the very modest collection of virtues that Mr. Hughes hopes to teach in this superior atmosphere. It may be worth while to quote it again, because it is almost exactly what a school that claims gentlemanliness rather than Christianity would say it was aiming at. "Thus we teach that lying and stealing and the exertion of an immoral influence are wrong: on the positive side we try to inculcate honesty, industry, consideration for others, loyalty

[1] Donald Hughes, *The Public Schools and the Future.*

in positions of responsibility and the ability to lose sportingly and to win gracefully."

But even schools that hold a very different status claim a closed society to teach their particular virtues. The Camp School for slum children must hold its pupils for two years, so that it may teach them to wash; and the children who attend a boarding-school specializing in ballet leave it with the traditions of the profession deeply ingrained.

Day schools, of course, have never made claims of this sort. They have hoped to influence children in certain directions but they have never had the child's world so completely under their control that they could propound ideas without fear of opposition. The child could always check the claims of the school against the common practices of his other environment; and the more senseless educational appeals, particularly those concerned with the "honour of the school", lost much of their persuasiveness when the criticism of parents was brought to bear. Moreover the momentous problems and decisions that burden the life of a boarding-school child lose their overwhelming significance when viewed from a mile away.

This idea of a closed society is very repugnant to many minds. Some of those who support the public-school system because it teaches so effectively the ideas and manners that in England characterize the "gentleman", expect the boy to leave with his intellectual ideas free and unfettered. There are others who, perhaps because they have suffered in their youth, claim for the child a complete freedom of thought. "Children", they say, "should be led to discuss everything, and to make up their own minds". This is simply phrase-making, and is dangerous because it is quite false to facts. In the first place the topics about which children are capable of making up their own minds are few and unimportant. They are handicapped mainly by ignorance: all the important questions, from politics and religion to the choice of a career, involve factors that are quite outside the grasp of their minds. They need very careful information and only gradually reach an age when they are able to judge of their own inclinations in the light of what they are told. In the second place, if they did really make up their own minds, and in a sense unfavourable to these broad-minded reformers, there would be horror and dismay. A child who decided to steal, to indulge in sexual perversion, and do no work, is not at all what they have in mind.

From the age of ten or eleven children become interested in the world beyond their own sphere. It is not a very intellectual interest and is often satisfied by looking at picture-books of foreign lands and being told stories of history or geography. It also leads them to read about murders in the newspapers and to try and collect facts about the "fastest train" or the "largest liner". The lessons that some schools give on "current events" are intended to use this interest in the best way, and a world war will enable the enterprising teacher to communicate a large number of facts to an unusually receptive class.

This teaching and this training, important as it is, is not the kind of discussion that leads to decisions on important matters. It is not till a child is fourteen or so that he normally becomes at all politically conscious. He will generally take sides in an election, but the side he takes is determined by emotion rather than any knowledge of the facts. He will either follow his family and side with the Conservatives, or revolt against his family and work for Labour. Political propaganda is seldom honest. It is largely an affair of slogans and appealing to the lowest intelligence of the group. At sixteen or so children enjoy discussing politics and if there are good books provided they may do so on a reasonably well informed level. As any political party is legal, and as, where morals are concerned, no great difference exists, no one much minds freedom of political discussion.

It is quite a different matter with religion. Although all faiths are legally indifferent, few teachers would be pleased if, as a result of their discussions, their children elected to join a faith alien to their own. All that free discussion means in these cases is that the child should be allowed to put a few objections which are promptly and effectively answered. In most cases even the idea of a discussion is frowned upon. The children are made to attend corporate worship till the forms are embedded in their minds, and it is assumed that this conditioning will make them proof against antagonistic ideas when they meet them later.

Still less is it contemplated that children should freely decide on sexual matters or obedience to the law.

If children are, in fact, to have the opportunity of decision they must be brought up in a mixed environment such as is supplied by home and school, by going to the theatre and cinema, by talk with various people and by reading books. But even those who believe in this kind of life for children expect that some element, usually the home, sometimes the school, will have a preponderant effect. The pattern of life that the child sees displayed somewhere should be the standard, and against it the other types of life are measured.

There are very few people who still believe in the fully closed environment as a place to rear children. The argument rather is about the degree of selection that should be exercised and the age at which a child's ideas are so firmly fixed as to be essentially unalterable. The selection of the environment generally goes on, for those who can afford it, till seventeen or eighteen; and it is then assumed that the character is stable. The selection of the environment is achieved at a very early age by the parents' choice of a nurse or of little friends, and later by the choice of a school.

This choice of a school is not an easy matter. Many people assume that a school environment will be good if the pupils come from well-to-do homes. They will then, it is felt, speak with an acceptable accent, and have the right ideas about private property. Other parents are concerned with the methods of teaching, the games, or other facilities. The private fee-paying school enables those parents who desire it to obtain for their children a special type of environment. The greater part of the nation, who send their children to the unselected state schools, have to trust to the overriding influence of home life to counteract any company or teaching they dislike. In most cases the home influence proves dominant.

The English parent appears to the intelligent foreigner singularly lax in the training of his children. The richer classes almost universally debarrass themselves of the supervision of their children's education. For the small children they employ nurses of another social class, and they send the babies to boarding-school—often at seven, occasionally at two—thus losing any real control of them or their ideas, for the rest of their lives. The artisan class takes more care, day schools are universal and the home influence is consequently of far greater importance. In Continental countries boarding-schools hardly exist, and even the rich are forced to do something personally about their child's training. The working man or the peasant prides himself greatly on his child's manners and development, and would never entrust so delicate a matter as his training to another. On the other hand the English boarding-school for the rich prevents the development of extreme forms of egoism. We said earlier that the story of *Kong at the Seaside* could never have had an English setting. The reason is that the child, boy or girl, at an English boarding-school, is never allowed to form quite such an egoistic pattern of life, or to think money of quite such importance. Whatever may happen to him in the world at large, while he is at school he must do his share of the jobs and be ordered about by older boys and masters just like anyone else. No amount of money will buy him exemption, or win him any prestige except with a small and unimportant minority.

The social as well as the educational teaching of virtue, in whatever way it is carried out, involves two elements—the presentation of ideas, and the provision of rewards and punishments according as the ideas are rightly selected and acted on. The punishments we have discussed. The rewards vary. They may be something material and intrinsically desirable, as a piece of chocolate for getting all one's sums right; but, on the whole, it is social prestige and approval that is offered. One of the determining characteristics of any society, whether a home, a school or a state, is the qualities that carry prestige and those that involve disgrace. Not only will individuals try to achieve the desired characteristics, but those in charge of a child will encourage his efforts. To take skill in athletics as an example. The captain of football and the members of the XV are great men, looked up to with veneration. The boy dreams of joining their Olympian ranks; and form master, house master, games captain and coach all help him to acquire the necessary skill. It is the same if the offspring of an academic family turns his mind to learning, or of a commercial group to money-making. Thus the qualities that society has chosen to honour become intensified, and, conversely, those disliked decrease. An example of how this social bias can affect a state is afforded by the contrasted culture of ancient Athens and Sparta, where, separated by, roughly, the distance between London and Bristol, two states of like race and environment developed intensely different and antagonistic ways of life. It also explains how an age can be dominated by a set of ideals and values quite different from the preceding period. The rigid morality and the importance of wealth and its evidences through most of the nineteenth century is an example of this, and certain groups in the community have retained till to-day similar standards of value. On the other hand a type of behaviour, previously unobjectionable, may acquire social disapproval and gradually cease to be usual. Drunkenness, universal in the eighteenth century among the gentry and the aristocracy, became much rarer in the nineteenth among the well-to-do. It was then the characteristic behaviour of labourers and soldiers. It has now become distasteful to all sections of the population, and the comparatively moderate potations of "bottle parties" now provoke censure from all but the few.

This general social pressure is reinforced by art and literature. To-day films are the most potent instrument for the dissemination of these standards, books next, plays and pictures last. When nations are conscious of their myth and are actively trying to establish it, the control of films, particularly of films for children, is one of their most definite steps. In England we have not yet organized special children's cinemas and the half-hearted attempt to keep those under sixteen from seeing most of the films shown is deplorable. It is ineffective and it brings the law into contempt. The attempt of some cinemas to organize children's film-going on Saturday morning, and then to provide a suitable programme, is an acknowledgment of a problem rather than its solution, though the programmes thus offered are good. The ideas that are displayed in a film are undoubtedly important for the effect they have on the minds of the audience, but no one has clearly and explicitly set out what ideas these should be. A film such as *Henry V* was the representation of the chivalrous tradition; a film in which a silly girl achieves success and money as a variety artist, crooning about love, while a chorus shows its legs, holds up a totally different set of values for contemplation. What does the intelligent and critical adolescent make of the morals of *The Merry Widow* (to digress for a moment from films), or of the comic in which people get into the wrong bedroom and fall into pails of whitewash?

This absence of standard, or rather the confusing of many standards, is inevitable when there is no acknowledged canon of morality. The conscientious director anxious to put the best before the world can only use his own judgment, and it depends on him what the result will be. In a country with a clear-cut idea of what ethic it wants, films as well as other forms of art can conform.

The danger of course is that the State-demanded ethic may be bad, or if not bad, static. The Nazi ideas were largely bad. If we look back to history we can see many types of ethic that, considered good in their day, we have outgrown; and official regulation might have perpetuated them. Further, each nation has its own ethic, and if there are to be cultural interchanges we must accept what others give. This is particularly true of England and America. There are very deep cleavages between certain points of ethic in the two countries, and these differences appear in their films. When each country is doing its best the differences are less, but the lower the standard falls the more they appear.

Literature is a field in which the myth is mirrored. Some books are definitely in the chivalric tradition, and can be used to teach it. There is an amusing account of the education of an Indian prince by an Englishman, and the way in which he tried to make *Rupert of Hentzau* serve as the basis for lessons in honour and truth. In one scene the villain, Rupert, and the hero, Rudolf, are confronting each other with loaded revolvers across a table. Rupert suggests a duel with swords as a more gentlemanly way of settling the affair, and Rudolf consents. Rupert continues:

" 'Put your revolver on the table, then, and I'll lay mine beside it."

" 'I beg your pardon', smiled Rudolf, 'but you must lay yours down first.'

" 'I am to trust you, it seems, but you won't trust me.'

" 'Precisely. You know you can trust me; you know that I can't trust you'."

The shining eyes of the young prince during this scrap of dialogue made the tutor feel that the romance was doing its work, and instilling an enthusiastic admiration for a high sense of honour. When finally "With an angry muttered oath Rupert flung his revolver on the table", the child could no longer restrain his excitement. "Now's his time!" he exclaimed, jumping up and down. "His time for what?" queried the puzzled tutor. "Why, to shoot him, of course", was the devastating reply.[1]

History is used extensively in this way and the selection of episodes for the qualities of heroism, courage, determination and so on has a considerable effect both on books and teaching. We have discussed elsewhere the Englishman's view of his own history, the imparting of this view is part of the process of education. How important the teaching of history can be in building the myth is shown by the eagerness of the Nazis to provide the nation with textbooks, and the zeal with which the Allies have collected and destroyed them. The new textbooks, still hardly written, must present facts in a very different selection.

The books written specifically for children, on the commercial level, or for adults whose mental processes have remained largely on the childish plane, are nearly all of the adventure type. In them virtue, truth, honour, are always triumphant. Cruelty brings its appropriate horror, and deceit and lying their due punishment. We have no fear of what the end will be. Were vice triumphant the book would be considered fit only for the sophisticated. This tradition, in England at least, is quite firm. It is maintained, largely for financial reasons, by publishers. As one once said, "We have so good a reputation that librarians put our books on the public shelves without reading them"; and therefore, for fear of offence, the author must cut out a passage relating to childbirth. This unofficial censorship is not resented, except by a very few, because it is, really, in accord with the general belief of the nation, and the hero of *Bar X Range* has gathered to himself the mantle and halo of Robin Hood and other national figures.

It is essential for this kind of moral teaching that prestige should attach to the character who shows the right qualities. The hero who is

[1] This story is taken from Edmund Campion's *Youth and the East*.

faithful and true, gallant and resourceful, must be successful, or if not must fail in such a way that he wins esteem. He must be some one who can act as a focus for the wish-fulfilment dreams of the audience, and the reward he receives must be one that they themselves desire. The success stories that end in wealth appeal to one type; those that bring marriage, to another; while a third may desire glory even with death. When later in this chapter we give examples of the moral story as written for didactic use, it will be seen that one of the great weaknesses of the type is that the heroes of the tales are so unlikely to cause any emotion or interest in the auditors. They could never identify themselves with these often anonymous and always insipid characters.

It is occasionally amusing to see the struggle between the customary morality and the belief that something salacious might be more profitable. There are various paper-covered books to be seen on station bookstalls on whose coloured cover a charming young woman poses in very revealing underwear. The titles are variations on the theme *Love and Women*, but the text within is a sexless adventure story of chivalry and ferocious fist-fights. Purchaser and publisher thus achieve a double satisfaction.

There is another type of literature to be considered, the comic and the cheap magazine. This is the normal mental fare of children up to about ten years of age and, in its more expensive form, of women of all ages. The child's comic, *Krazy Kids* or *Jumbo Jim*, is a harmless affair of simple tricks such as a child dreams of playing on his elders, but does not, because, really, he wishes them well. Perhaps psychologically it is the least dangerous way of working off the mild feelings of frustration and annoyance that must afflict any child, however well treated. The schoolboy stories that follow in age often concern mutinies at school, or the selection of boys for the cricket team on their power to hit, rather than their science. There are boy detectives who unravel dreadful mysteries, and prigs and swanks come to a deserved downfall. It is all simple and innocent wish-fulfilment. It is not particularly moral, but it is certainly not immoral, and probably conduces to morality by offering a harmless outlet for slightly anti-social emotions. George Orwell in his essay on boys' periodicals is more interested in the social content of these stories; in their attempt to initiate the less privileged boy into the mysteries of the public school, which for them, as for Humbert Wolfe, had the halo of glory. It is quite probable that this prestige value attaches to an expensive type of education, but it is also possible that the real charm of these stories, as of the school stories that girls read, is the escape from family life, the gang spirit of being all children together, and the presentation of the minor peccadilloes of childhood, that the intelligent parent waves aside, as matters of extreme importance on which great issues hang.

At a later stage the papers provided by a commercial press for girls and women are almost exclusively sexual. The morality, the technical morality, is always impeccable. Life is seen through a wedding ring. It is always the good, quiet girl who gets her man. At the same time by presenting a woman's life as first a hunt for a mate, and then a struggle to keep him, and lastly the surge up the financial steps to a better house and garden, they are unduly limiting the scope of women's activity and indirectly discouraging them from trying to acquire skill and a personal position. The girl who refuses to learn a trade and waits, doing the simplest and stupidest tasks, "till I marry", has been affected for the worse by the myth that these stories embody; as has also her employer, who refuses to consider her as a valuable individual and give her the training she needs.

The school moralist too often ignores this vast pressure of social standards. He is concerned, as was the worthy Gustav Spiller, with the formal lesson attempting to teach some moral idea in isolation from the environment. The ineffectiveness of the Japanese pictures recommending peace, could be adequately matched by the failure of sermons on the beauties of poverty and humility preached to certain congregations. Even if such sermons are listened to, and they generally are not, ideas which fall on stony ground wither away.

The moral story intended for the edification of the young tends to provoke our disgusted mirth. Some of the virtues recommended are very admirable, occasionally a story has a certain charm, but in the main the soul sickens at their unreality, and can imagine no healthy child feeling anything but scorn at such a picture of life. We naturally feel most hostile to those moral stories that recommend virtues we have ceased to esteem, or hold up rewards that we no longer think worthy, however convenient it might be to receive them. The saintly children who died beautifully, after having learned all the collects for the year, fill us with a furious pity; the industrious apprentices whose wages are always being doubled, though less painful, fail to stir us to emulation. What would a child think of the following story?[1]

"A few years ago the owner of a large drug store advertised for a boy. The next day the store was thronged with boys applying for the place. Among them was a queer-looking little fellow accompanied by his aunt. 'Can't take him', said the gentleman, 'he's too small'.

[1] Adams, *Exposition and Illustration in Teaching.*

" 'I know he's small', said the aunt, 'but he's prompt and faithful'.

"After some consultation the boy was set to work. Not long after a call was made on the boys for some one to stay in the store all night. The other boys seemed reluctant to offer their sevices. But this boy promptly said,'I'll stay, sir.'

"In the middle of the night the merchant went into the store to see that all was right, and found the boy busy at work cutting labels. 'What are you doing, my boy', he said. 'I did not tell you to work all night.'

" 'I know you didn't sir, but I thought I might as well be doing something useful.'

"The next day the cashier was told to 'double that boy's wages for he is prompt and industrious'.

"Not many weeks after this, a show with wild beasts was passing through the streets, and naturally enough all the hands in the store rushed out to see them. A thief saw his opportunity and entered by the back door to steal something. But this prompt boy had stayed behind. He seized the thief, and after a short struggle captured him. Not only was a robbery prevented, but valuable articles stolen from other stores were recovered.

" 'Why did you stay behind', asked the merchant of the boy, 'when all the others went to see the show?'

" 'Because, sir, you told me never to leave the store when the others were absent, so I thought I'd stay.'

"Orders were once more given, 'double that boy's wages, for not only is he prompt and industrious, but also faithful'. That boy is now getting a salary of twenty-five-hundred dollars a year, and before long he will become a member of the firm. He was following Elijah's model of promptness, and it helped to make his fortune."

Quite apart from the absurdities in the story, the lack of interest or the impossibility of feeling any identification with the hero, the rewards, and the motives, seem to us at a very low level. Slightly more exalted is the morality and expression of a book of poems originally published in 1813, on which one of the authors was nurtured from the age of three years.

INDUSTRIOUS RICHARD.

When Richard left his humble home,
 A simple cottage boy,
Friendless from town to town to roam
 In search of some employ;

How many obstacles combined
    To check his honest zeal,
To damp the ardour of his mind
    And make him timid feel.

But perseverance led him on,
    And hope threw out her light;
Till difficulties, one by one,
    Vanished, and all went right.

Years of success now rolled away
    His every wish to crown;
And Richard lived to see the day
    That brought him just renown.

While thus among the wealthy classed,
    His thoughts would often rove
To those with whom his youth was passed—
    Objects of early love.

Again the rural scene he sought,
    The dear, though poor abode;
Where, to the pious lesson taught,
    The present good he owed.

His grateful bosom proudly beat
    To think he now could share
The wealth by industry made sweet,
    With those oppressed by care.

His parents his first thoughts engage,
    The next his friends demand;
But all had claims, whose feeble age
    Needed his fostering hand.

And thus the humble cottage boy
    Strewed charity's fair seed;
While grateful thanks and heartfelt joy
    Repaid each generous deed.

This is a far pleasanter picture to hold before the mind of the child than that of becoming a "member of the firm", but how far it would influence behaviour is uncertain. As the author has never achieved Richard's financial success its effect on her cannot be known.

As late as 1913 industrious authors were struggling to put more truth, more matter, more topicality, into the moral story. The following is quoted, not in derision of its teaching, which is excellent, but of its form, which is deplorable[1]:

"A lesson may be prepared with a view to showing, in a series of simple contrasts and stories, the quality of various kinds of power, the intention being to lead to the conclusion that the power of Love is the most noble. Let us suppose that the teacher tries to show how this power may be exercised by a corporate and social agency as well as by individuals. Here is a programme of the lesson, graduated to such a climax and yet avoiding a didactic termination.

"Lightning striking a tree: power of Nature creating fear.

"Story from Herodotus of the African tribe who fought the South Wind and were defeated. The power of Nature feared and resisted, though unsuccessfully.

"A glimpse of Hindus throwing gift of propitiation into the mouth of the Ganges: worship of Nature, half in fear, half in affection.

"The power of Nature, then, appears greater than that of Man.

"Here is the sea; it has power. On the sea is a liner, which also has power. The power of Man's mind has conquered Nature.

"Let us study powers:

"(1) A performer at a circus lifts enormous weights: Physical power.

"(2) You children can answer the questions I put to you. (Here ask some question.) You are able to reply. You have Mental power.

"(3) A man who is caretaker of a girls' school, removes a lunatic who has broken in and terrified the scholars. The illustration combines Physical Power and Duty.

"(4) Mr. Luke Fields's picture of 'The Doctor' portrays the physician exercising thought as well as doing his duty: Mental Power and Duty."

So far the teacher has been developing the conception of various forms of power and has attached them in the latter examples to moral ideas of Duty. He has thus sought to make Duty admirable and also to make Power admirable when harnessed to Duty's chariot.

"The final story may be given in full:

"(5) Saint Cecilia played the organ with so fine a touch that the power of the melody drew an angel down, and the angel

[1] F. J. Gould, *Moral Instruction*.

bore a red and white rose as a gift. Stronger yet than her music was another power in Cecilia's nature. She journeyed along the banks of the Rhine, and crowds attended her steps. Among the people was a man with a deaf-and-dumb boy whom he wished the saint to heal. Cecilia took the child in her arms.

[Deaf-and-dumb children are taught in the schools of London and other cities to-day and with great pains teachers teach them to pronounce words aloud instead of the old plan of the finger alphabet. What feelings move the teachers' hearts? Pity? Yes. Kindness? Yes. Love? Yes, the Power of Love. But who chooses the teachers? The citizens. Would it be true then to say that the power of Love for their unfortunate childrens moves the people's hearts? It would be true: and it is wonderful to see how much is done in the world by persons and cities, through the Power of Love. Of all the powers we have considered—Nature, Physical, Mental, Duty, Love—which is the noblest? Love.]

"But Cecilia took the deaf-and-dumb boy in her arms (so says the old Catholic legend) and in her eyes was a look of love. She kissed the unhappy child and looking up into her gentle face he burst into speech and praised Cecilia's goodness.

"That is all. You may safely leave the children with a subtly-minded memory picture in which the figures of the public-school teacher and St. Cecilia are almost blended."

It would, however, be wrong to think that the moral story has no place in education; but if the story is to be a success it must have its own artistic value and charm, quite apart from its morals. The auditor must enter into the story, feel himself part of it, identify himself with the hero, react to it, in fact, as a work of art before it is likely to have any serious effect on his thought and conduct. The literary artist is the first ally of the moralist. It is also important that the story should be in accordance with the thought of the day. The confused, unanalysed morality of the market-place can and should be pointed by the concrete and perfect example. A recent film, *The Last Chance*, presented in the most vivid form a story of self-sacrifice and heroism that was all the more moving because the people for whom the gallant young soldier died were so much his inferiors. The characteristics of the modern saint, of whom perhaps Mme Curie is the best example, are well marked. Scientific curiosity, intellectual integrity, personal endeavour, a neglect of financial gain or even of public recognition, and a faithfulness to death if need be, combine in the characters of such men as Scott of the Antarctic,

Leigh and Mallory on Mount Everest, or Pasteur. These heroes of the modern world expected no reward on earth or in heaven. There was no martyr's crown or handsome bank-balance to shine before their eyes. They are an interesting illustration of Kipling's dictum that "the best work is done freely and for devilment", and are a long way from the apprentice who was always having his wages doubled.

At the same time these examples are not for direct imitation. As the hymn says,

> We may not die, we cannot fight . . .

and neither mountains nor ice-floes are in our back yard. It is an attitude of mind that the stories wish to impart, not an immediate course of action.

> A servant with this clause
> Makes drudgery divine;
> Who sweeps a room, as for Thy laws,
> Makes that and the action fine.

The motive to which the moral story appeals is an important part of the moral effect. Is a boy to be good and industrious in the hope of having his salary doubled, or of returning to his village to do good, or, to take a motive more insisted on to-day, to make his contribution to the export drive and the financial recovery of his State. The first motive is purely egoistic and in addition depends on the conventional value of money. The second is partially egoistic in the gratified self-assertion of benevolent power, but it also concerns itself with other individuals, the third is devoted to an abstract entity, the State, and demands a higher level of education and a sense of impersonal public duty. It will only be effective at the end of a training in ideas.

This choice of motive is a function of the whole social environment. Certain motives, common in one age, lose their effectiveness in another. The sanctities of royalty have been largely stripped away. The impediments to marriage that royal birth imposes, and the noble sacrifices that arise from them, seem to us silly. The exclusively financial motive seems ignoble. Devotion to the abstract State would have seemed a curious idea to an industrialist of another age; while the scientist, who labours he knows not why, is an object of mild derision to all but believers. There must then be a choice of motives in the moral tale, as well as art in the telling, if it is to be successful.

There is in many people a genuine readiness, almost a desire, for self-sacrifice. Whatever the ultimate psychological basis of this attitude may be, it appears as an emotional eagerness to sink one's own interest in that of the group, and a willingness to suffer so long as one suffers with and for others. The moral story therefore is often most powerful when the conduct it describes or commends contains this element of sacrifice and effort. Part of the power of the Nazi and Fascist movements was their ability to give to their members this experience of self-sacrifice. One of the most popular Nazi songs specially intended for the use of young girls ran:

> Many must fall and sink into the grave,
> Before our goal is reached and our banners in victory wave
> You who are left behind are branded with the sign of death,
> You will have to learn that happiness and bliss
> You only can earn,
> If you bleed and die and leave your life behind.

When the school is part of the general culture round about it, we are apt to overlook the specific contribution that it makes to this teaching of ideas. We can realize it a little more clearly when we consider minorities anxious to preserve their specific traditions and ideals in a partially alien culture. In many countries various religious and ethnic groups maintain their own schools, supplementary to the state schools of the country, which the children also attend, so that they can teach their own tradition and often their own language. This is particularly common among the Jews, but in America Greeks, Armenians, Russians, Poles and others all maintain their own schools.[1] The aim of the schools is to teach the national language, history, religion and general ideas. Where the aim of the school is mainly religious, Jewish or Catholic, the attempt is often made to give a complete education which takes the place of the education given in the state schools. In other cases the school is conducted in the evening and the children sent there after their normal day's work, like the hard-working little boys in Louis Golding's novel, *Magnolia Street*.

The Jewish schools are probably the most numerous and the strongest. One of the fundamental tasks of the school is to teach Hebrew, and from this knowledge of the language springs a knowledge of Jewish literature and thought. This is how an American Jew expressed it: "Hebrew and the sacred cultural tradition which it frames remains the integrative basis for a social system which lacks the fundamental basis of a single geographical *locus*. Without the Hebrew language we shall become severed from the great tree which is life unto those that cling to it: Hellenistic Judaism is the only one

[1] Warner and Srole, *Social System of American Ethnic Groups*.

in history which dared to make the experiment of dispensing with the sacred language. The result was death. It withered away and terminated in total and wholesale apostasy from Judaism. Let us not deceive ourselves. There is no future in this country for any Jew that resists either the English or the Hebrew language."

Through Hebrew the child should come to a direct appreciation of Jewish literature, the Bible and more modern writings. Through these again he learns the peculiarly significant concepts of the race, and the social and religious ideas that the race has evolved. We, in so far as our culture is founded on the Jewish Bible, share some of these concepts and beliefs, but they have not for us the urgency and intimacy that they have for the Jew, nor do they serve as the basis of racial cohesion. For the Jew, these beliefs, this history, are the bonds that hold the past to the present, and unite Jews in different lands. With this sense of cultural unity comes a desire to assist in the solution of Jewish problems, and to bring about a development of Jewish culture.

Without these schools, although the child would be exposed to the ideas of his home, the larger environment would probably prove dominant, and the racial and cultural peculiarities largely disappear. This the more race-conscious parents resist. As one father said, "I want my children to say, 'We are Greeks and our fathers were Greeks', even though they have been in this country three or four generations". Those who resist minority schools do so because they object to this form of double patriotism.

It is this sense that the language must be taught to children as the vehicle of the literature, history and specific ideas of the race, that makes the minority demand for special schools so urgent; and causes so violent a reaction when they are suppressed by a dominant and intolerant majority. It is the same spirit which makes the Welsh or the Dutch South Africans insist on bilingualism in the schools, or the Irish resurrect a dead language to become the vehicle of national ideas. The schools which are working with, instead of parallel to, the dominant culture play an equally important part in the general development of social ideas though this part is less noticed in the common culture.

Nations that have been keenly conscious of the myth they wish to teach, especially when the myth has been a rather narrow one, have developed methods of education that illustrate most clearly the points we have been making. Their methods seem strange to members of less definite cultures, or perhaps we should say, to members of cultures which deliberately leave a wider field for variations.

Ancient Sparta was an example in the past of this type of culture, and Nazi Germany in our own day. Both states were organized explicitly for war, and the whole economy of the state, as well as the education, directed to that end. In both states children grew up with a definite path before them, the boys were taught to think of themselves as soldiers and the girls as the mothers of warriors. Both states pursued their aims by much the same methods; a closing of the state to ideas from without, a continual repetition of the ideas within, harsh discipline with much calculated physical suffering, and the maltreatment of an inferior race. There was in both a break-up of family life, and though the German child was not removed to barracks soon after the age of seven, as was the Spartan boy, he was largely taken from the control of his parents. In both nations the girls received careful physical training to fit them for their part as mothers of sturdy sons. The brutality and intolerance of Spartan life has been well described in some stories by Naomi Mitchison, the stages of German education in a book by an American, Gregor Ziemer. The greatest difference between Sparta and Germany is the heavy weight of sentimentality that lies over the German mind. The Spartan boy suffered for the state and his ideal of himself, the German for the Fuehrer, and this monstrous figure making its appeal to all the sentiments of herd loyalty, of uncritical devotion, towered over everything, and turned what should have been honest patriotism into unnatural adoration.

The German child by the age of four was expected to have military ambitions and to aspire to be a storm-troop leader and shoot "ugly Poles". He grew up with the same words, the same ideas, continually repeated. All the prizes of life lay in the gift of the party, any deviation from the code brought down on him punishment from elders or equals. He was compelled to physical exhaustion and suffering in the name of the Fuehrer, and his life was punctuated by ceremonies, made as emotional and impressive as possible, to imprint his destiny on his mind. The following is an account of the initiation of the little boys of ten into the higher order of the Jungvolk:

"Half an hour before sunrise, 19th April, I hurried up to the Marksburg, best preserved medieval castle on the Rhine. Even before I reached the pinnacle I realized that two hundred youngsters, aged ten, who would that morning be promoted from Pimpf to Jungvolk, had got up long before me.

"The lads had walked a distance of fifty miles and had arrived at the village of Braubach at the foot of the castle the night before. I saw them come stumbling in, weary and hungry, but singing their songs with high piping voices thin with fatigue. This march was their last test of endurance, a sort of final examination in leg prowess before their graduation. They were now standing in rank and file in the largest of the Marksburg courtyards.

"A sharp command stopped all movement and noise down in the castle yard. Hundreds of youthful heels clicked to attention.

"The officer in charge introduced the guest of honour, a high official of the Hitler Youth.

"The sun came crawling over the wall, and its rays stopped some of the youngsters from shivering quite so much. Here are excerpts from his address:

" 'You boys must be hard—hard as iron; the Fuehrer has demanded it. You must be loyal; the Fuehrer has demanded it. But above all you must be ready and willing to give up your lives for the Fuehrer; he has demanded that, too.

" 'On you rests the future of Germany. Our Fatherland needs you. Germany will one day be ruler of the world. Our Nordic culture will go out and cover the earth. The youth of other lands does not understand us. Democracies point fingers at us. They say we are making you boys into soldiers. Do we deny it? Certainly not. They do not realize how eager and ready German boys are to become soldiers for Hitler. We will make Germany what she should always have been, a force to be reckoned with. . . .'

"A penetrating fanfare of trumpets from the tower of the castle sent a flock of pigeons wheeling towards the valley. I could see tenseness in the young ranks. A short silence. Somebody gave a note on a pitch pipe and the boys burst forth with *Deutschland, Deutschland über alles*. . . .

"Another fanfare shattered the spring morning. A sharp command, and from round the corner of the castle came a colour guard of three boys. The one in the centre carried an old tattered swastika flag on a tall pole.

" 'The Blutsfahne', said the commander. 'The bloodflag. And look, the sun casting its rays right on it'.

"This was one of the flags from the fighting days of the Party. Some of the members of the squad to which it belonged had been killed by the Communists.

"Quickly the boys formed a circle. Some were pale, some were flushed. The most momentous occasion of their young lives was approaching.

" 'Raise your hands', came the command. 'Repeat after me'.

"And the spring sunshine and the Rhine and the medieval castle heard the following:

" 'In the presence of the bloodflag which represents our Fuehrer, I swear to devote all my energies, all my strength, to the saviour of our country, Adolf Hitler. I am ready and willing to give up my life for him, so help me God. One Nation, One People; one Nation, one Fuehrer'."

This kind of ceremony, repeated sometimes at midnight, sometimes in the blaze of noon at each stage of a boy's progress towards manhood, must have a very deep effect. When the lessons it teaches are repeated daily, in school and out, they rapidly acquire the character of self-evident truths. When this system of ideas is strengthened by the emotional, non-rational teaching of adults, when the martyrs of the cause are the heroes of the story-books, and when ordinary impulses of sex or leadership are caught up and given the special twist that the myth demands, then it is easy to see why the teaching is so successful.

Such teaching has its parallels among other nations, but it can only be carried out fully when the myth is simple and unitary, and when the whole state is organized in accordance with it. A complicated code, one that allowed for individual judgment or one that differed in different sections of the community, could not be taught in this way. Above all the whole state must be consistent. If the child is taught one thing in school, and looks out and sees a society organized on a different principle, then he is presented with a choice and he is most likely to follow the custom of the larger, more powerful, group.

Nazi Germany, preaching the duty of being "hard" and "ruthless", ready for blood, could show in many ways a consistent front. The child witnessing carefully staged scenes of persecution and violence, being himself expected to suffer for the cause, could not be conscious of an opposition between principle and practice. It is harder for a state that preaches justice and mercy to show them in universal operation. As we shall see later in our chapter on the arts, it is essential for full success in teaching that adult life should manifestly demand the qualities and knowledge that are taught in school, and should be organized on the same fundamental principles.

Moreover precept and emotion in Nazi Germany could find a real outlet in action. It is characteristic of emotion that if it does not issue in action it becomes sentimentality, and is indulged in for its own sake and weakens subsequent impulses to action. In Germany there was always something to do, some service demanded in the name of duty. Every moment of emotion was made use of, the child was never allowed to lose himself in romantic emotion and feel that no more was required. In England we shrink both from arousing this emotion and employing it. Our very half-hearted celebration of such an occasion as Empire Day is an example of this unwillingness. Even those people who think such a celebration ought to be held will not take the trouble to make it truly impressive, and do not accompany the flag-waving with any thought-out scheme of activity into which the emotion could be turned. The Nazi Party had made the raising of emotion a matter of great skill, and they never wasted the emotion they generated.

The Nazi child knew what he was working for: the liquidation of the Jews, the destruction of the Poles; England must be humbled, France crushed. The ring of enemies round Germany must be broken through and Germany attain her rightful position as lord of the world. Each thing that the boy did had its place in the scheme. When he learned trigonometry, that was for gunnery. When he went camping or marching or on manoeuvres, that was a part. When he studied a foreign language, or travelled on the borders of Germany, that was another part. Everything had an aim, and everything was clearly explained.

For the educationalist, even one trying to teach a very different myth, this system has its lessons. We may not wish to close our children's minds to all ideas but our own, but there are undoubtedly some ideas that we think supremely important, and that we really wish to have absorbed. If we can formulate them clearly, express them simply, and translate them into everyday activities, we shall have done much towards teaching them. If we can show them in operation in the world about us, and, above all, if we can invest them with emotional significance, we shall have done more. English children are no less ready than German to feel the emotions of patriotism or self-sacrifice; they are probably more ready to feel benevolence and loving care. When these emotions are aroused and the children are shown how to express their emotion in action they will gladly make their contribution. The moral teaching of the past far too seldom showed the path of action. We learn to love by service, and a community that makes too few demands gets neither the love nor the service it might easily have.

Books referred to in this chapter:

H. O. Evennett, *The Catholic Schools of England and Wales.*

H. C. Barnard, *The Little Schools of Port Royal; The Port Royalists on Education.*

Donald Hughes, *The Public Schools and the Future.*

Edmund Candler, *Youth and the East.*
Anthony Hope, *Rupert of Hentzau.*
George Orwell, *In the Belly of the Whale.*
John Adams, *Exposition and Illustration in Teaching.*
J.E.M., *The Rose; The Crocus.*
F. J. Gould, *Moral Instruction.*
Warner and Srole, *Social System of American Ethnic Groups.*
Gregor Ziemer, *Education for Death.*
Naomi Mitchison, *Black Sparta.*
Louis Golding, *Magnolia Street.*

Chapter 4

ART AND MORALITY

ALL learning has a moral effect, but there is one field in which the moral element is particularly important—the arts. Through them, and the closely allied crafts, both the child and the adult gain valuable experiences which, for most people, are not given by other forms of activity. For the child the arts, especially drawing, provide a most valuable emotional release, through them and the crafts he gains a sense of power and dignity. The adult gains an interest, a subject of thought, a realm of life where he is lord, which all set him apart from his daily work, and enable him to endure with calmness much that would be intolerable without this other type of experience.

It is important to realize that art can be pursued at various levels. There is a great difference between the amateur and the professional artist. At the professional level the arts are a field in which competition is very keen. Too many people have coveted the supposed ease, excitement and distinction of an artistic life. Poverty, jealousy, self-advertisement, have been characteristic of too many practitioners. *La Vie de Bohème* with all its sordid emotionalism is one result of this, and the temperamental prima donna another. The amateur, on the other hand, is a much happier person. He is not dependent on his art, and therefore does not have to strive. He suffers artistic agonies, but they are not the agonies of competition. His music and his painting, his verse and his model ships, are a solace and a delight rather than a nervous drive to fierce activity. Society has need of a finite number of professionals, but every one can be an amateur and do nothing but good. In discussing the arts, therefore, we are not thinking much of the training of professionals. Their training is laborious, and it is only by continual close application that they can keep their skill at the necessary pitch. We are thinking rather of the man who comes home from work and plays in a band, or who, in the N.F.S., paints pictures and constructs toys, or the woman who weaves scarves or makes leather bags. Such activities, started in childhood, can add value to a whole life; and though, judged by an external standard, the performance may not be of a very high order, from the experience of the performer it is of the utmost importance. As one sage remarked, "If a thing is worth doing, it is worth doing badly".

The first thing to realize about art, both as an individual activity and as a social expression, is that it is natural. Before industrialism laid its hand on urban life, the community provided beauty as a social service. Most of the smaller towns in England, or in almost any other European country, are adorned with churches, pleasing avenues, market-places with gracious houses surrounding them. Towns such as those in the Cotswolds have achieved fame simply because they have remained largely untouched by modern developments. There are races in Europe, especially the Celts, who appear to have little sense of visual beauty in architecture, but they are much in the minority, and they find their art-expression in other forms such as music. Not only was this visual beauty part of social life, there were also music, dancing, and such crafts as weaving and embroidery. Among simpler people art always has a place, so long as the society is healthy. Different communities have of course reached different levels of accomplishment, but hardly any are quite without some art-forms.

The extraordinary thing is the extent to which art disappeared with the advent of the industrial age. While the country town retained its charm, the great cities grew up without the smallest element of beauty. The thinkers of the early nineteenth century saw society divided between the working and enjoying classes, and did not imagine that the needs of the two could be in any way similar. The wealthy enjoyed leisure, beauty and the contemplation of music and painting. Their houses were elegant and their stretching parks set with noble trees. They read the best literature and wore silk dresses. The worker lived in filth and degradation. Whilst a delicate sensibility was part of the traditional equipment of the wealthy, it was almost impious to suggest that a love of beauty might be shared by the lower orders. A walk to-day in the slums of any industrial city, improved though they now are, shows how unnecessary beauty seemed to the builders. The streets may be wide, the houses decent. There is no object of beauty anywhere, unless, perchance, a sunset cloud shines above the grey rooftops. The very people, the children, the dogs, the shop windows, all alike lack aesthetic charm.

This denial of beauty was part of the commercialism of the nineteenth century; it was also a legacy of Puritanism. Presbyterianism came to us from Scotland, and possessed some of the characteristics of the country of its origin. The dress of the Roundheads deliberately eliminated the beauties of colour and texture, as well as the follies of foppery. The Puritan tradition passed to the industrialists, and if Whitefield could say that there "was scarcely any recreation that could be called innocent" and Wesley could prohibit all "play" at Kingswood, the employers were quite ready to agree with them—for commercial as well as moral reasons. Even when the alternative to recreation was not work but idleness, as at Mansfield Park, theatricals were still condemned. In this condemnation of art and beauty the artist was naturally included, and art, in itself, became immoral. The only form in which it was reasonably respectable was as the accomplishments of the Victorian Miss.

It has been suggested that it was the very rapid growth of population during the nineteenth century that destroyed the tradition of art, because, with the excessive number of children, there were not enough teachers. In England the situation was made worse by the employment of women and children in factories. Women then could not educate their own children, and the children were not free to learn. Even when children were sent to school, social meanness kept the supply of teachers inadequate and the classes too large for effective teaching. Thus the urban proletariat was cut off from traditional culture, which should have been given in the home, and failed to get the formal teaching which a good school might have given.

In addition, the whole concept of school education was hostile to art. The essential psychological experience of artistic creation is freedom, individuality and personal power. The methods of class teaching, which required seventy children all to do the same thing at the same moment in the same way, made art impossible. Moreover the children were expected, for the most part, to sit still in their desks, and thus many forms of art were barred. Drawing was possible, in the form of copying "freehand models" or working from "groups". This was not art, and was properly hated by most children. In music the class could sing, and sing they did with some enjoyment, but that was all. Such an absence of art in school, combined with a life outside in which there was neither recreation nor beauty, naturally produced people who were insensitive, and who thought of art as something quite alien to themselves.

The change from this state of mind has come slowly. The admission that aesthetic enjoyment was one of the pleasures of life that could be enjoyed by all was only made with great difficulty. That all men had a *right* to this enjoyment was made more slowly still—in many towns it has not been made yet. The slum school, ugly, dirty, unpainted, with heaps of decaying coke in the playground, is still with us, and no one in authority seems to think it an outrage. The planting of urban trees stops short with the wealthier streets, though many in the poorer districts are quite wide enough to have them. Music and the drama are concentrated in the centre of the town, thus putting them beyond the reach of the dweller in the outer districts. Only the cinema goes to the people, and thus becomes one of the most important factors in life.

The particular nature of aesthetic enjoyment makes it easy for a certain type of thinker to ignore it. When social theory was dominated by the myth of the economic man the pleasures of art were naturally ignored. They could not be assessed in terms of money, they were not necessary for survival. Those who have never experienced them, be they rich or poor, cannot imagine what they have lost; and, unless some training is given in youth, many grow up insensitive. Those who have never known them, and yet by nature desire them, cannot diagnose the sickness in their souls. The way these pleasures permeate and colour the whole of life is well suggested in the speech of Pericles that he made in praise of one of the loveliest cities of all time:

"And we have not forgotten to provide for our weary spirits many relaxations from toil; we have regular games and sacrifices throughout the year; our homes are beautiful and elegant; and the delight which we daily feel in these things helps to banish melancholy."

The failure to recognize art as a natural social service accompanied a failure to realize that art was a natural part of the individual life. The artist, or the child of artistic tastes, became a curiosity, something to be wondered at or tormented as a freak. The child had either to contend with parents who tried to display him as a prodigy, or else with hostile companions who derided his tastes and tore up his pictures. Even to-day it is hard for the artistically minded to receive the right sort of treatment. Persecution has softened to a vague contempt. One university lecturer on being told that a young man painted and wrote verse remarked consolingly, "I expect he will grow out of it". Faced with this attitude, many, naturally, do. The amazed and adoring parent is an even greater danger. Children are taught to regard themselves as remarkable, and, instead of cultivating a moderate talent, they lose heart as soon as they find how great is the labour involved, and how small a figure they cut in comparison with others. If art seemed as natural, or more natural than arithmetic, which it is, both these attitudes would disappear.

Apart from their value for enjoyment the arts provide the ideal occupation. Proverbial wisdom knows that idleness or lack of amusement is productive of ill, and that the way to keep people good is to provide them with harmless occupation. How innocent and inoffensive mankind can be when placed in suitable surroundings is demonstrated every day in August at a popular seaside resort. Rows of families sit in deckchairs, children run about with buckets and spades, and lads and lasses splash in the surf. There is amiability everywhere, and a flock of sheep could not be more harmless. The complete contrast is the unoccupied gang of the city streets, seeking any mischief to lend excitement to dirt and ugliness.

The condition of the past, when work and exhaustion made up man's day, has departed, and there are now for most people many hours a day that have to be filled. The provision of suitable occupations has thus become of great importance. There are at least two classes of enjoyments—the passive and the active. The cinema, the wireless, a concert, a football match, a novel, need varying degrees of attention and response from those that enjoy them; but in none of these is the audience taking any active personal part in the action. Of commercially provided amusements the majority belong to this class, or to the rather closely connected group where the activity is on the lowest level—pin tables, dodgem cars or merry-go-rounds.

The genuinely active amusements can cover the whole field of social life. The least obvious are the intellectual amusements, but for many people they are real and important. A miner who, after a day's work, will walk five miles there and back to attend a class on psychology, obviously thinks he is getting something worth while. Even in a town a class will flourish in spite of the competing attractions of the cinema. The more obvious fields of active amusement are athletics and the arts and crafts.

The importance of athletics for young people has been realized, and the provision of facilities for certain sports has become a public duty. Playing-fields, swimming-pools, youth hostels, are a recognized part of our civilization, and the young themselves organize cycling clubs and various kinds of sports. All this by providing occupation and giving an outlet for energy makes for virtue; and the regular cure for the bad boy is to teach him to box or to take him camping.

The place of the arts in virtue is far less generally realized and this blindness is perhaps greatest at the top. A public school in which games are compulsory for every boy charges large extra fees for instruction in music and drawing, and often arranges the lessons at times like Sunday afternoon when there is naturally the greatest unwillingness to attend. To tell such a headmaster that art is as important as cricket or football would be to utter a heresy so black as to be incredible.

The importance of art as part of a scheme for virtue lies largely in the psychological experience that it provides. In the first place the practice of art demands full concentration, and very considerable mental effort. In this lies much of its charm. This concentration shuts the mind to all else. For the time being, the artist passes to the blessed world where all is at one and where mundane troubles and worries cannot come. Thus he is happy while he works, and returns to common things refreshed spiritually. In the second place the artist is free to be himself and to enjoy the god-like pleasures of creation. In almost all types of work, particularly factory work, humiliations and boredom are inevitable. Official superiors are unreasonable; there is fear, monotony, fatigue; the sense that one is working for another and to another's plan. This is often true of children at school, and even in all but the best homes. An art sets one free. Power flows from the artist to his work. He conceives it, carries it out, it is his. Particularly when work is monotonous and requires little thought it is invaluable to the worker to have some matter that can occupy his mind during his hours of employment. The picture one is going to paint, the model one is making, are far better topics of thought than one's grievances or the daydreams of wealth based on a football coupon.

Some art is solitary, such as painting, some communal, as is music or to some extent dancing. Both of these forms of art have their own importance. The solitary arts give a sense of independence and can be pursued without reference to others at the artist's own convenience. The communal arts give an emotional release, through the experience of the group, that is most important. Communal art binds men together in a particularly satisfying way. It gives a sense of personal value and of belonging to the group, of order and free co-operation. The ritual dance which is so important a part of life among simple people is an emotional experience of great power and significance. Nothing that the ordinary civilized man does is quite so vivid and exciting. The dance combines personal activity of a highly skilled kind with membership of a team, and the whole is done before an appreciative crowd for an important social and religious purpose. Even the onlooker, coming from another civilization, unable to share fully in the ideas expressed, is carried away by the communal emotion. As he sits in the dust by the side of the dance-floor he feels himself caught up in the vast tide of emotion. The members of an orchestra must enjoy something of the same experience.

In modern European civilization the art of dancing has been degraded. It has lost its religious significance, and has nearly lost its social. It has become largely a form of sexual amusement or a vehicle of sexual display. The boy and girl at a hot and tawdry dance-hall, the chorus showing its legs before the eager eyes in the stalls, have degraded dancing to a very low level. Even dancing on the level of the ballet is no longer an important social function. The dancer has become so specialized that he can do nothing but his art, and his performance is solely for the temporary pleasure of the spectators. Some attempts are being made to re-introduce dancing as an art that gives the individual performer a mode of expression; it seems unlikely, at present, that it will once more attain social importance as an expression of group purposes. It will need more than the present activities of folk-dance societies to revive the ideas on which communal dancing was based and give us back anything like the Greek chorus, or the boys and girls round a maypole. Yet, for all its degradation, dancing remains one of the most popular activities of the young. It is the lack of social setting, as we shall say again about the other arts, that has deprived dancing of the opportunity of becoming once again an important part of social life.

The more uniform a culture is, the fewer opportunities it gives for individual variation and distinction, the more important the arts become as a form of psychic release. For a large part of the population life is unendurably monotonous and lacking in the possibilities of successful self-assertion. It is here that the arts play their part. In Samoa there is the strongest emphasis on conformity, and strict social subordination of the young to their superiors in status. The one outlet for all the repressed emotions is dancing.

"The attitude of the elders towards precocity in singing, leading the dancing or singing, is in striking contrast to their attitude to every other form of precocity. On the dance-floor the dreaded accusation, 'You are presuming above your age', is never heard. Little boys who would be rebuked or possibly whipped for such behaviour on many other occasions, are allowed to preen themselves, to swagger and bluster, to take the limelight without a word of reproach. The relatives crow with delight over a precocity for which they would hide their heads in shame were it displayed in any other sphere. . . . The ubiquitous ascendancy af age is somewhat relaxed in the interest of greater proficiency—each child is a person with a definite contribution to make, regardless of age and sex. The emphasis on individuality is carried to limits which seriously mar the dance as an aesthetic whole. Each dancer moves in a glorious individualistic oblivion of the others—there is no pretence of co-ordination or of subordinating the wings to the centre of the line. . . . The strong emphasis on dancing does not discriminate against the physically defective. Instead, every defect is capitalized in the form of the dance, or compensated for by the perfection of the dance. I saw one badly hunchbacked boy who had worked out a most ingenious imitation of a turtle and also a combination dance with another boy in which the other supported him on his back. . . . So every defect, every handicap, is included in this universal, specialized exploitation of personality"[1].

The freedom from the conventionally imposed trammels of personality mentioned here is a most important function of art. Society has its pattern, work its organization, and the individual must conform or lose his usefulness in many ways. There is no place for the assertion of originality. Art can move on a different plane, and give the potential rebel a delightful outlet for his feelings of difference.

The power and dignity that art can give enables young people to realize themselves on a higher plane than the customary one. A. E. Morgan in *Young Citizens* has written of the ambitions of the less-cultivated adolescent:

"To be able to buy cigarettes and to spit, to cheek the foreman (at least once), to dream of the backchat you'll give the boss, to go to the pictures with your own money, to swagger in the monkey parade with your own girl."

These are the desires of one who has no resources or powers of his own. They are all directed to impressing others, and they are largely hostile self-assertion against the environment. The satisfactions, the dignity that art gives, are of a very different kind. Often no audience is required. The craftsman works alone, and may only show his work formally at the end. The artist, too, is friendly to the universe. His appreciation of beauty in the world, and his attempt to express it in his work, are movements outward to embrace existence, to make himself part of a superior unity.

Children especially find in art or craft occupation and emotional satisfaction. We have said that the normal child seems to find it much easier to discover amusement than the neurotic, and that homes for difficult children seem to depend for their success largely on the number of occupations they provide for the children. Arts occupy minds and hands, and when the artist is not working he is often thinking about his work. But children need to be taught these amusements. They may discover a few for themselves, but, for the majority, they need training. In games and athletics this is very clear. No one expects a boy to play cricket by nature, and much time and effort are spent on training him. The tradition once established is handed on from one to another, and even where there is no formal coaching

[1] Margaret Mead, *Coming of Age in Samoa.*

the big boys teach the little ones in the backyard. Training is reinforced by social custom. The successful athlete has great prestige. The football teams, the county cricketers are known to every boy. A lad may not actually aspire to national fame, but the mere thought that through cricket a man may become a hero adds zest to his street game. Sports receive public encouragement. There are fields set aside for them, grown-ups play them, the eager crowd surges to see the cup-tie. The child, when he puts on his football boots, joins in a nation-wide activity. It is very different with the arts. For most children the teaching is perfunctory, there is no provision after they leave school, or indeed outside it while they are still young. No artist's name is honoured as is a famous footballer's and no adults seem to find in it a dignified amusement. If the arts are to form an important part of life, all this must be changed.

In the first place the teaching in school must have a new quality. Teachers must be eager, active practitioners of the arts they teach, and must see them as a real part of life. The present system in primary schools, by which every teacher is expected to teach everything, makes this requirement nonsense. The teacher is not gifted and the formal lessons required of him are given without conviction. There are many of the better teachers, particularly in the primary schools, who feel that their pupils' highest role in art is as a passive audience, admiring the more talented, and not being themselves executants. This is particularly true in music. Custom decrees that children should sing, but apart from that all teachers claim to do is to give their children a taste for *listening* to music; they never imagine that they themselves might make it. In the same way custom gives a child a pencil or an indifferent brush and some powder colours for painting, but this activity is only a school occupation, valueless, even in the teachers' own view. If the children can be taught to go to the art gallery once or twice in a life-time, that is all they are capable of. The idea that they should themselves really paint pictures never occurs to any one. Even in craft the same thing holds. The activity is not thought of as being real and a real part of life. For example the "book crafts" which are taught year after year in many schools leave the child quite unable even to bind a book. He certainly has never imagined printing one, still less writing it.

This attitude is due to quite a number of things, apart from the actual inability of the teachers to teach well. It is cheaper and easier to try and train children to be part of a concert audience, or to drift into a picture gallery, than to attempt to make them executants. The mass passive instruction that schools still in their hearts prefer, is quite in accord with lessons in "musical appreciation". To teach a child an instrument requires much more effort and expense. So, too, if a child is to paint after he leaves school, he must be given better materials, more careful instruction, and more true freedom.

There must also be a bridge between the school and the world. One of the inventions of Soviet Russia is the Children's Palace. There selected children are taught arts and crafts by experts in conditions of dignity. The Children's Palace is part of the extension of the school. Every large school has a member of the staff whose duty it is to organize out-of-school activities, athletic and artistic. In summer the children are taken bathing, in winter skating. There are classes for arts and crafts and indoor games. The ablest are recommended for the palaces and can there be taught by acknowledged experts, who receive honours for this teaching. There is thus a sense of importance, and facilities quite beyond those of any ordinary school. The Russians claim that, before the war disrupted their arrangements, they had reduced juvenile delinquency to unimportance by giving children these alternative occupations.

If this training in the arts is to be fully effective the community as a whole must confer honour on the artist. This happens, of course, with film-stars, and some broadcasters, and in consequence thousands of young people every year crave a stage career and try to force their way into this already overcrowded profession. Other thousands content themselves with amateur acting, and draw their enthusiasm for their art largely from the sense of social importance. More still would probably act if they had any place to do it in. For the community must give the artist facilities for his art, at least equivalent to those given to athletics. He must feel that through his art he has a place in the community. For example in painting: the amateur artist needs somewhere to work, other than the kitchen, and a chance to show his pictures. If a community centre put a studio room at the service of the neighbourhood, if it held six-monthly exhibitions, if it allowed some local man to fresco the wall of the hall, most unexpected talent would be displayed. If, in addition, it held every alternate six months an exhibition of professional art, and occasionally provided a sympathetic teacher, taste and knowledge would grow at a remarkable rate. The child at school would realize that his classroom occupation bore a relation to the world outside; he would see interest and honour before him; and adults would use their influence to encourage the child to do his best.

In the same way with music: school and community must work together if it is ever to become a reality in life. The schools must teach better. They should realize that the world does not sing folk-songs exclusively, and that many children could be trained to play an instrument. It has recently been discovered that the violin can be taught to a class of children together, and some authorities provide the teaching. But in this there is a lack of clear thinking. The world to-day does not want many indifferent fiddlers. In the past, before the advent of the cheap church harmonium, fiddlers supplied the music, and in *Under the Greenwood Tree* we have a picture of the village orchestra as a vital part of village life. Now society longs for jazz bands, brass bands, and those who play the banjo and the accordion. One of the supreme pleasures of music is the opportunities it gives for corporate activities, and children should be taught to play in an orchestra, of whatever instruments it might be composed. Even the unmusical will derive great pleasure from being allowed to bang a tambourine in the school jazz band. Music can exist at many levels and a great number of people can find amusement at each. Then when the children leave school, the community should find uses for their powers. In the parts of England where brass bands are a popular form of activity, children in school learn to play the instruments with eagerness. They see how their present activity fits into their future life. So, too, cadets spend their Sunday mornings learning to play in their corps band because of the glory of the march past. The child repeating the same folk-song, that he never hears outside the school walls, or listlessly scraping the fiddle, which is a very difficult instrument, and for which he sees no social use, is an example of theory which ignores the actual facts of existence.

In the past religion has been one of the chief social activities that have made use of music. This music of the Church has had a very important effect on European culture. To it we owe much of our best work. In the general decay of the arts, when all else was denied to the poor, religion, within or without the Churches, kept some tradition of music alive. In Wales the Chapels played the most important part in the musical life of the nation. To-day many a poor boy finds his only hope of real musical training as a choirboy. The Churches with this tradition behind them could do much to stimulate the musical life of the nation. But Church music is only one aspect of music, and for music really to take its place in society, much more is needed. The choral society, the amateur opera, the dance band, the works concert, all have their place. Society must provide opportunity, teaching, encouragement and honour.

Of all the arts literature fares worst. The field for the amateur is, in fact, very restricted. The Roman senator who recited his epic at the baths has left no successor. Apart from the shy poet, most people who write want their works published; and to achieve that means entering the ranks of professionals. A few authors print their own poems and give them as presents to their friends, but a printing-press is rather an expensive luxury for an individual, and takes up more space than most can spare. The schools, too, teach in such a way that few children even imagine that there is any connection between "literature" and the miserable compositions they are asked to write. Half a laborious page about "My Dog" or the "History of a Shilling" provides no training even in the fluency necessary for writing a letter. There are hardly any schools in which the children write verse or stories or plays with a sense that they are creating something. The literature which is laboriously explained or "appreciated" in lessons is never presented as something that was really written by a man like any other, and was adapted to a certain purpose. Yet when children receive encouragement and freedom, when they are taught literature as an art that they might themselves practise, they write very well.[1] If there is a school printing-press, and if schoolbooks are bound; if plays written in school are acted, if even the school magazine becomes vital and interesting; then there is an opportunity for a real development of an art.

Unfortunately, as we have said, it is harder to carry literature over from the school to adult life, but it can be done, and the amateur dramatist can be of great use to a Church or a community centre.

The Christian Church has been the patron of all the other arts. Architecture, painting, sculpture, metalwork, enamelling, weaving, embroidery, printing, so long as the Church held a place of supreme social importance it lent dignity to the arts and crafts of the community. The craftsman or artist worked not only for a living but in the service of God. His work was not only part of the setting for the community, but added to its spiritual glory. Now that the Church has ceased to dominate society in the old manner, no alternative focus for the arts has been provided.

The Welsh eisteddfod is a solution to many of these problems in the communal encouragement of the arts. The Welsh amateur poet can read his works, is brought into contact with others, and introduced to a field of study in classical Welsh verse. Music has its official place, and the whole celebration is strongly national, and directs the thoughts of the participants to the ancient glories of the race. There is pageantry, symbolism and sufficient formality to make it seem important. There are titles and robes. If other sections of the

[1] Caldwell Cook, *Perse Play Books*.

British Isles, if counties and districts, organized something similar, the arts would grow in importance, and find a vital place in school and later life.

Not all people are capable of the effort that the arts require, but almost all are capable of taking an interest in various crafts. The crafts differ very greatly in difficulty, and most of them can be pursued at different levels, and produce satisfactory psychological results. Silversmithing, for example, is a very difficult craft which requires long training and probably some special aptitude. Vitreous enamelling is easier; weaving is easier again; basketry is simple. There are occupations for completely different types, from the woman who wishes to make soft toys to the man who wants to do elaborate repairs to antique furniture. In schools the crafts should be well taught with consideration of a child's aptitudes as well as his desires. They should be taught with his future life in view, not merely because there is a fashion for that type of school activity. Outside school the community can do much by providing workshops and teaching and by organizing exhibitions and sales of work. Nothing is more beneficial for example than for a group of men to combine to make toys for a school or a children's hospital. If when the toys are made they are exhibited and suitable honour given to the craftsmen, the maximum encouragement is received. In the same way a group of people could undertake the furnishing of a church, or of a new clubroom. If work of this sort is well organized, and facilities are given for its accomplishment, the curse of isolation which afflicts so much of our present way of life is removed.

The amateur artist needs the professional to set a standard and to give him the supreme enjoyment that goes beyond his own power to provide. The amateur is a far more enthusiastic and critical auditor than the man who has no technical knowledge of the art, and merely enjoys it. Thus the more amateurs there are the larger and more enthusiastic will be the audiences the professional can command. The better and more numerous the professionals the more people will become interested and anxious to take part in the activity. There is thus no opposition between the two classes; they work together. The training of the professional is long and arduous, and in England it is often very difficult to acquire. This is particularly so in music. There is over most of the country no national provision by which a poor child can acquire a musical training. In the very few schools where a training in instrumental playing is given it is very slight, and hardly sufficient to pick out a talented child. In some towns there may be scholarships for music, but in the great majority there are none, and no adequate instruction even if the child could pay for it. The richer not only must pay considerable fees, but must contend with the apathy and hostility of schools, who think a child would be better employed "getting on with his work". The system of homework makes it very difficult for a child to practise without being overworked, and the actual hour a week for a lesson may be very difficult to find. The case of the child who wants to paint is slightly better, because municipalities do, in many cases, provide art schools where a reasonable standard of technique can be acquired cheaply. The would-be dancer is the most unfortunate. Unless she spends all her out-of-school hours at one of the academies which train children for the pantomime, or pays large fees at one of the residential boarding-schools which give a specialized training, she encounters nothing but hostility and frustration. The artistic child exposed to this either loses heart, or forms too high an opinion of his abilities; or he represses his powers to avoid appearing exceptional. He does not get either the teaching or the company he needs. The Russians have realized this problem and solved it by having special schools for talented children in which they collect the future professionals in the arts, and train them together under the care of the best teachers. The child thus gets what he needs, teaching, stimulus, and the companionship of his peers.

Under this treatment children develop remarkably quickly, and show powers far in advance of those usually expected of children. Whether in the end such children are superior to the artists of other lands is as yet uncertain; but at least potential artists have not been lost by neglect or discouragement.

One of the results in England of the recent war is an increased interest in the arts, and it has become the duty of the Arts Council to foster this interest. With the aid of public money the Council has increased the number of concerts and music groups, provided traveling exhibitions of pictures and organized companies of actors. All this is excellent. It gives pleasure, it introduces many to the arts who before had no opportunity of contact with them, and it should act as a stimulus to individual effort. The best fruit of a strong professional art is an even stronger growth of amateurs who themselves participate in the pleasures of artistic creation. When this happens, and when it is realized how large a part of the population is capable of some artistic activity and how important it is that they should have facilities for this work, then a great step will have been taken towards social virtue.

Books referred to in this chapter:

Murger, *La Vie de Bohème.*
Thucydides, *Histories*, Bk. II, Funeral Speech.
Margaret Mead, *Coming of Age in Samoa.*
A. E. Morgan, *Young Citizens.*
Caldwell Cook, *Perse Play Books.*
Herbert Read, *Art in Education.*

Chapter 5

RELIGIOUS EDUCATION OF CHILDREN

Jesus spoke first and foremost of God, and only in the second place of man and his conduct. No Christian would regard his faith merely as the sanction and guarantee of an accepted morality. The tree may be known by its fruits, but we do not gather fruits except from trees, and they have roots and branches as well. Christian education cannot here be treated in all its bearings, but only in relation to the theme of this book. From all that has gone before in this book it must be clear that we regard the religious education of children as of the highest importance. It is not only that Christianity is so closely connected with ethics, and is for many people the supreme validation of ethical behaviour, it is also that the whole fabric of our state is Christian. We have perhaps left behind "the age of faith" in which all the functions of social life took place in a religious setting, and in which craftsmen, merchants and princes alike felt themselves in direct contact with God in the details of their daily work; but even to-day we can only fully understand our civilization if the religious background is also fully known.

Organized Christianity, in one form or another, is an established part of the cultural pattern of all European nations and those who derive their ideas from them. The forms of our buildings, the arrangements of our towns, our history, our ideas, all have a profoundly religious element. In England and other Protestant countries the Bible was for centuries the chief reading of the people, and formed a bond between all classes. The polite literature of the day was often, in the periods before literacy became general, confined to a very small educated class, but, if all could not read, all could hear the Bible read, and the ideas it contained, the language and the images, were known to all. The writers of English, as well as the non-literary, were influenced by the Bible, and English literature, however secular it might seem to be, was the product of men who had learnt their ideas from this source.

In the same way European history is a record of men acting in the framework of Christian ideas. It is true that many ignored these ideas, and that men, naturally, interpreted them in terms of their own day. In many respects ideas change with each generation and its special problems, but, nevertheless, there runs from age to age so large a common body of ideas that it is not too difficult to understand the past by reference to our own beliefs. The Knights of the Round Table were faithful sons of the Church, the Crusades were a great corporate expression of a common faith, even Ferdinand and Isabella driving the Moors from Spain and persecuting the remnants, were bound to us by many common ideas. The great humanitarian reformers of more recent years, Elizabeth Fry[1], Lord Shaftesbury, or Wilberforce and the anti-slavery party,[2] were deeply religious Christians who were trying to translate their beliefs into action. Many of our soldiers, from Oliver Cromwell to Durand of the North-West Frontier, tried to live their military lives with the help of divine guidance. They did what they believed to be right, sometimes in the face of military expediency. For example Durand refused to accept command of the group charged with the task of blowing up the Kabul gate at Ghazni because he thought it wrong to hurt the feelings of his incompetent superior officer.

However much some people may deplore the decay of faith in England, Christianity is about us everywhere. We do not notice it in our everyday life because we are so habituated to the forms and customs. If one travels through the smaller towns of the prairie states in the U.S.A. there is never a church to be seen. It is only when one enters Utah that the village church reappears, and one realizes what had been lacking. In the same way one only realizes the all-pervasiveness of Christian forms in Europe when one visits another continent where the tradition is different. Even our commonest political appeals to justice, consideration for others, and mercy, have a religious background, and contrast with, say, the Nazi demand that men should be "hard" and relentlessly pursue the interest of the stronger. That being so, to put the matter on the lowest level, no English child can enter into the cultural heritage of his nation unless he has the Christian knowledge on which the state is built.

In fact, English education has always been religious. With the final break-up of the Roman Empire the Christian Church assumed responsibility for handing on the traditions of Greco-Roman civilization, and for many centuries provided the literary education of Europe. There existed, of course, the mainly non-literary education of the noble or the craftsman; but the Church was the body which gave a training in language, letters and philosophical ideas. Then in centuries

[1] Janet Whitney, *Elizabeth Fry*.

[2] John A. Patten, *These Remarkable Men*.

when kings could not read, the offices of state were filled by clerics, and England did not have a lay Lord Chancellor till the reign of Queen Elizabeth.

The new schools which followed the break-up of the monastic education were deeply religious. In St. Paul's of Milton's day, founded by Colet with the support of Erasmus, the figure of Christ with his hands raised and the motto, *Hear ye Him*, looked down on the classroom. The founder had left it as a charge to the scholars to "lift up their little white hands and pray for him". Beneath the statue to Christ Alexander Gill taught English verse, and expounded his views on phonetics and classical metres. If Gill had his "flogging fits", and if he taught Latin and Greek for themselves rather than for their pious content, he still felt his work to be the service of God. This combination of godliness and sound learning is typical of the Reformation grammar schools, and might be illustrated from the records of many less famous than St. Paul's. For instance Dr. Samuel Harsnet founded a free grammar school at Chigwell, where he was for eight years rector, and resembles Colet in his insistence on pure Latinity. Like him also, he emphasizes the importance of the teaching of religion and morality. "I charge my school masters respectively", he wrote, "as they will answer for it to God and good men, that they bring up their scholars in the fear of God and reverence toward all men; and that they teach them obedience to their parents, observance to their betters, and ingenuity in all their carriages; and above all that they chastize them severely for three vices—lying, swearing and filthy speaking."

A hundred years or so later, when the public schools had fallen into mental stagnation and the grip of the Church of England, the Dissenting Academies combined educational progress with religious teaching. Later still the beginnings of elementary education were due to the activities of the religious bodies and we have the remnants of the system in the Church schools to-day.

It is probable that most people desire their children to receive a religious education. If they do not themselves practise their religion, they at least think they believe it, and feel strongly that religion is good for the young. They also have considerable confidence in the ministers of religion. They think that the chaplains in the Forces can give great help to the young soldier, and, though they may never appeal to him themselves, derive a certain consolation from the Vicar. This attitude, which characterizes a large part of the British public, is felt as indifference by the more devout, but it is something real and quite strong, and can be brought into action by any event that is felt to threaten the religious fabric of the nation. Thus the great public interest felt in such a topic as the reform of the prayer-book, and the heat that the discussion caused. Thus, too, historical anomalies, such as tithe, survive in spite of protests, because any attempt to end them would evoke so much opposition.

The latest Education Act has realized the strength of this feeling. The Butler Act has laid it down that religious observances shall be a part of the life of every school. There shall be a daily "act of worship", and there shall be definite instruction in the Bible according to an agreed syllabus drawn up by representatives of the Church of England, the Free Churches, the education authorities and the teaching profession. The intention, clearly, is to bring back religious teaching as a serious part of school work, and to make it possible for the state schools to make a real contribution to the religious life of the nation.

But this intention cannot be carried out quite as simply as the framer of the act seemed to think. The present position in schools in regard to religion is the result of historical causes, and corresponds to a real feeling in many people. In England there are many sects of the Christian religion, and there are, in addition, many people who are not Christians, either because they belong to some other religion or because they disapprove of any organized religion. So long as a man's beliefs do not lead him to break any law of the land, his religion is officially indifferent. He can believe, or not believe, as he likes; and a religious test is not enforced as a preliminary to office. We have learned to ignore a man's religion and concentrate on his other characteristics. But this is quite a new attitude. When Florence Nightingale was recruiting nurses a friend asked if she objected to Presbyterian nurses. She said she did not care to what denomination they belonged, so long as they were good nurses and did not weigh more than fourteen stone, as the beds were not strong enough. So modern a view took a long time to penetrate into education, and the early history of elementary education is dominated by the denominational controversy. The two religious bodies which organized schools with the aid of an ever-increasing government grant naturally taught in accordance with their own convictions, and when the school boards added schools that belonged to neither body, it was felt necessary to make special arrangements. The Cowper Temple Clause forbade the teaching of any formula characteristic of any particular denomination, while the "conscience clause", till recently "prominently displayed" in every school hall, declared that any child might be "withdrawn" from religious instruction without suffering any penalty; and, to make this easy, religious instruction must be given at the beginning or end of the session.[1] In addition inspectors were not allowed to visit scripture lessons or to include religious knowledge in their examinations. In consequence of the first part of this arrangement scripture was regularly taken first thing in the morning, and of the second part it was felt to be quite unimportant. When an inspector came and judged the school, even assessed the teacher's salary, on success in reading or arithmetic, no one was going to waste any effort on scripture which was not even looked at. Thus when schools became more and more places where children paid milk money, got dinners, amassed war savings, and received extra clothing coupons, it was only natural to use a convenient period at the beginning of the morning for transacting such business, and, that over, one could settle down with a calm mind to the serious business of arithmetic. Teachers are not at all anxious to have this arrangement disturbed.

But there is a far more serious objection. It is felt by many that the imposition of an "act of worship" as a compulsory part of the day's routine will, in fact, constitute a religious test for the teachers, in particular for the head. As it is at present illegal for religious considerations to enter into the appointment of a teacher, except in certain special cases, it is quite possible that the headship of a school may be held by a Catholic, a Jew, or an atheist, and they must now face a situation that had never been contemplated. Moreover young teachers who have a conscientious objection to forms of religion will feel themselves debarred from seeking promotion.

There are also a number of parents, though not perhaps a large number, who strongly object to their children receiving any religious instruction, or to contributing to the cost of others receiving it. This objection is not frivolous but is based on deep moral conviction, as a letter to the Journal of Education shows:

"The whole point I sought to make in my former letter was that this is not a question which ought to be settled by the exclusive concourse of Christians at all. It is a question involving the expenditure of public money and must be settled by the citizens as such, irrespective of whether they "profess and call themselves Christians" or not.

"To me, teaching children to believe the Apostles' Creed is intellectually and morally equivalent to belief in the truth of the ancient Greek myths of Zeus or Apollo, or the old Norse legends of Odin and Thor. If people want their children taught religion, then I feel that the least they can do is to be prepared to pay the appropriate religious body to do so for them. I fail to see why as a taxpayer and ratepayer, I should be expected to contribute money for the purpose

[1] Spencer Leeson, *Christian Education.* (See useful appendix on legislation.)

of teaching other people's children what I believe to be myths and legends.

"Nothing less than secular education is therefore my own attitude. To-day we have a compromise between secular education and full-blooded religious and doctrinal teaching. It is war-time, and we are as far as circumstances allow setting aside our pre-war controversies and disputes in order to stop the Nazis from herding us all, Christians and agnostics alike, into their concentration camps. With that peculiar blindness of which only ecclesiastics are capable, a number of Church leaders are now seizing the opportunity to press forward the institutional claims of their Churches for the purpose of extending clerical control over our educational system, and of allowing even greater opportunities of using the schools for religious propaganda.

"Let it be made clear to all concerned that any attempt to force through the recent proposals of the archbishops and their colleagues will lead to embittered controversy and confusion of counsel, for those who think as I do will fight such proposals to the bitter end."

The argument of these and many others is that large numbers of specially trained men are supported by the community in order that they should give religious instruction and consolation to those who want it, and that there is one day in seven set apart for them. If they cannot do their own teaching then they must be incompetent, and if they cannot collect an audience they cannot be offering something the nation really wants. It is wrong to make the schools work to cover up their inefficiency, and to use the law's power to compel children to attend school to save priests and parents the trouble of ensuring that children are taught on Sundays. It follows from this that the schools should give a purely secular education as happens in some European countries and the U.S.A. and leave parents and religious bodies to arrange for religious teaching either in schools that are outside the state system, or on Sundays, or in the evenings for those children who attend state schools.

The number of really convinced atheists is small, but a very much larger group of people distrust the control of education by any organized Church. There have been so many occasions in history when the control of an educational institution by a specific religious body has been associated with its decadence. On the whole, English education has been comparatively free from ecclesiastical domination. Before the Reformation the Church was never in control, because it was early decided by the High Court that teaching children was a virtuous act which ought not to be restrained, and that therefore the teacher did *not* need the bishop's licence before setting up school. After the

Reformation there was an even greater diversity and no attempt was made to prevent any one who wished from teaching, until the Schism Act of 1714 attempted to exterminate Dissent in the next generation by suppressing the many excellent schools and academies maintained by the Nonconformists, and handing the education of their children to teachers licensed by Anglican bishops. The historian, G. M. Trevelyan, after stigmatizing the act as "this peculiarly odious and unnatural form of religious persecution" says that it rendered the downfall of the Tory party "a pre-condition of religious freedom in England".[1] At present no qualification, religious, academic, or personal, is required for opening a school; our laxity in the matter is a subject of endless surprise to the traveller from more organized lands. This diversity has allowed experiment and the development of many different types of school so that in many ways England has given the lead to the world. On the other hand the period when Oxford was the closed preserve of the Church of England was the most degraded in its history. The opponents of religion in schools, who are yet themselves Christians, fear that it may lead to a growth of Church control which may bring a narrowing of ideas, and the exclusion of teachers who do not profess a definite creed. In the past this fear was well founded. Religious intolerance ran high and there were very close political affinities between the different religious sects and the political parties. There is still some basis for it to-day, but though some types of organized religion may still be a threat to human liberty, the real danger lies in the totalitarian philosophies which are hostile to Christianity as a whole. The Christian faith teaches doctrines completely opposed to Nazism or Fascism or Communism. The essential equality of man is opposed to the belief in a Master Race, the belief in the value of the individual to the claim that all must be subordinate to the state. As has happened before in history, Christianity to-day stands for liberty, equality and fraternity, and is opposed to a state that denies all three. Moreover the Christian claims to criticize the state by standards that lie outside it. The "will of God" is superior to every other consideration, and the "welfare of the State" must give way if the two conflict. The Christian, like Sir Thomas More, always claims to be "the King's good servant; but God's first". Thus the conscientious objector to war would rather that the state should perish than that it should engage in the activity of war which he thinks totally wrong. The totalitarian state could

[1] G. M. Trevelyan, *History of England*, p. 500. (The earlier Five Mile Act had applied to schoolmasters as well as to clergymen, and done grave injury to Dissenting education, but was not directly aimed at making it impossible.)

never tolerate such an attitude. Thus the opponent of religious teaching who feels that it is a threat to individual liberty is, to-day, out of date. The world has changed round him, and the slogans of 1900 have lost their meaning. It is only a little curious that the state should be so anxious to teach a religion that claims the right of criticism and of resistance, if necessary, to all state acts. It may be sinister, if we suspect that it is the aim of the state to secure that this religion shall be taught in such a form as to provide a spiritual sanction for itself. This is what happened in Fascist Italy, as we can see if we put side by side two utterances of Mussolini in May, 1929: "These children must be instructed in our faith, but *we need to integrate this faith*, to give these young people the sense of virility, power, conquest". "The Fascist State fully affirms its ethical character; it is Catholic, but it is Fascist, indeed above all *exclusively*, essentially Fascist" (italics mine). Many of the distinguished men who support the Secular Education League do so on the grounds that the state control of religion is disastrous to the religious life of the nation.

If we are to have religious teaching in schools, of what should it consist? We can roughly divide the teaching into three parts. There is a general act of worship, an acknowledgment of God, and an attempt to show our relation to him. There is definite teaching of the historical facts of the Christian religion and of the ideas which it contains. Lastly there are the practices and observances of a particular sect. The first two parts constitute the undenominational religion which it is intended to establish in schools, the last is the concern of individual religious bodies.

In the "act of worship" laid down by the Butler Act we must imagine the reciting of such prayers, and the singing of such hymns as, whatever their authorship, are now acknowledged to be the common heritage of Christians, and the reading of passages from the Bible. If this "worship" is to remain "undenominational" there is little else that can be done. So much is already common in many schools, but to make it compulsory by law will give offence to many groups of people, and there may be many children whose parents will object to their taking part in it. What effect the service will have on those who do attend it depends entirely on the manner in which it is conducted. It could be a valuable part in the children's lives, and it could be a few minutes spent in an activity that no one even pretends to think important. The *spirit* in which such an "act of worship" is conducted can never be made the subject of a regulation, and those schools which did it with belief before will continue to do so, and the others will do it in any manner they think fit.

The teaching of the Bible, the historical facts of the Christian religion and the ideas associated with it, involve less metaphysical problems than worship, but it is still far from easy. How are we to teach the Bible? What are we to say if a child asks if the Creation story is *true*? The Spens report has a sentence on this subject which appears excellent till we try to interpret it in practice. "The approach to the study of Scripture which we had in mind is historical and objective, the temper and method of the teaching being such that the teacher's primary purpose will have been attained when he or she has made the pupil understand the meaning of the book which is being studied, and by "meaning" is to be understood the meaning, so far as it can be understood, of those who wrote the book and those for whom it was written."

Considered in the light of practice this sentence is nonsense. Did the framers of it think they themselves understood the meaning of those who wrote the Epistle to the Ephesians, or the Book of Daniel, or the meaning these works had for those for whom they were written? Did they expect a teacher in Birmingham in 1947 to convey to her children an imaginative understanding, for they ask no less, of even so comparatively simple a thing as the first chapters of St. Matthew. What of the casting out of the money-changers from the Temple, or Christ's verbal exchanges with the Pharisees? To understand what these meant to the writers or their first readers is to become a denizen of another historical age and clime, and is achieved only after years of study and great intellectual effort. To claim that such an understanding should be given in the classroom is to lose all sense of school reality. Equally difficult is the question of *truth*. When a child says "Is this *true*?" he means, "Did it really happen just like that. Can I lie in bed and think of Noah really getting all those animals into the Ark". He does not mean is the story true as an allegory, or true in a spiritual sense, or true as being an attempt to rationalize observed facts of nature. Stella Benson amusingly sets out her difficulties with scripture when she tried in Hong Kong to instruct a class of fifty boys, with an age-range of ten to twenty-three, "in all branches of human knowledge".

"Scripture was my worst subject since, to my generation, I think, the Bible is rather a sentiment than a conviction. Most of the stories are hard to teach, from the school point of view, as *true*, and still harder to show from my own point of view as the wild and lovely things they are to me. And it is difficult to cope conscientiously with the response of fifty sceptics to Western 'superstition'.

" 'But that only superstition, ma'rm, didn't it?'

"Such a fire can dry up the watery triumphs of Noah, and, in the clear light of that fire, all glories and stories—the angelic checking of Abraham's fanatic knife above the neck of little Isaac; the excellent close bargaining with God for the preservation of Sodom and Gomorrah; David's schoolboy victory over Goliath; the business successes of the dreamer, Joseph; and the poetry that intrudes into the efforts of the prosy Moses—all these look cold and lifeless."

Many other teachers have the same difficulty. There are very few who can honestly assert the literal truth of much of the Old Testament, and there are many who would hesitate to say that the shepherds and Wise Men behaved in exactly that way on Christmas morning, much as they would like them to have done so. However much a Scripture syllabus may be "agreed" these difficulties remain.

There is also the point that a knowledge of Scripture does not necessarily imply belief. It is possible to be a profound Biblical scholar, and yet remain an atheist; just as it was said of one classicist that "he knew far more about Zeus than Aeschylus ever did, but unfortunately he had never worshipped him". It is to be assumed that a study of Scripture is intended to lead to belief. It may have this result, but it is quite possible that it may not. In recent years some of the most energetic opponents of Christianity have been men with a very wide knowledge of the Scriptures. The difficulty is greatest with stories from the Old Testament. Many of the ideas and standards are so alien to our own, that the thoughtful child regards them with puzzled discomfort. They are apparently offered for admiration, and yet outrage all the principles he is taught to apply in his daily life. The New Testament is far more closely bound up with our thought to-day, and if we do not accept every detail our unbelief has that enchanted quality that is the effect of all great imaginings. In any case, however, whether the Bible stories are believed to be literally true or not, a knowledge of them does serve the purpose mentioned earlier, that of helping the child to enter into the cultural heritage of his nation.

The ideas and stories of the Bible do not exhaust all that Christianity has to offer us. The history of Christendom contains many figures that are very well suited to be illustrations of the myth. The lives of the saints, whether they be saints of the Middle Ages, such as Elizabeth of Hungary, or of modern times, such as Albert Schweitzer, serve as examples of virtues that we need. The story of a saint can have the same effect on daily life that poetry has—it raises the trivial and customary to the realm of the universal and significant. The child who scatters crumbs for the birds and thinks of St. Francis has a richer and more valuable experience than one who just scatters. In the same way many of our daily acts can gain by being given an ideal prototype, and one of the ways in which religious teaching can help a child is by making him feel that, in his own way, he is part of a long and noble tradition of kindness and virtue.

The actual study of ecclesiastical institutions is less rewarding for children. Church history, during many periods, has been a melancholy chronicle of quarrels and persecutions. Though the adult should know it, partly to enable him to understand the world as he finds it, partly as a dreadful warning, it is hardly suitable for the young.

The third element in religious education is the habituation of the child in the observances characteristic of one sect. It is not proposed that this should form part of the teaching given in state schools. It could not in fact be done without rousing the greatest hostility and opposition. Most Christian sects believe in the importance of corporate worship in accordance with their own forms, and also think that it is dangerous or even wrong for children to take part in services conducted in a different manner. The state schools must therefore confine themselves to such observances as are common to as large a number of sects as possible, and more specific rituals must be an affair of out-of-school teaching. Fee-paying schools, particularly boarding-schools, normally make attendance at some form of organized worship a part of the training, but these schools have been chosen by the parents because they give the kind of training approved.

Even those who most ardently desire definite religious teaching in schools would not claim that such instruction was sufficient in itself. The phrase "Christian way of life" frequently occurs in their writings, and though no exact account is ever given of what this way is, as opposed to the way of life of an enlightened humanist, it is possible to imagine something of what is meant. Probably it is easier to say what is excluded. Eton in the days of Keate gave religious instruction, had "acts of worship" and insisted on attendance at Chapel, and yet there is no father to-day who would condemn his son to such a hell. There have been many practising Christians who have been so narrow-minded in their fervour, so restrictive in their pursuit of godliness, that no child could develop normally under their control. Our Christianity in education must be different from theirs. When we speak of a "Christian way of life" we are really asking for moral excellence, tolerance, humanity, lovingkindness, in a framework of religion. The ethical virtues we ask for could quite well exist without the Christian religion. Paganism has shown many such, and many modern examples of excellence have not been Christians. It is a common mistake to think that such virtues only flourish in Christian soil; they can grow in any religious garden or in none. When we demand in our educators a Christian way of life we are making a demand for two qualities, not necessarily conjoined; ethical excellence and Christian belief. If this is clearly understood we can at any moment decide which is the more important, supposing them dissevered. It may well be that in the education of a child we prefer culture, humanity and kindness to doctrinal respectability when joined to less attractive moral qualities. We must understand what exactly we are seeking, and not allow the use of a word to blind us to the facts. But we may equally be blinded by academic, or unreal, or arbitrary disjunctions; A. E. Housman wittily parodied a whole class of profitless discussions in the question: "Which weighs most, a fat man or a tall man?" If it is true that "an Englishman can now live decently without the religious call and the religious help because he is living on the moral capital of the country's Christian past", then it is also true that "the problem to be thought out here is whether, and if so for how long, our nation can hold on to Christian standards of everyday morality after neglecting its religious basis".[1]

There is also another school problem. How far is it possible to isolate the subjects of the curriculum one from another? Does our religious teaching end, intellectually, with the scripture lesson, or does a religious outlook imply a certain method of approaching the whole world and the study of it? There is in some writers a quantity of vague talk about unifying the whole curriculum around religion so as to avoid the dangers of a "divided mind". It was a favourite device a century ago to set sums in simple arithmetic on the numbers of the Israelites, just as the Nazis used to set problems in ballistics and the endurance of submarines. Mr. Spiller saw in all the subjects of the curriculum a field for moral teaching. Modern writers mean something rather more subtle, and the two subjects which cause them most concern are probably science and history. Both these subjects introduce the child to a range of ideas and facts which will contribute to his conception of the kind of world in which he finds himself, and his philosophy of life, or religion. The danger is that they may be regarded as in themselves providing an adequate basis for the complete interpretation of experience.

"To mistake each aspect of reality for the whole of reality, or to regard each system which we can isolate for scientific purposes as closed, self-contained and independent, are the most obvious and

[1] T. E. Jessop, *Evangelism and Education.*

most recurrent mistakes in philosophy. Every form of naturalism and materialism is but a dreary repetition of these patent illusions, and one of the most grievous defects of higher education is its failure to save those who are exposed to it from confusing the fallacies of naturalism with the findings of science."[1]

In the last century this difficulty was felt most strongly over geology and biology, to-day it is history. Fossils and the development of species were felt to threaten religious belief, as astronomy had earlier. We have already discussed in another chapter the way in which facts will affect beliefs, and said that religion has already decided that the ammonite offers no threat to man's belief in God. But at the end of the last century science offered a new interpretation of the universe. It was trying to show that man and the universe were alike controlled by mechanical laws, and that in this ordered machine there was no need and no room for a divine personal principle. Order and predictability reigned, religion and its fantasies of free will, of heaven and hell, of divine decree, could all be swept away, and man set free from fear, and, probably, from hope. In so far as this attitude removed the need for thought, or took from the world its beauty and mystery, it was anti-religious. When it was used to claim the right to free questioning, to proof and to criticism, it was a protest against the unholy alliance between organized religion and political and social conservatism. It was then valuable as part of the liberating movement of fifty years ago.

To-day both science and religion have changed. Science uncovers new mysteries with every advance, and is farther than ever from reducing the universe to a tidy self-actuated machine. The intricacies of the physical world leave reason helpless and dazzled. The commonest event, when viewed intelligently, becomes a marvel. No baby can be born, no flower grow, no ant die, without the scientific onlooker being amazed at the depth of the mystery. It is, of course, possible to teach science, as one can teach anything else, without due understanding. It can become dull, the formal repetition of explanations that leave everything unrevealed; but the good scientist is not opposed to religion, he knows himself at but the edge of the mystery, and the better he teaches the more he excites the reverence and wonder of his class.

To-day, History, rather than science, is made the servant of anti-religious propaganda. It has been discovered that in the presentation of history almost any political or religious idea can be embodied. The histories of the world, as seen by opposing ideologies, have

[1] H. G. Wood, *Christianity and the Nature of History*.

strangely little in common. There are of course a number of events in history that still stir emotion in some men's hearts. The Reformation, the treatment of Catholics under Elizabeth, the Battle of the Boyne or Cromwell's behaviour at Drogheda still have considerable emotional significance. Any teacher dealing with them cannot fail to imply that one side or the other had the better right. Charles I has had Anglican churches dedicated to him. Many now, as then, think him a foolish and obstinate man who resisted changes that were inevitable and desirable. It is not really possible to present both points of view. There is no dispute over facts, those are fairly certain. The difference lies in the interpretation; and nothing will bring Milton's tracts and the *Ikon Basilike* into the same picture.

According to the historian, A. L. Rowse, the result of modern methods of historical study is that "The idea of God has been rendered superfluous."[1]

More important by far than these differences in the presentation of individual facts is the general pattern of history as it appears to different types of thought. History can be seen as the lives of great men each striving as an individual, and making his individual contribution to the events of his age, or it can be represented as the blind play of impersonal economic forces which, with the crushing weight of millstones, grind out the course of man's destiny. The individual is lost and worthless in the general movement of events. It is possible also to represent the history of a nation as God's dealings with his chosen people, or as the random batterings of chance. Clearly it matters considerably to a child, both what interpretation is put on particular events and also what is represented as the general nature of mans' story. It is vain for a teacher to claim that he is only presenting facts and doing no more. The truth of a past epoch is so incredibly complicated, so infinitely hard to discover, that we must always have simplification and selection, emphasis and conjecture. The myopically industrious historian who is always analysing medieval washing lists or records of courts leet can best claim that objective truth alone is his concern, but even he can hardly avoid some general conclusions. He must at times let his mind stray over the general field, and form some glimmerings of a philosophy of history. The ordinary teacher, far less well equipped, must deal with wider issues, and cannot fail to give to his pupil some picture of man's destiny. There is no doubt, as we have said earlier, that man's mental picture is projected outward on the world. The man whose impulses are well harmonized and who enjoys peace of mind feels benevolence somewhere supreme

[1] A. L. Rowse, *The Use of History*.

in the universe. It is the emotionally unhappy who find the devil or blind chance in control of man's destiny. If history is to reinforce religious teaching, those who advocate this must think out their philosophy of history. They must not merely offer an interpretation of certain facts in accordance with their convictions. That is easy enough. They must see the whole course of history as a manifestation of certain principles that are themselves religious, and they must do this in such a way that truth is still preserved.

There is one last question before we close this chapter; what do we expect to be the result of our religious teaching? Mr. Donald Hughes suggests the difficulty of the answer:

"If we were really as interested in the religious results of the school as we are in the results which it obtains in the School Certificate or in the inter-school matches, we should not tolerate for another term the tepid substitute for enthusiasm which most boys experience at the thought of anything connected with the religious life of the school."

This is a typical passage from this class of writing, but it is interesting for the blatancy of its assumptions and the confusion of thought. Can a school achieve "religious results" in any way comparable to its success in inter-school matches or in the School Certificate? Of course it cannot. Nor can a school be conducted in the emotional atmosphere of a revivalist meeting. The religion that we are trying to impart should be a settled attitude to life, and it will not affect all children in the same way, nor will its effects be fully apparent during the school years. Religious feelings, like all others, must change and develop with life. The majority of children will, we hope, leave school with peace in their hearts, a sense of security, and a belief in the possibility and duty of making a contribution to the welfare of those about them. They should be happy, and eager to share that happiness with others. They should have a sense of the wonder and beauty of the world and a reverence for the personalities of others. None of these characteristics are necessarily religious. They become so when they are united by a sense of the divine within and about us; they become Christian when they are enriched by the "knowledge of the glory of God in the face of Jesus Christ", and stabilized by loyalty to the heavenly vision. This may be called the religion of the many. There should also be the religion of the few who through thought, and often suffering, have come to accept definite principles of conduct, have tried to mould their lives on a pattern nearer the divine and who are prepared to transmit to others the ideals that are truly Christian.

Books referred to in this chapter:

A National Basic Outline of Religious Instruction. (Obtainable from the N.U.T.)
T. F. Kinloch, *Pioneers of Religious Education.*
Spencer Leeson, *Christian Education.*
Reeves and Drewett, *What is Christian Education.*
H. P. Wood, *Christianity and the Nature of History.*
A. L. Rowse, *The Use of History.*
Janet Whitney, *Elizabeth Fry.*
John A. Patten, *These Remarkable Men.*
M. V. C. Jeffreys, *Education, Pagan or Christian?*
T. E. Jessop, *Evangelism and Education.*
Stella Benson, *The Little World.*

INDEX

For Product Safety Concerns and Information please contact our EU representative GPSR@taylorandfrancis.com Taylor & Francis Verlag GmbH, Kaufingerstraße 24, 80331 München, Germany

Batch number: 10725800

Printed by Printforce, the Netherlands